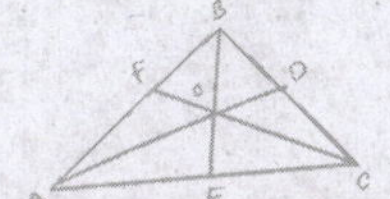

THE MATHS MAGIC

Formulae & Concepts

SSC (CGL, CHSL, CPO, MTS), Railway (NTPC, Group D, ALP, JE)
Banking, CDS, DSSSB, CUET, CTET, State Police
UPSC एवं State PSCs के लिए समान रूप के उपयोगी

BILINGUAL

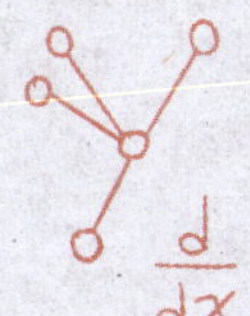

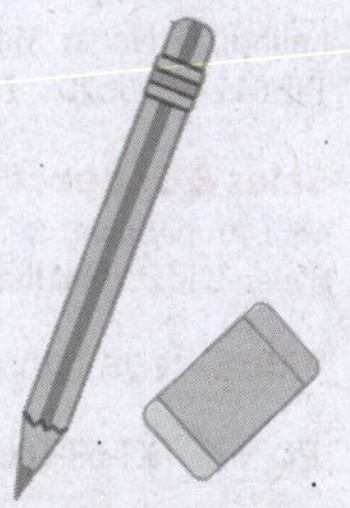

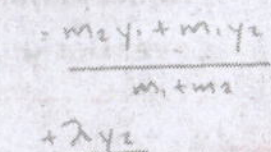

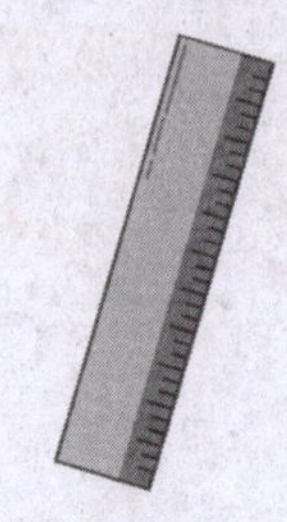

ARIHANT PUBLICATIONS (INDIA) LIMITED

Administrative & Production Offices

Regd. Office

'Ramchhaya' 4577/15, Agarwal Road, Darya Ganj, New Delhi -110002
Tele: 011- 47630600, 43518550

Head Office

Kalindi, TP Nagar, Meerut (UP) - 250002
Tel: 0121-7156203, 7156204

Sales & Support Offices

Agra, Ahmedabad, Bengaluru, Bareilly, Chennai, Delhi, Guwahati, Hyderabad, Jaipur, Jhansi, Kolkata, Lucknow, Nagpur & Pune.

PRICE ₹ 265.00

PO No. : TXT-59-T065278-05-25

Published by Arihant Publications (India) Ltd.

Publishing Manager	:	Mahendar Singh Rawat	**Layout & Design**	:	Ankit Tyagi
Project Manager	:	Shahid Saifi	**Front Page Designer**	:	Ravindra Kumar
Cover Design	:	Bilal Hashmi	**Proof Readers**	:	Rashid, Sagar, Ravi

For further information about the books published by Arihant, log on to www.arihantbooks.com or e-mail at info@arihantbooks.com

Follow us on   

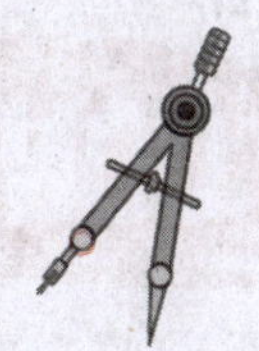

दो शब्द...

वर्तमान समय में किसी भी प्रतियोगी परीक्षा में गणित (Maths) विषय का एक महत्त्वपूर्ण स्थान है तथा परीक्षा पैटर्न में तीव्र गति से आ रहे बदलाव को देखते हुए हमने इस पुस्तक 'The Maths Magic' को प्रस्तुत किया है।

इस पुस्तक को प्रकाशित करने का मुख्य उद्देश्य (Main purpose) अभ्यर्थियों को सभी प्रतियोगी परीक्षाओं (Competitive Exams) में सरलता से बहुत कम समय में (In short time) सफलता दिलाना है।

यह पुस्तक इस प्रकार तैयार की गई है कि प्रत्येक विद्यार्थी बिना किसी कठिनाई के प्रत्येक प्रश्न में प्रयोग होने वाली अवधारणा (Concept), सूत्र (Formula) या ट्रिक (Trick) को समझ सके। इस पुस्तक में थ्योरी और प्रश्नों को द्विभाषी (Bilingual) दिया गया है, जिससे दोनों माध्यम के अभ्यर्थियों को समझने में कोई कठिनाई न हो सके।

इस पुस्तक में कुछ ऐसे सूत्र/ट्रिक दिए गए हैं, जिनके प्रयोग से कमजोर अभ्यर्थी भी किसी परीक्षा में गणित के प्रश्नों को सरलता से हल कर सकता है।

पुस्तक की विशेषताएँ

इस पुस्तक के प्रत्येक अध्याय को निम्नलिखित आकर्षण बिन्दु के अनुसार तैयार किया गया है

- सम्पूर्ण थ्योरी (Theory)/ परिभाषाएँ (Definitions)
- सभी महत्त्वपूर्ण कॉन्सेप्ट्स उदाहरण सहित
- सभी सामान्य सूत्र (All General Formulae)
- ट्रिक सूत्र (Tricks formulae)
- विगत परीक्षाओं के हल किए गए उदाहरण

हम आशा करते हैं कि "अरिहन्त पब्लिकेशन्स रहेंगे" द्वारा प्रस्तुत यह पुस्तक आपको परीक्षा में अवश्य सफलता दिलाएगी।

आपके सुझाव के लिए सदैव आभारी रहेंगे।

धन्यवाद (Thanks)!

लेखकगण
AJAB SINGH
RANJAN KUMAR
DEVESH SONKAR

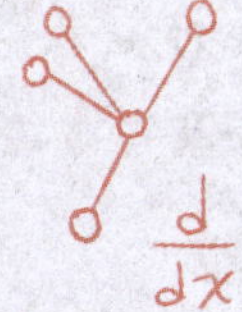

CONTENTS

अध्याय 01

Line and Angle

रेखा और कोण

Point बिन्दु

- A dimension-less figure or a circle with zero radius, is called a point.
  एक आयाम-रहित आकृति अथवा एक वृत्त जिसकी त्रिज्या शून्य हो, बिन्दु कहलाता है।

Types of Points बिन्दुओं के प्रकार

(i) **Collinear Points :** If three or more than three points are situated on a straight line, then these points are called collinear points.
संरेख बिन्दु : यदि तीन या तीन से अधिक बिन्दु एक सरल रेखा पर स्थित हों, तब ये बिन्दु संरेख बिन्दु कहलाते हैं।

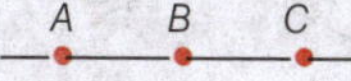

Here, point A, B and C are collinear. /यहाँ, बिन्दु A, B और C संरेख हैं।

(ii) **Non-Collinear Points :** If three or more than three points are not situated on a straight line, then these points are called non-collinear points.
असंरेख बिन्दु : यदि तीन या तीन से अधिक बिन्दु किसी एक सरल रेखा पर स्थित नहीं होते हैं, तब ये बिन्दु असंरेख बिन्दु कहलाते हैं।

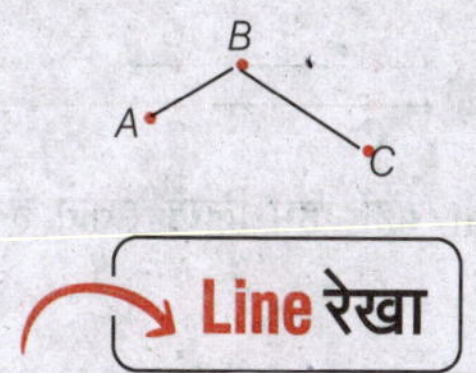

Line रेखा

- A **line** is a one-dimensional figure, which has length but no width and no end points. It can be extended infinitely in both directions.
  एक **रेखा**, एक-आयामी आकृति है, जिसकी लम्बाई तो होती है लेकिन कोई चौड़ाई तथा अन्त बिन्दु नहीं होते हैं। इसे दोनों दिशाओं में अनन्त तक बढ़ाया जा सकता है।

- **Ray** A part of line with one end point, is called a ray. It can be extended only in one direction.
  किरण : एक अन्त बिन्दु वाला रेखा का एक भाग, किरण कहलाता है। इसे केवल एक दिशा में बढ़ाया जा सकता है।

A ———— B →

- **Line Segment** : A line with a fixed length, is called a line segment.
रेखाखण्ड : एक निश्चित लम्बाई वाली रेखा, रेखाखण्ड कहलाती है।

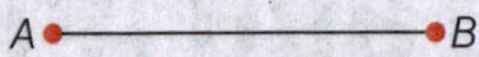

- **Curve Line** : A line which is not a straight line and changes its direction, called curve line.
वक्र रेखा : ऐसी रेखा जो सीधी रेखा नहीं होती और अपनी दिशा बदलती है, वक्र रेखा कहलाती है।

Types of Lines रेखाओं के प्रकार

(i) **Intersecting Lines :** If two or more than two lines intersect each other (or meet each-other at any point), then these are called intersecting lines.
प्रतिच्छेदी रेखाएँ : यदि दो या दो से अधिक रेखाएँ एक-दूसरे को प्रतिच्छेद करती हैं (या एक-दूसरे से किसी बिन्दु पर मिलती हैं), तब ये प्रतिच्छेदी रेखाएँ कहलाती हैं।

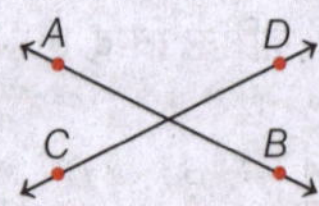

(ii) **Parallel Lines :** When two or more than two lines never intersect each other, then they are called parallel lines. Parallel lines are denoted by symbol '||'.
समान्तर रेखाएँ : जब दो या दो से अधिक रेखाएँ एक-दूसरे को कभी प्रतिच्छेद नहीं करती हैं, समान्तर रेखाएँ कहलाती हैं। समान्तर रेखाओं को चिन्ह '||' से प्रदर्शित करते हैं।

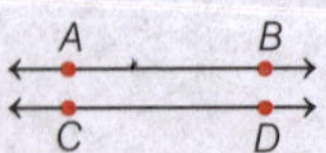

Here, we can write that/यहाँ, हम लिख सकते हैं कि

$$AB \parallel CD$$

(iii) **Perpendicular Lines :** If two lines intersect each other at right angles, then these are called perpendicular lines. Perpendicular lines are denoted by symbol '⊥'.
लम्बवत् रेखाएँ : यदि दो रेखाएँ एक-दूसरे को समकोण पर प्रतिच्छेद करती हैं, तब ये लम्बवत् रेखाएँ कहलाती हैं। लम्बवत् रेखाओं को चिन्ह '⊥' द्वारा प्रदर्शित करते हैं।

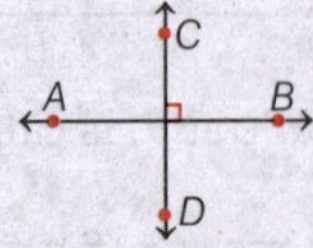

Here, we can write that/यहाँ, हम लिख सकते हैं कि

$$AB \perp CD \text{ or } CD \perp AB$$

(iv) **Concurrent Lines** : If three or more than three lines passes through the same point, then they are called concurrent lines and the point through which all these lines pass is called **point of concurrency**.
संगामी रेखाएँ : यदि तीन या तीन से अधिक रेखाएँ एक ही बिन्दु से होकर गुजरती हैं तब वे संगामी रेखाएँ कहलाती हैं और जिस बिन्दु से होकर ये सभी रेखाएँ गुजरती हैं, उसे **संगामी बिन्दु** कहते हैं।

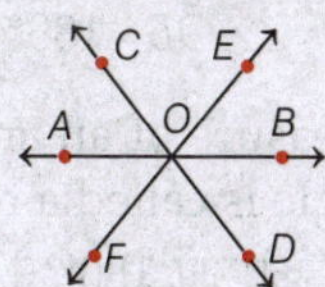

(v) **Transversal Line** : A line, which intersects two or more than two given lines at distinct points is called a transversal line.
तिर्यक रेखा : वह रेखा, जो दो या दो से अधिक दी गई रेखाओं को अलग-अलग बिन्दुओं पर काटती है, तिर्यक रेखा कहलाती है।
Here, LM is a transversal line/यहाँ, LM एक तिर्यक रेखा है।

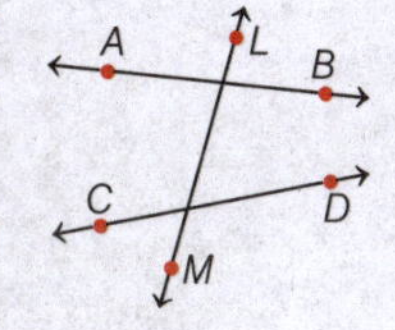

- Inclination between two rays is called an angle.
दो किरणों के बीच के झुकाव को कोण कहते हैं।

$$\angle AOB = \theta$$

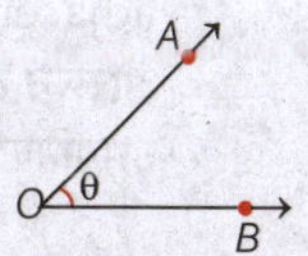

Types of Angles कोणों के प्रकार

(i) **Acute Angle** : If the measure of an angle is in between 0° to 90°, then this angle is called an acute angle.
न्यून कोण : यदि किसी कोण की माप 0° से 90° के मध्य है, तब यह कोण, न्यून कोण कहलाता है।

$$0° < \angle AOB < 90°$$

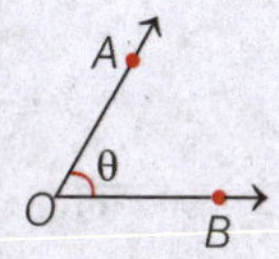

(ii) **Right Angle** : If measure of an angle is equal to 90°, then this angle is called a right angle.
समकोण : यदि किसी कोण की माप 90° के बराबर हो, तब यह कोण, समकोण कहलाता है।

$$\angle AOB = 90°$$

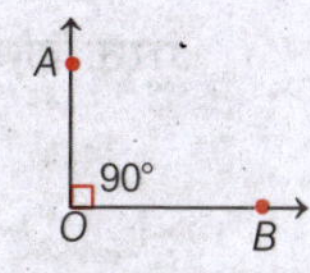

(iii) **Obtuse Angle** : If the measure of an angle is more than 90° but less than 180°, then this angle is called an obtuse angle.
अधिक कोण : यदि किसी कोण की माप 90° से अधिक परन्तु 180° से कम हो, तब यह कोण, अधिक कोण कहलाता है।

$$90° < \angle AOB < 180°$$

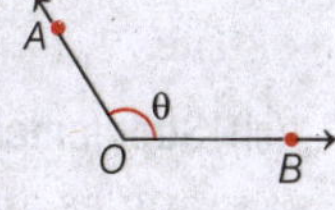

(iv) **Straight Angle :** If the measure of an angle is equal to 180°, then this angle is called straight angle.

सरल कोण या ऋजु कोण : यदि किसी कोण की माप 180° के बराबर हो, तब यह कोण, सरल कोण कहलाता है।

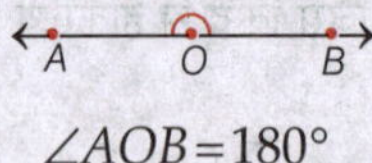

$$\angle AOB = 180°$$

(v) **Reflex Angle :** If the measure of an angle is more than 180° but less than 360°, then this angle is called a reflex angle.

वृहत कोण : यदि किसी कोण की माप 180° से अधिक परन्तु 360° से कम हो, तब यह कोण, वृहत कोण कहलाता है।

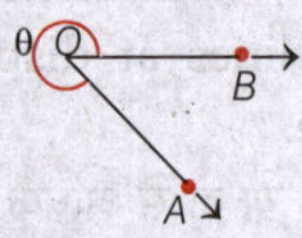

$$180° < \angle AOB < 360°$$

Pairs of Angles कोणों के युग्म

(i) **Adjacent Angles :** Two angles are called adjacent angles, if they have the same vertex and a common arm. Here, $\angle AOC$ and $\angle BOC$ are adjacent angles.

आसन्न कोण : दो कोण आसन्न कोण कहलाते हैं, यदि उनके शीर्ष समान हों तथा एक भुजा उभयनिष्ठ हो।

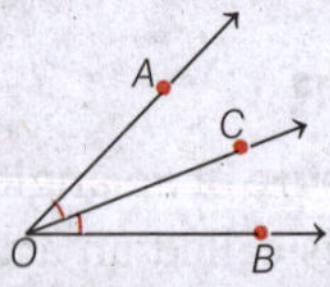

यहाँ, $\angle AOC$ और $\angle BOC$ आसन्न कोण हैं।

(ii) **Complementary Angles :** If the sum of two angles is equal to 90°, then these angles are called complementary angles.

पूरक कोण : यदि दो कोणों का योग 90° हो, तब ये कोण पूरक कोण कहलाते हैं।

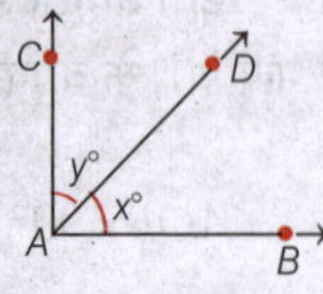

i.e., $x + y = 90°$

(iii) **Supplementary Angles :** If sum of two angles is equal to 180°, then these angles are called supplementary angle.

सम्पूरक कोण : यदि दो कोणों का योग 180° हो, तब ये कोण सम्पूरक कोण कहलाते हैं।

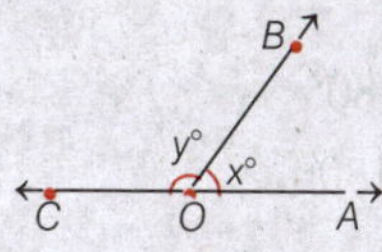

i.e., $x + y = 180°$

(iv) **Linear Pair of Angles** : Two adjacent angles are said to be a linear pair of angles, if their non-common arms are two opposite rays. The sum of two linear pair of angles is 180°.

कोणों का रैखिक युग्म : दो आसन्न कोण, कोणों का रैखिक युग्म कहलाते हैं, यदि उनकी गैर-उभयनिष्ठ भुजाएँ दो विपरीत किरण हों। दो रैखिक युग्म कोणों का योग 180° होता है।

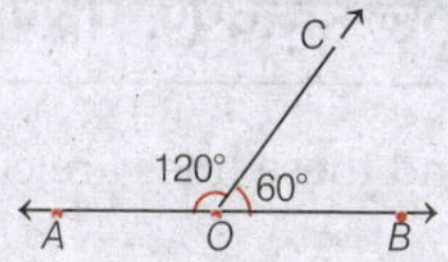

(v) **Vertically Opposite Angles :** When two lines intersect each other, the opposite angles are called vertically opposite angles. Vertically opposite angles are equal to each other.

शीर्षाभिमुख कोण : जब दो रेखाएँ एक-दूसरे को प्रतिच्छेदित करती हैं, तो आमने-सामने के कोण शीर्षाभिमुख कोण कहलाते हैं। शीर्षाभिमुख कोण एक-दूसरे के बराबर होते हैं।

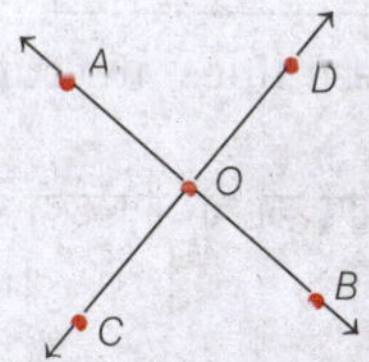

Here, $\angle AOD$ and $\angle BOC$, $\angle AOC$ and $\angle BOD$ are vertically opposite angles.

यहाँ, $\angle AOD$ और $\angle BOC$, $\angle AOC$ और $\angle BOD$ शीर्षाभिमुख कोण हैं।

$$\therefore \quad \boxed{\angle AOD = \angle BOC, \quad \angle AOC = \angle BOD}$$

- When two parallel lines l and m are intersected by a transversal line (n), then

  जब दो समान्तर रेखाओं l और m को एक तिर्यक रेखा (n) प्रतिच्छेद करती है, तब

  (i) **Corresponding Angles**/संगत कोण

  $$\angle 1 = \angle 5, \angle 4 = \angle 8$$
  $$\angle 3 = \angle 7, \angle 2 = \angle 6$$

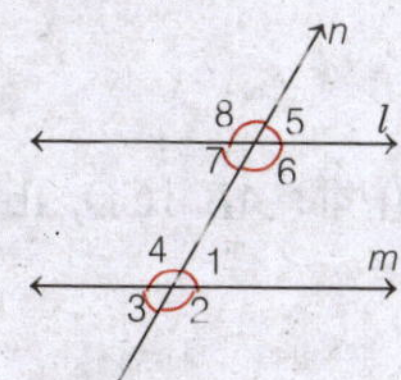

  (ii) **Alternate Interior Angles**/एकान्तर अन्तः कोण

  $$\angle 6 = \angle 4, \angle 7 = \angle 1$$

  (iii) **Alternate Exterior Angles**/एकान्तर बाह्य कोण

  $$\angle 5 = \angle 3, \angle 8 = \angle 2$$

(iv) **Consecutive Interior Angles**/क्रमागत आन्तरिक कोण

$$\angle 6 + \angle 1 = 180°$$
$$\angle 7 + \angle 4 = 180°$$

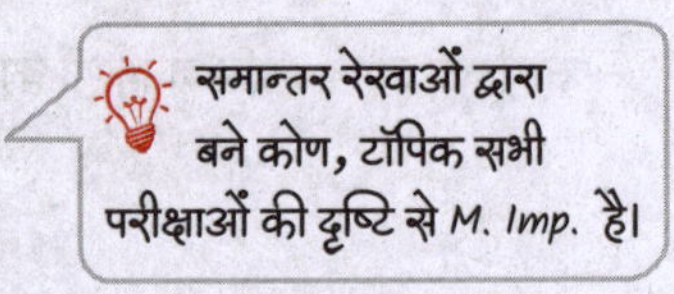

(v) **Vertically Opposite Angles**/शीर्षाभिमुख कोण

$$\angle 1 = \angle 3,\ \angle 2 = \angle 4$$
$$\angle 5 = \angle 7,\ \angle 6 = \angle 8$$

- When $AE \parallel BF \parallel CG \parallel DH$ and these lines are cut by two transversal lines.
  यदि $AE \parallel BF \parallel CG \parallel DH$ हों तथा ये रेखाएँ दो प्रतिच्छेदी रेखाओं को काटती हों।
  Then, /तब, $a:b:c = p:q:r$

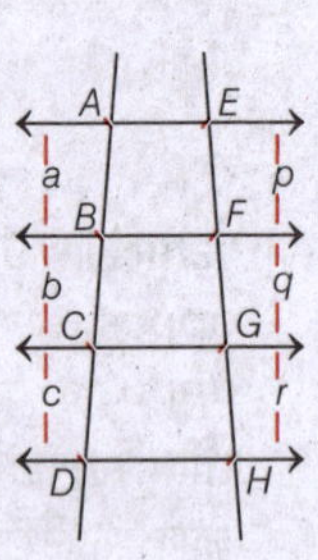

$$\boxed{\frac{a}{a+b+c} = \frac{p}{p+q+r}}$$

- When $AD \parallel BE \parallel CF$ and these lines are cut by two transversal lines.
  जब $AD \parallel BE \parallel CF$ हों तथा इन रेखाएँ को दो प्रतिच्छेदी रेखा काटती हों।
  Then, /तब,

$$\frac{AB}{BC} = \frac{DE}{EF} = \frac{m}{n}$$

$$\boxed{BE = \frac{an+bm}{m+n}}$$

- If/यदि $P \parallel Q$, then/तब $\theta = \alpha + \beta$

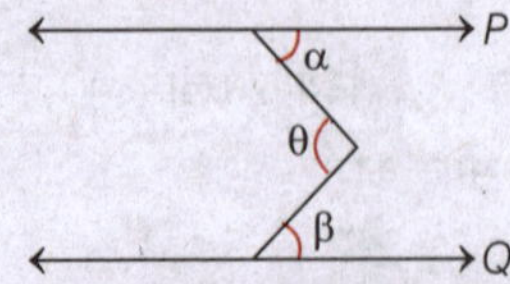

- If/यदि $AB \parallel CD$, then/तब $\mathbf{x + y + z = 360°}$

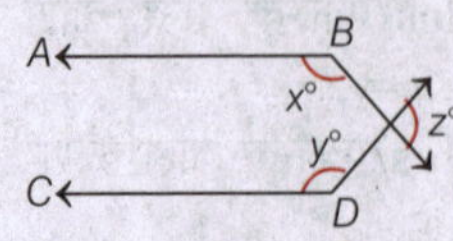

- In such figure/इस आकृति में, $\theta = x + y + z$

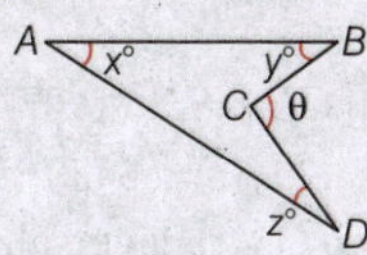

रेखा और कोण पर आधारित मिश्रित उदाहरण निम्नलिखित हैं

Example **In the given figure, line l is parallel to m and p is the transversal. Find the measure of angle 1.** *RRB NTPC 2022*

दी गई आकृति में, रेखा 'l', m के समान्तर है तथा 'p' एक प्रतिच्छेदी रेखा है। कोण '1' की माप बताएँ।

Sol. As $l \parallel m$ and p is a transversal.

$\therefore \quad x + 80° = 2x + 10°$ [∵ corresponding angles]

$\Rightarrow \quad x = 70°$

As, $\quad x + 80° = \angle 1$ [∵ vertically opposite angle]

$\Rightarrow \quad \angle 1 = 70° + 80° = 150°$

Example **In the given figure, LOM is a straight line. What is the value of $x°$?**

दी गई आकृति में, LOM एक सरल रेखा है। x का मान क्या है?

Sol. ∵ LOM is a straight line

$\therefore \quad \angle LOQ + \angle QOP + \angle POM = 180°$

$\therefore \quad (x + 20)° + 50° + (x - 10°) = 180°$

$\Rightarrow \quad x = 60°$

Example **In the given figure, $l_1 \parallel l_2$ and p is their transversal with $\angle 4 = 5x + 35°$ and $\angle 7 = 45°$. Find the value of x.**

दी गई आकृति में, $l_1 \parallel l_2$ तथा 'p' उनकी प्रतिच्छेदी रेखा है, जिसमें $\angle 4 = 5x + 35°$ तथा $\angle 7 = 45°$ है। x का मान बताएँ।

Sol. As $l_1 \parallel l_2$ and 'p' is a transversal.

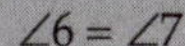

$\angle 6 = \angle 7$ [∵ vertically opposite angle]

$\therefore \quad \angle 4 + \angle 6 = 180°$ [consecutive interior angle]

$5x + 35° + 45° = 180°$ [∵ $\angle 6 = \angle 7$]

$\Rightarrow \quad x = 20°$

Example **In the given figure, $AB \parallel CD \parallel EF$, if $\angle GHE = 100°$, then find $\angle AGD$.**

दी गई आकृति में, $AB \parallel CD \parallel EF$ है, यदि $\angle GHE = 100°$ है, तब $\angle AGD$ की माप बताएँ।

Sol. As $CD \parallel EF$ and GH is transversal.

$\therefore \quad \angle CGH + \angle GHE = 180°$ [co- interior angles]

$\Rightarrow \quad \angle CGH = 80°$

As, $\angle CGH = \angle AGD = 80°$ [vertically opposite angles]

$\therefore \quad \angle AGD = 80°$

Example **In the given figure, $AC \parallel BD$ and $AE \parallel BF$. What is the value of $\angle x$?** **CDS 2022**

दी गई आकृति में, $AC \parallel BD$ तथा $AE \parallel BF$ है। $\angle x$ का मान क्या है?

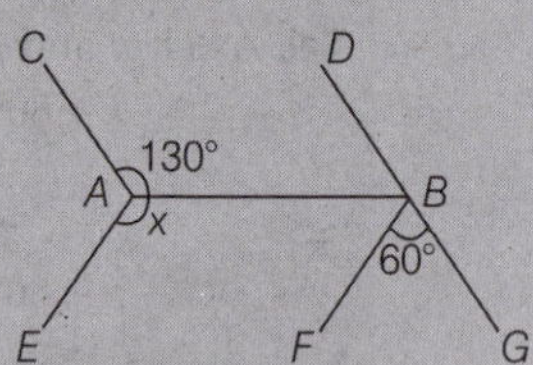

Sol. Since $AC \parallel BD$

$\therefore \quad \angle DBA = 180° - 130° = 50°$ [co- interior angle]

As, DBG is a straight line.

$\therefore \quad \angle ABF = 180° - 50° - 60° = 70°$

Since $AE \parallel BF$

$\therefore \quad x = 180° - \angle ABF$ [co- interior angle]

$\Rightarrow \quad x = 180° - 70°$

$\Rightarrow \quad x = 110°$

अध्याय 02

Triangle त्रिभुज

Triangle त्रिभुज

- A closed figure bounded by three line segments, is called a triangle. It is denoted by Δ.
  तीन रेखाखण्डों से घिरी एक बन्द आकृति, त्रिभुज कहलाती है। इसे Δ द्वारा प्रदर्शित करते हैं।
  In ΔABC,
  ΔABC में,
  (i) Three vertices, namely A, B and C.
  तीन शीर्ष, जो A, B और C हैं।
  (ii) Three sides, namely AB, BC and CA.
  तीन भुजाएँ, जो AB, BC और CA हैं।
  (iii) Three angles, namely $\angle A$, $\angle B$ and $\angle C$.
  तीन कोण, जो $\angle A$, $\angle B$ और $\angle C$ हैं।
  (iv) Three heights, namely h_1, h_2 and h_3.
  तीन ऊँचाइयाँ, जो h_1, h_2 और h_3 हैं।
  (v) Opposite sides of $\angle A$, $\angle B$ and $\angle C$ are a, b and c, respectively.
  $\angle A$, $\angle B$ और $\angle C$ की सम्मुख भुजाएँ क्रमशः a, b और c हैं।

Types of Triangle त्रिभुज के प्रकार

On the Basis of Angles कोणों के आधार पर

- **Acute Angle Triangle/न्यूनकोण त्रिभुज**
  Its all three angles are less than 90°.
  इसके तीनों कोण 90° से कम होते हैं।
  $\angle A, \angle B, \angle C < 90°$

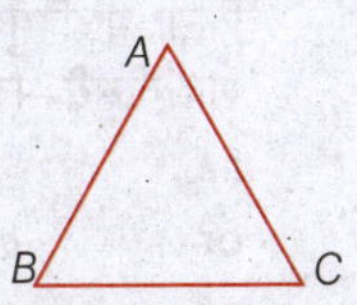

- **Right Angle Triangle/समकोण त्रिभुज**
  Its one angle is 90°.
  इसका एक कोण 90° का होता है।
  $\angle B = 90°$ and/और $\angle A + \angle C = 90°$

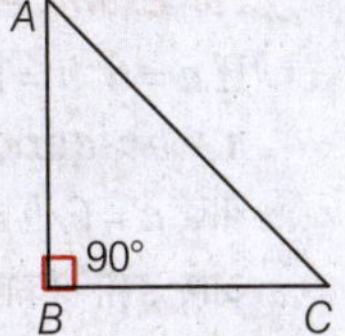

- **Obtuse Angle Triangle/अधिक कोण त्रिभुज**

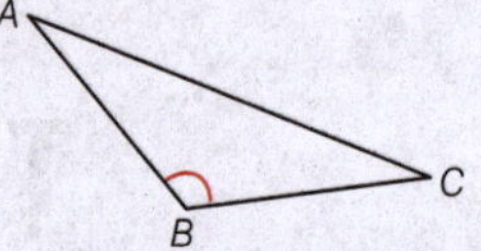

Its one angle is greater than 90°.
इसका एक कोण 90° से अधिक होता है।

$$\angle B > 90°$$

On the Basis of Sides/भुजाओं के आधार पर

- **Equilateral Triangle/समबाहु त्रिभुज** Equilateral triangle has 3 equal sides and each angle is 60°.
समबाहु त्रिभुज में 3 बराबर भुजाएँ होती हैं तथा प्रत्येक कोण 60° होता है।

$$AB = BC = CA\ ,$$
$$\angle A = \angle B = \angle C = 60°$$

- **Isosceles Triangle/समद्विबाहु त्रिभुज** Isosceles triangle has two equal sides and opposite angles of equal sides are also equal.

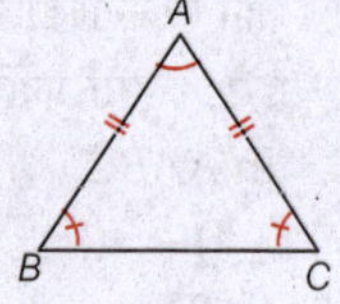

समद्विबाहु त्रिभुज में दो भुजाएँ बराबर होती हैं। समान भुजाओं के सम्मुख कोण भी बराबर होते हैं।

$$AB = AC$$
$$\angle C = \angle B$$

- **Scalene Triangle/विषमबाहु त्रिभुज** Scalene triangle has three unequal sides and three unequal angles.
विषमबाहु त्रिभुज में तीन असमान भुजाएँ और तीन असमान कोण होते हैं।

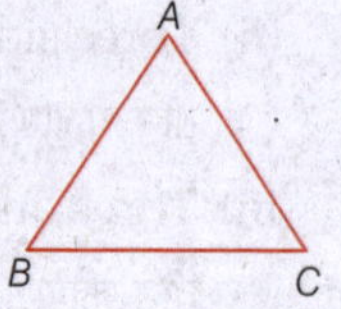

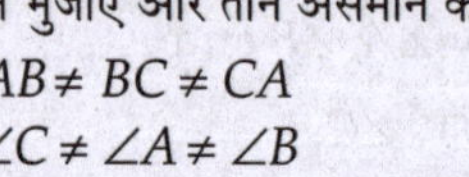

$$AB \neq BC \neq CA$$
$$\angle C \neq \angle A \neq \angle B$$

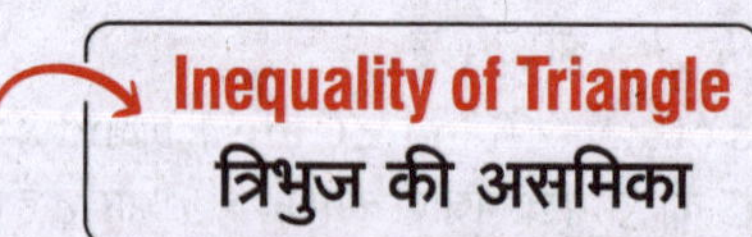

Inequality of Triangle
त्रिभुज की असमिका

(i) For any triangle the sum of the lengths of any two sides must be greater than the length of the remaining third side.
किसी भी त्रिभुज के लिए किन्हीं भी दो भुजाओं की लम्बाई का योग, शेष तीसरी भुजा की लम्बाई से अधिक होना चाहिए।

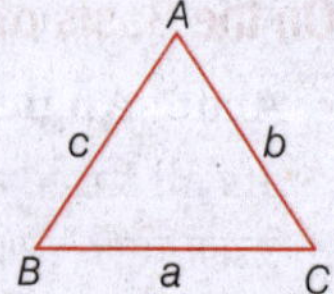

i.e., $a + b > c$
or $b + c > a$
or $c + a > b$

For Example

If $a = 6$, $b = 11$, $c = 15$, then triangle is possible, because $6 + 11 > 15$ *i.e.*, $a + b > c$ and so on.
यदि $a = 6$, $b = 11$, $c = 15$, तब त्रिभुज सम्भव है, क्योंकि $6 + 11 > 15$ अर्थात् $a + b > c$ और इसी प्रकार आगे भी।

(ii) For any triangle the difference of lengths of any two sides is always less than the length of third side.
किसी भी त्रिभुज के लिए किन्हीं दो भुजाओं की लम्बाई का अन्तर सदैव तीसरी भुजा की लम्बाई से कम होता है।
i.e., $|b-c|<a$ or $|a-c|<b$ or $|a-b|<c$

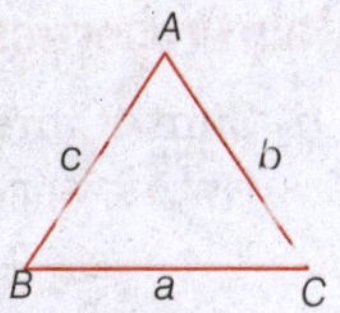

For Example If $a=6$, $b=11$ and $c=15$, then triangle is possible, because $15-11<6$ *i.e.*, $c-b<a$.

यदि $a=6$, $b=11$ और $c=15$, तब त्रिभुज सम्भव है, क्योंकि $15-11<6$ अर्थात् $c-b<a$।

(iii) If a, b and x are sides of a triangle, where x is unknown and a and b are known and also $a<b<x$.
Then, number of possible value of $x=2\times$ Small side -1
यदि a, b और x त्रिभुज की भुजाएँ हैं, जहाँ x अज्ञात तथा a और b ज्ञात हैं और साथ ही $a<b<x$ तब, x के सम्भावित मान की संख्या $=2\times$ छोटी भुजा -1

For Example If 9, 16 and x are sides of a triangle, where x is an integer, then how many values of x are. (Where, $x>16$)
$\therefore$ Possible values of $x=2\times9-1=17$
यदि 9, 16 और x एक त्रिभुज की भुजाएँ हैं, जहाँ x एक पूर्णांक है।
तब, x के कितने सम्भव मान हैं। (जहाँ, $x>16$)
$\therefore$ x के सम्भावित मान $=2\times9-1=17$

Identification of Triangle त्रिभुज की पहचान

If a, b and c are the sides of a triangle with also $a<b<c$; then
(i) If $c^2<a^2+b^2$, then triangle is an acute angle triangle.
(ii) If $c^2=a^2+b^2$, then triangle is a right angle triangle.
(iii) If $c^2>a^2+b^2$, then triangle is an obtuse angle triangle.
यदि a, b और c त्रिभुज की भुजाएँ हैं अर्थात् $a<b<c$; तब
(i) यदि $c^2<a^2+b^2$, तब त्रिभुज न्यूनकोणीय होगा।
(ii) यदि $c^2=a^2+b^2$, तब त्रिभुज समकोण होगा।
(iii) यदि $c^2>a^2+b^2$, तब त्रिभुज अधिक कोणीय होगा।

त्रिभुज की पहचान पर आधारित उदाहरण निम्न है

Example **If sides of triangle are 7, 12 and 15, then which type of triangle it is?**
यदि त्रिभुज की भुजाएँ क्रमशः 7, 12 और 15 है, तो यह किस प्रकार का त्रिभुज है?
Sol. Let a, b, c are 7, 12 and 15, respectively.
Then, $a^2=7^2=49$; $b^2=12^2=144$; $c^2=15^2=225$
Here, $49+144<c^2$ $193<225$, $(\because a^2+b^2<c^2)$
Then, it is obtuse angle triangle.

Main Properties of Triangle त्रिभुज के मुख्य गुण

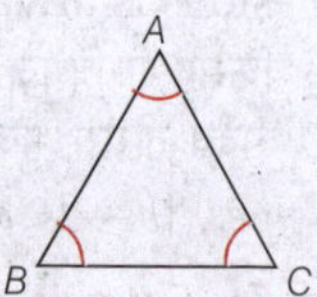

(i) Sum of three angles of a triangle is 180°
त्रिभुज के तीनों कोणों का योग 180° होता है।

$$\angle A + \angle B + \angle C = 180°$$

(ii) If a side of triangle is produced, then the exterior angle so formed is equal to the sum of two interior opposite angles.
यदि एक त्रिभुज की किसी एक भुजा को बढ़ाया जाता है, तब इस प्रकार बना बाह्य कोण दो विपरीत आन्तरिक कोणों के योगफल के बराबर होता है।

$$\angle ACD = \angle CAB + \angle ABC \text{ or } \angle 4 = \angle 1 + \angle 2$$

(iii) In a triangle the angle between internal bisector of a base angle and external bisector of the other base angle is half of the third vertex angle.

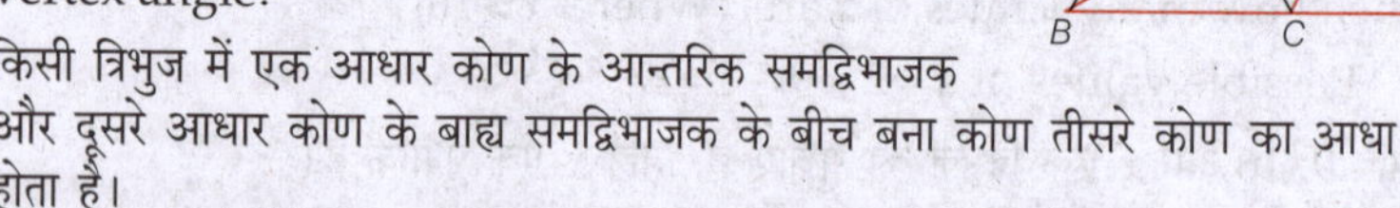

किसी त्रिभुज में एक आधार कोण के आन्तरिक समद्विभाजक और दूसरे आधार कोण के बाह्य समद्विभाजक के बीच बना कोण तीसरे कोण का आधा होता है।

$$\angle BEC = \frac{\angle A}{2}$$

त्रिभुज के कोण समद्विभाजक गुण पर आधारित उदाहरण निम्न है

Example **In the given figure, QT and RT are angle bisector of $\angle PQR$ and $\angle PRS$ respectively and $\angle QPR = 100°$, then find $\angle QTR$.**

दी गई आकृति में QT और RT क्रमशः $\angle PQR$ और $\angle PRS$ के कोण समद्विभाजक हैं और $\angle QPR = 100°$, तब $\angle QTR$ ज्ञात करें।

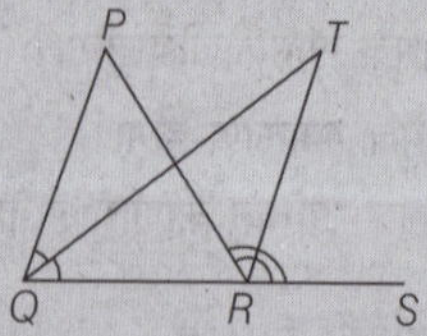

Sol. In given figure, $\angle PQT = \angle TQR$ and $\angle PRT = \angle TRS$

So, from formula $\angle QTR = \frac{\angle QPR}{2}$

$$\angle QTR = \frac{100}{2} = 50°$$

(iv) In ΔABC, the side BC is produced to D and angle bisector of $\angle A$ meets BC at E, then

$$\angle ABC + \angle ACD = 2\angle AEC$$

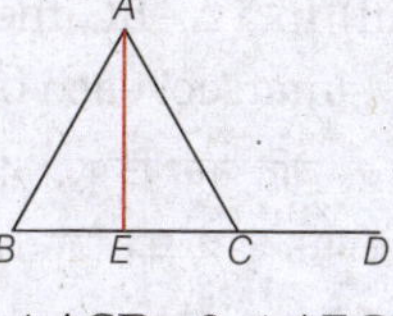

ΔABC में, भुजा BC को D तक बढ़ाया गया है और $\angle A$ का समद्विभाजक भुजा BC से बिन्दु E पर मिलता है, तब $\angle ABC + \angle ACD = 2\angle AEC$

(v) The angle between perpendicular drawn by a vertex to opposite side and angle bisector of the vertex angle, is half of difference between the other two remaining vertex angles.

शीर्ष से विपरीत भुजा पर डाले गए लम्ब और शीर्ष कोण के समद्विभाजक के बीच का कोण, शेष दो शीर्ष कोणों के अन्तर का आधा होता है।

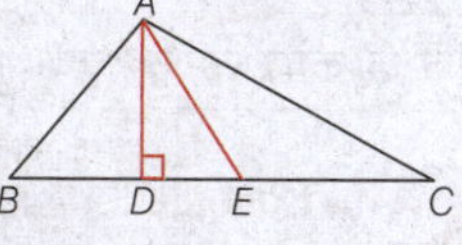

$$\angle DAE = \left|\frac{\angle B - \angle C}{2}\right|$$

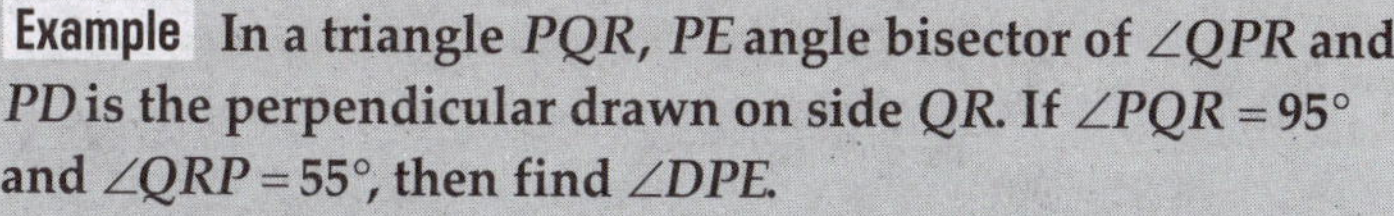
शीर्ष कोण के समद्विभाजक और लम्ब के बीच के कोण पर आधारित प्रमेय पर उदाहरण निम्न हैं

Example **In a triangle PQR, PE angle bisector of $\angle QPR$ and PD is the perpendicular drawn on side QR. If $\angle PQR = 95°$ and $\angle QRP = 55°$, then find $\angle DPE$.**

त्रिभुज PQR में, $PE, \Delta QPR$ का कोण समद्विभाजक है और PD भुजा QR पर खींचा गया लम्ब है। यदि $\angle PQR = 95°$ और $\angle QRP = 55°$ तो $\angle DPE$ ज्ञात करें।

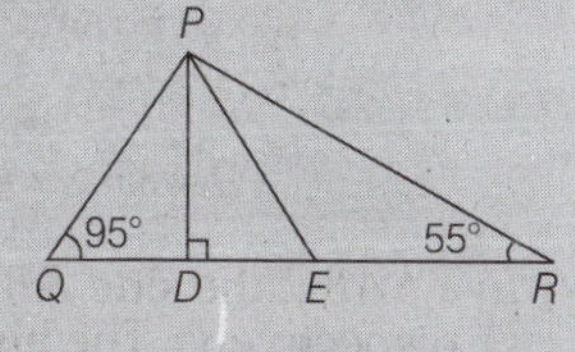

Sol. $\angle DPE = \dfrac{\angle PQR - \angle PRQ}{2}$

$$= \frac{95° - 55°}{2} = \frac{40°}{2} = 20°$$

(vi) The sum of three reflex angles of three angles of a triangle is 900°.

किसी त्रिभुज के तीनों कोणों के वृहत कोणों का योग 900° होता है।

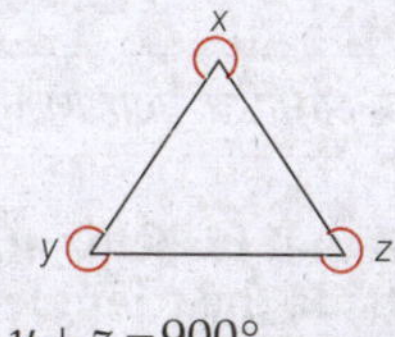

$$x + y + z = 900°$$

(vii) If angles of a triangle are in AP, then middle angle is always 60°.

यदि किसी त्रिभुज के कोण समान्तर श्रेणी में हैं, तब मध्य का कोण सदैव 60° होता है।

For Example In ΔABC, $\angle A, \angle B, \angle C$ are in AP, then $\angle B = 60°$

उदाहरणार्थ ΔABC में, $\angle A, \angle B, \angle C$, AP में हैं, तब $\angle B = 60°$

(viii) In a ΔABC, the bisector of $\angle B$ and $\angle C$ intersect each other at a point O, then

यदि एक त्रिभुज ABC में, $\angle B$ तथा $\angle C$ के समद्विभाजक एक-दूसरे को बिन्दु O पर प्रतिच्छेद करते हैं, तब

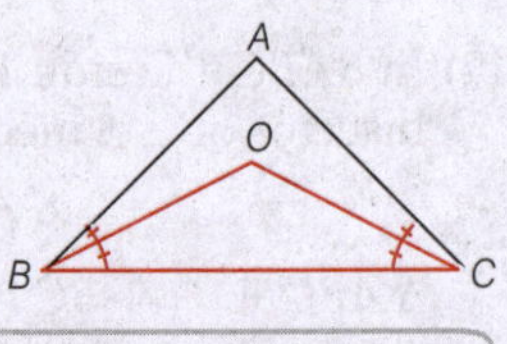

$$\angle BOC = 90^\circ + \frac{\angle A}{2}$$

परीक्षा ब्रह्मास्त्र
त्रिभुज का कोण समद्विभाजक गुण SSC की सभी परीक्षाओं के लिए *M. Imp.* है।

त्रिभुज के कोण समद्विभाजक गुण पर आधारित उदाहरण निम्न है

Example 3 **In a triangle PQR, the bisector of $\angle P$ and $\angle R$ meet at a point M inside the triangle. If the measurement of $\angle PMR = 127^\circ$, then measurement of $\angle Q$.**

त्रिभुज PQR में, $\angle P$ और $\angle R$ का समद्विभाजक बिन्दु M पर त्रिभुज के अन्दर मिलता है। यदि $\angle PMR = 127^\circ$, तब $\angle Q$ की माप ज्ञात कीजिए। *SSC (10+2) 2023*

Sol. $\angle PMR = 90^\circ + \frac{\angle Q}{2}$

$$127^\circ = 90^\circ + \frac{\angle Q}{2}$$

$$\frac{\angle Q}{2} = 127^\circ - 90^\circ = 37^\circ$$

$$\angle Q = 37^\circ \times 2 = 74^\circ$$

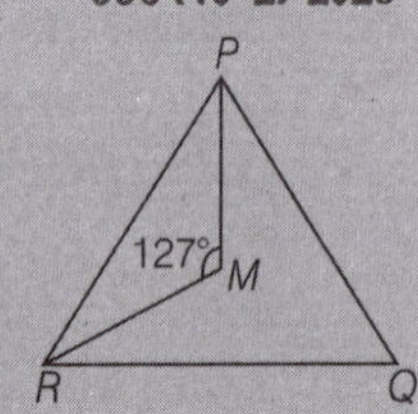

(ix) In a ΔABC, the side AB and AC are produced to P and Q, respectively. The bisectors of $\angle PBC$ and $\angle QCB$ intersect at a point O, then

एक त्रिभुज ABC में, भुजा AB और AC को क्रमशः P और Q तक बढ़ाया जाता है। $\angle PBC$ और $\angle QCB$ के समद्विभाजक एक बिन्दु O पर प्रतिच्छेद करते हैं, तब $\angle BOC = 90^\circ - \frac{\angle A}{2}$

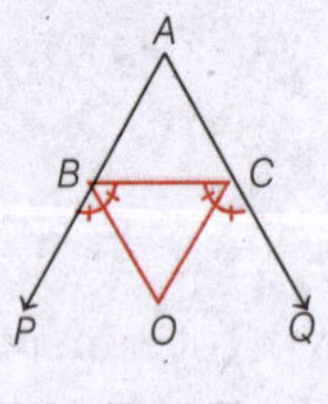

त्रिभुज के बाह्य कोण समद्विभाजक गुण पर आधारित उदाहरण निम्न है

Example **In ΔABC, external bisector of angle $\angle B$ and $\angle C$ C meets at point P outside the triangle. If angle $\angle BAC = 60^\circ$, then find the measurement of $\angle BPC$.**

ΔABC में, कोण B और कोण C का बाह्य समद्विभाजक त्रिभुज के बाहर बिन्दु P पर मिलता है। यदि $\angle BAC = 60^\circ$, तो $\angle BPC$ का माप ज्ञात करें।

Sol. $\therefore\ \angle BPC = 90^\circ - \frac{A}{2} = 90^\circ - \frac{60^\circ}{2} = 90^\circ - 30^\circ$

$$\angle BPC = 60^\circ$$

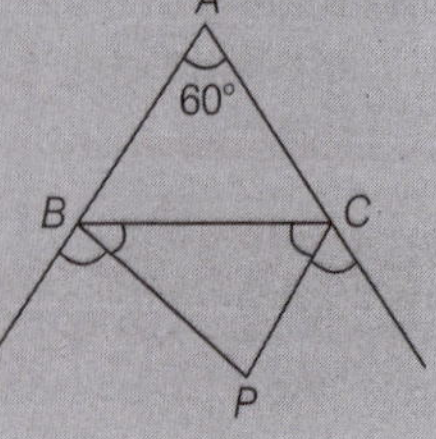

Congruency of Triangle
त्रिभुज की सर्वांगसमता

Two triangles are said to be congruent, if all three corresponding sides are equal and all three corresponding angles are equal. It is denoted by symbol '≅'.

दो त्रिभुज सर्वांगसम कहलाते हैं, यदि तीनों संगत भुजाएँ और तीनों संगत कोण बराबर हों। इसे चिह्न '≅' से प्रदर्शित करते हैं।

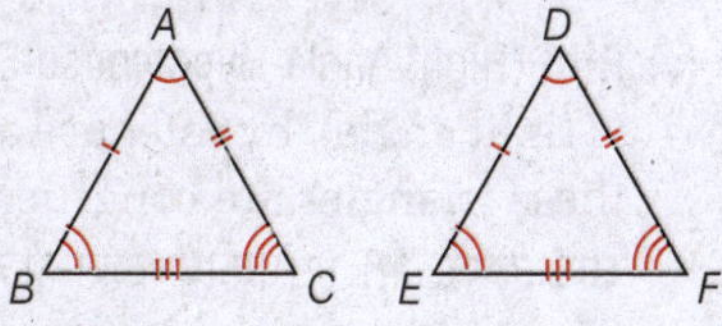

$AB = DE, AC = DF, BC = EF;$

$\angle A = \angle D, \angle B = \angle E, \angle C = \angle F$

Then/ तब, $\Delta ABC \cong \Delta DEF$.

Condition of Congruency सर्वांगसमता की शर्त

(i) **SSS** (Side-Side-Side) : If all three corresponding sides are equal in two triangles, then these triangles are congruent.

भु.भु.भु. (भुजा-भुजा-भुजा) : यदि दो त्रिभुजों में सभी तीनों भुजाएँ बराबर हैं, तब ये त्रिभुज सर्वांगसम होते हैं।

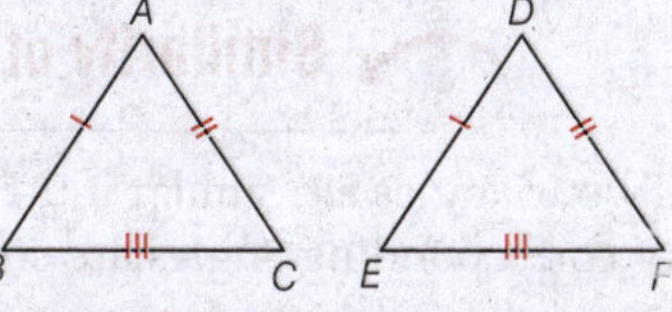

$AB = DE, AC = DF, BC = EF$

Then/तब, $\Delta ABC \cong \Delta DEF$

(ii) **SAS** (Side-Angle-Side) : If in two triangles, two sides and angle between them are equal, then these triangles are congruent.

भु.को.भु. (भुजा-कोण-भुजा) : यदि दो त्रिभुजों में, दो भुजाएँ तथा उनके बीच का कोण बराबर हो, तब ये त्रिभुज सर्वांगसम होते हैं।

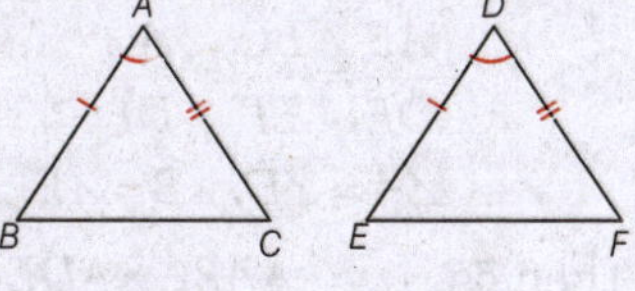

$AB = DE, AC = DF, \angle A = \angle D$

Then/तब, $\Delta ABC \cong \Delta DEF$

(iii) **ASA** (Angle-Side-Angle) : If in two triangles, two angles and side between them are equal, then these triangles are congruent.

को.भु.को. (कोण-भुजा-कोण) : यदि दो त्रिभुजों में, दो कोण तथा उनके बीच की भुजा बराबर हो, तब ये त्रिभुज सर्वांगसम होते हैं।

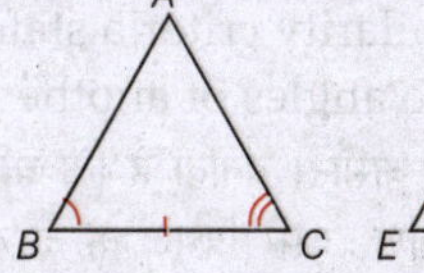

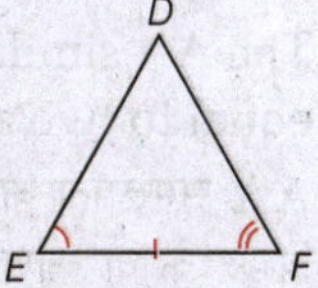

$\angle B = \angle E, \angle C = \angle F, BC = EF,$

then/तब $\Delta ABC \cong \Delta DEF$

(iv) **AAS** (Angle-Angle-Side) When two angles and non-included side of triangle are equal to the corresponding angles and sides of another triangle, then triangles are said to be congruent by AAS.

को. को. भु. (कोण-कोण-भुजा) : जब किसी त्रिभुज के दो कोण और असम्मिलित भुजा किसी अन्य त्रिभुज के संगत कोणों और भुजा के समान हों, तब त्रिभुज AAS द्वारा सर्वांगसम होता है।

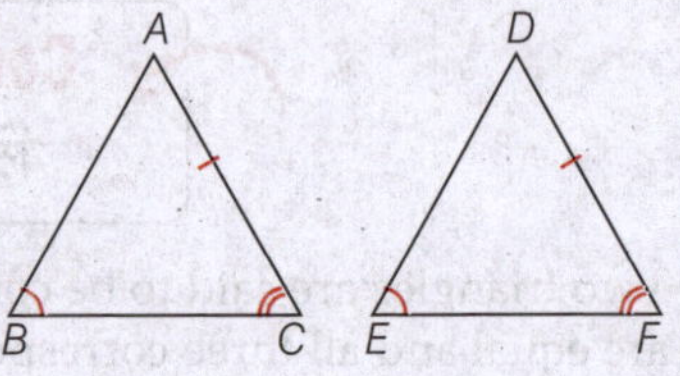

$\angle B = \angle E, \angle ACB = \angle DFE, AC = DF$

$\therefore \quad \Delta ABC \cong \Delta DEF$

(v) **RHS** (Right Angle-Hypotenuse-Side) : If in two triangles, one angle (which is right angle), hypotenuse and one corresponding side are equal, then these triangles are congruent.

सम. कर्ण. भु. (समकोण-कर्ण-भुजा) : यदि दो त्रिभुजों में, एक कोण (जो समकोण है), कर्ण तथा एक संगत भुजा बराबर हैं, तो ये त्रिभुज सर्वांगसम होते हैं।

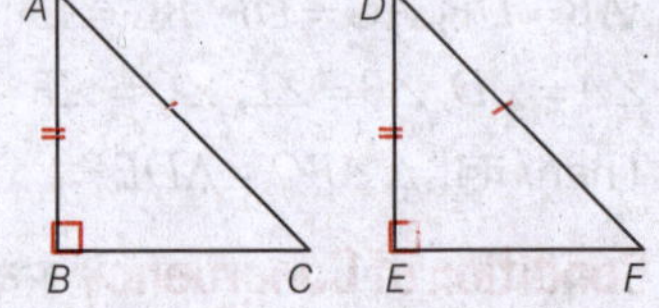

$\angle B = \angle E = 90°, AC = DF, AB = DE,$

Then/तब, $\quad \Delta ABC \cong \Delta DEF$

Similarity of Triangles त्रिभुज की समरूपता

Two triangles are similar if the ratio of their corresponding sides and pair of corresponding angles are equal. It is denoted by symbol '~'.

दो त्रिभुज समरूप होते हैं, यदि उनकी संगत भुजाओं का अनुपात और संगत कोणों के युग्म समान हों। इसे प्रतीक चिन्ह '~' से प्रदर्शित करते हैं।

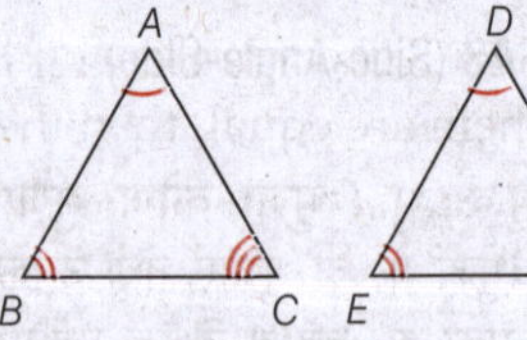

$$\frac{AB}{DE} = \frac{BC}{EF} = \frac{AC}{DF}$$

$$\angle A = \angle D, \angle B = \angle E, \angle C = \angle F$$

Then/तब, $\quad \Delta ABC \sim \Delta DEF$

Condition of Similarity समरूपता की शर्त

1. AA Similarity को. को. समरूपता

The AA similarity criteria states that if two angles of one triangle are equal to two angles of another triangle, then the triangles will be similar.

AA समरूपता कसौटी बताता है कि यदि एक त्रिभुज के दो कोण, दूसरे त्रिभुज के दो कोणों के बराबर हैं, तो त्रिभुज समरूप होंगे।

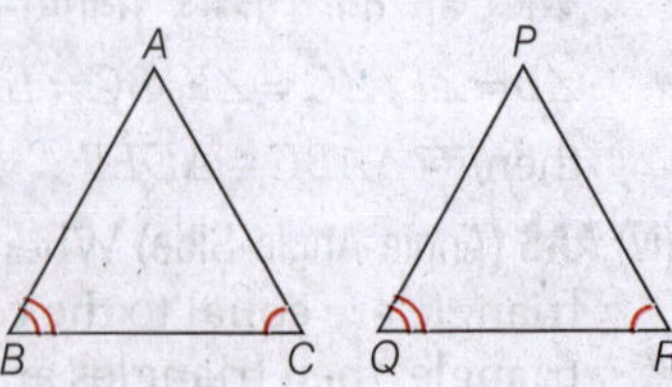

In the given figure/दिए गए चित्र में,

$\angle B = \angle Q$ and/और $\angle C = \angle R$

$\therefore \quad \Delta ABC \sim \Delta PQR$

2. SSS Similarity / भु.भु.भु. समरूपता

If the corresponding sides of two triangles are proportional, then the two triangles will be similar. It is also called side-side-side similarity theorem.

यदि दो त्रिभुजों की संगत भुजाएँ समानुपाती हों, तो दोनों त्रिभुज समरूप होंगे। इसे भुजा-भुजा-भुजा समरूपता प्रमेय भी कहा जाता है।

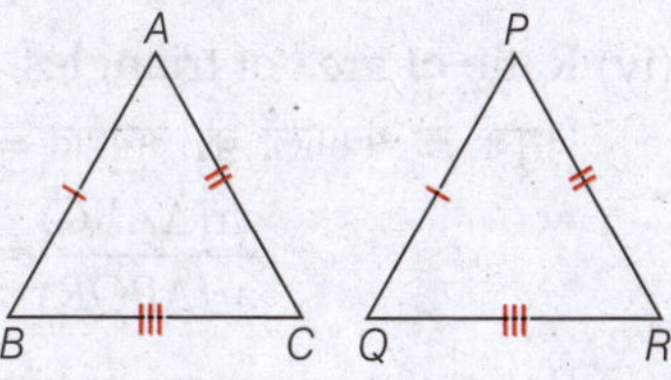

In the given figure/दिए गए चित्र में,

$$\frac{AB}{PQ} = \frac{BC}{QR} = \frac{AC}{PR}$$

$$\therefore \quad \Delta ABC \sim \Delta PQR$$

3. SAS Similarity / भु.को.भु. समरूपता

If in two triangles, two pairs of corresponding sides are proportional and the angle between these two sides are equal, then the two triangles will be similar. It is also called side-angle-side similarity theorem.

यदि दो त्रिभुजों में, परस्पर संगत भुजाओं के दो जोड़े समानुपाती हों और इन दोनों भुजाओं के बीच का कोण बराबर हो, तो दोनों त्रिभुज समरूप होंगे। इसे भुजा-कोण-भुजा समरूपता प्रमेय भी कहा जाता है।

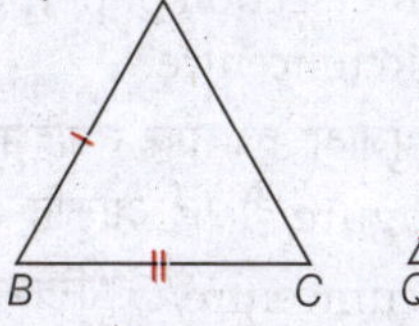

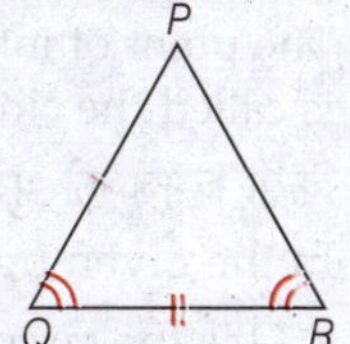

$$\angle B = \angle Q, \; AB = PQ, \; BC = QR$$

Areas of Similar Triangles/ समरूप त्रिभुजों के क्षेत्रफल

- If ΔABC and ΔPQR are similar triangles
  यदि ΔABC और ΔPQR समरूप त्रिभुज हैं।

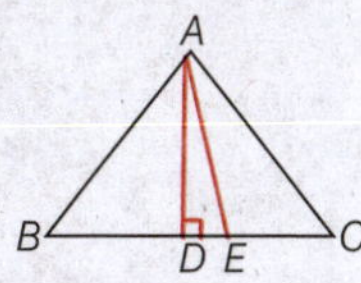

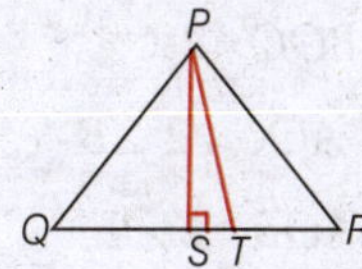

Then/तब,

(i) Ratio of area of triangles = (Ratio of corresponding sides)^2
त्रिभुजों के क्षेत्रफलों का अनुपात = (संगत भुजाओं का अनुपात)^2

$$\frac{ar(\Delta ABC)}{ar(\Delta PQR)} = \frac{AB^2}{PQ^2} = \frac{BC^2}{QR^2} = \frac{AC^2}{PR^2}$$

(ii) Ratio of area of triangles = (Ratio of corresponding altitudes)^2
त्रिभुजों के क्षेत्रफलों का अनुपात = (संगत ऊँचाइयों का अनुपात)^2

$$\frac{ar(\Delta ABC)}{ar(\Delta PQR)} = \frac{AD^2}{PS^2}$$

(iii) Ratio of area of triangles = (Ratio of corresponding medians)^2

त्रिभुजों के क्षेत्रफलों का अनुपात = (संगत माध्यिकाओं का अनुपात)^2

$$\frac{\text{ar}(\Delta ABC)}{\text{ar}(\Delta PQR)} = \frac{AE^2}{PT^2}$$

(iv) Ratio of area of triangles = (Ratio of perimeters of triangles)^2

त्रिभुजों के क्षेत्रफलों का अनुपात = (त्रिभुजों के परिमापों का अनुपात)^2

$$\frac{\text{ar}(\Delta ABC)}{\text{ar}(\Delta PQR)} = \frac{(AB+BC+CA)^2}{(PQ+QR+RP)^2}$$

(v) Ratio of perimeters of triangles/त्रिभुजों के परिमापों का अनुपात

$$\frac{AB+BC+CA}{PQ+QR+RA} = \frac{AB}{PQ} = \frac{AD}{PS} = \frac{AE}{PT}$$

Centres of Triangle त्रिभुज के केन्द्र

1. Circumcentre/ परिकेन्द्र

The point of intersection of perpendicular bisectors of the sides of a triangle is called the circumcentre.

किसी त्रिभुज की भुजाओं के लम्ब समद्विभाजकों के प्रतिच्छेद बिन्दु को परिकेन्द्र कहते हैं।

In the given figure/दी गई आकृति में,

- O is the circumcentre/O परिकेन्द्र है।
- O is equidistance from all three vertices of $\angle ABC$.

  O, $\angle ABC$ के तीनों शीर्षों से समदूरस्थ है।

  $AO = BO = CO = R$ (Circumradius/परित्रिज्या)

  i.e. $OA = OB = OC = R$

  where R is circumradius

$$\angle BOC = 2A$$

$$\angle AOC = 2\angle B$$

$$\angle AOB = 2\angle C$$

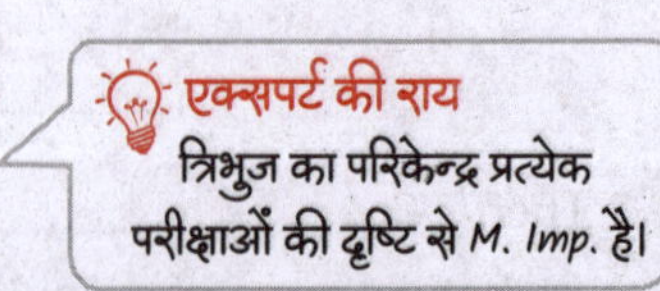

and $$R = \frac{abc}{4\Delta}$$

[where/जहाँ Δ = area of triangle/त्रिभुज का क्षेत्रफल]

त्रिभुज के परिकेन्द्र पर आधारित उदाहरण निम्न है

Example **In ΔPQR, perpendicular bisector of all sides of triangle intersect at point O. If $\angle B = 55°$, find $\angle AOC$.**

त्रिभुज PQR में, त्रिभुज की सभी भुजाओं के लम्ब समद्विभाजक बिन्दु O पर प्रतिच्छेद करते हैं। यदि $\angle B = 55°$ है, तो $\angle AOC$ का मान ज्ञात कीजिए।

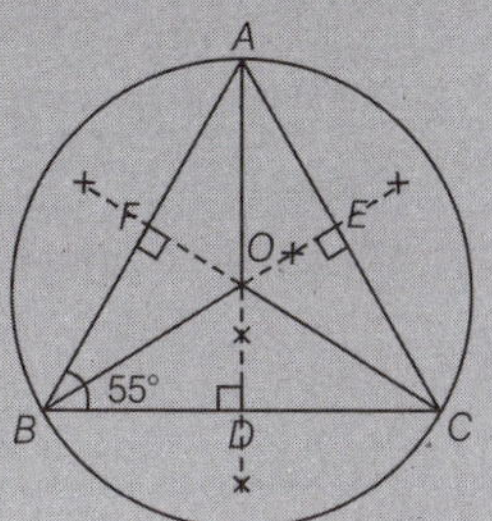

Sol. From figure, $\angle B = 55°$

$\therefore O$ is the circumcentre of triangle.

Then, $\angle AOC = 2\angle B = 2 \times 55°$

$\angle AOC = 110°$

Position of Circumcentre परिकेन्द्र की स्थिति

(i) In acute-angled triangle, circumcentre lies inside the triangle
न्यूनकोणीय त्रिभुज में, परिकेन्द्र त्रिभुज के अन्दर होता है।

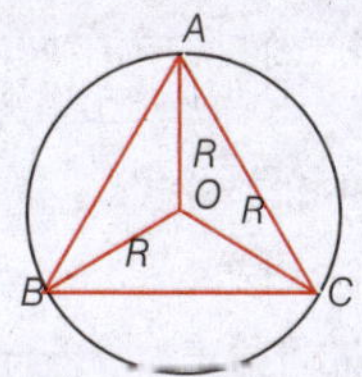

(ii) In a right-angled triangle, the circumcentre lies at the mid-point of the hypotenuse.
किसी समकोण त्रिभुज में, परिकेन्द्र सदैव कर्ण का मध्य-बिन्दु होता है।

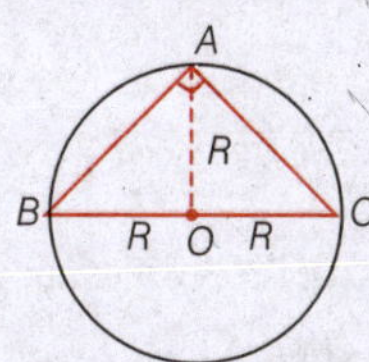

(iii) In an obtuse-angled triangle, the circumcentre is always outside the triangle and will be at the front of the angle which is obtuse.
किसी अधिक कोण त्रिभुज में, परिकेन्द्र सदैव त्रिभुज के बाहर और अधिक कोण के सामने होता है।

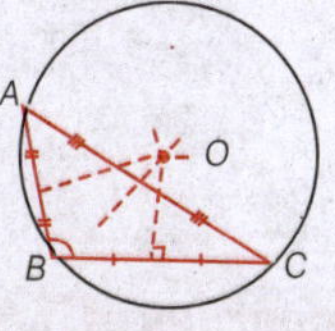

Note The circumradius of an equilateral triangle

$$R = \frac{\text{Side}}{\sqrt{3}} = \frac{a}{\sqrt{3}}$$ ($\because$ a is the side of equilateral triangle)

समबाहु त्रिभुज की परित्रिज्या, $R = \frac{\text{भुजा}}{\sqrt{3}} = \frac{a}{\sqrt{3}}$

($\because$ a समबाहु त्रिभुज की भुजा है)

त्रिभुज की परित्रिज्या पर आधारित उदाहरण निम्न है

Example ***ABC* is an equilateral triangle having side 12 cm. Find circumradius of triangle.**

ABC एक समबाहु त्रिभुज है, जिसकी भुजाएँ 12 सेमी हैं। त्रिभुज की परित्रिज्या ज्ञात करें।

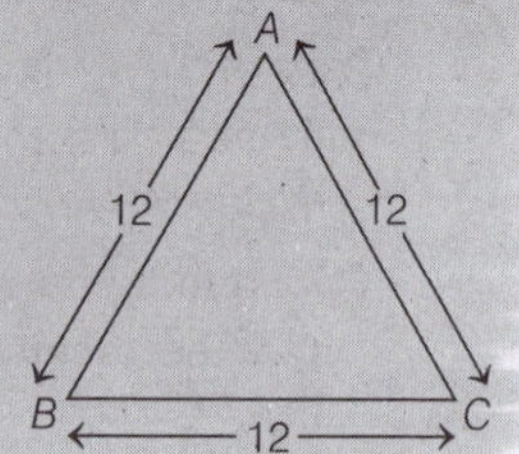

Sol. Here, $a = 12$ cm

Then, circumradius of equilateral triangle

$$= \frac{\text{Side}}{\sqrt{3}} = \frac{12}{\sqrt{3}} = \frac{12\sqrt{3}}{3} = 4\sqrt{3} \text{ cm}$$

2. Incentre अन्तःकेन्द्र

The point of intersection of the internal bisector of the angles of a triangle, is called the incentre.

In the given figure, *I* is the incentre and *r* is inradius.

त्रिभुज के आन्तरिक कोणों के समद्विभाजकों का प्रतिच्छेद बिन्दु, अन्तः केन्द्र कहलाता है।

दी गई आकृति में, *I* अन्तःकेन्द्र तथा *r* अन्तःत्रिज्या है।

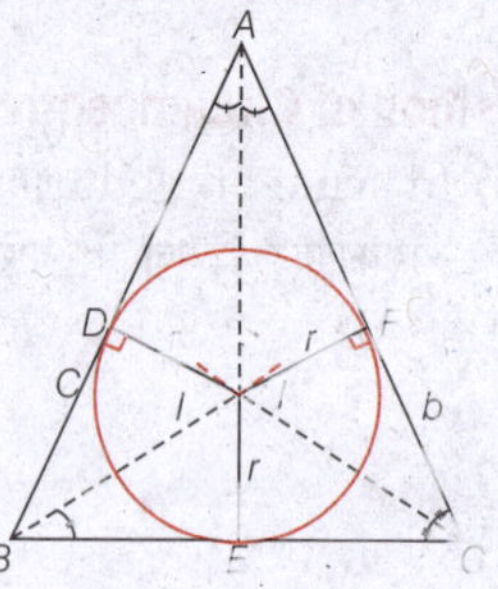

Inradius of triangles/त्रिभुज की अन्तःत्रिज्या

$$= r = \frac{\Delta}{s} = \frac{\text{Area of triangle / त्रिभुज का क्षेत्रफल}}{\text{Semi-perimeter / अर्द्ध-परिमाप}}$$

Note Inradius of equilateral/समबाहु त्रिभुज की अन्तःत्रिज्या $(r) = \frac{\text{side / भुजा}}{2\sqrt{3}}$

त्रिभुज की अन्तःत्रिज्या पर आधारित उदाहरण निम्न है

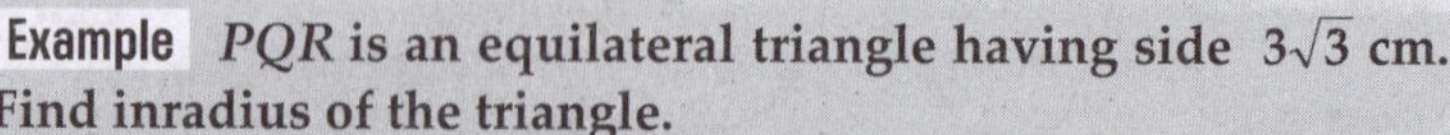

Example ***PQR* is an equilateral triangle having side $3\sqrt{3}$ cm. Find inradius of the triangle.**

PQR एक समबाहु त्रिभुज है। जिसकी भुजाएँ $3\sqrt{3}$ सेमी हैं, तो त्रिभुज की अन्तःत्रिज्या ज्ञात करें।

Sol. Inradius of equilateral triangle $= \frac{\text{Side}}{2\sqrt{3}} = \frac{3\sqrt{3}}{2\sqrt{3}} = \frac{3}{2} = 1.5$ cm

Properties of Incentre अन्तःकेन्द्र के गुण

- Incentre always lies inside the triangle.
  अन्तःकेन्द्र सदैव त्रिभुज के अन्दर स्थित होता है।
- Incentre is equidistant from all three sides of triangle.
  अन्तःकेन्द्र त्रिभुजों की तीनों भुजाओं से समदूरस्थ होता है।
- In the given figure, *I* is the incentre, then
  दी गई आकृति में, *I* अन्तःकेन्द्र है, तब

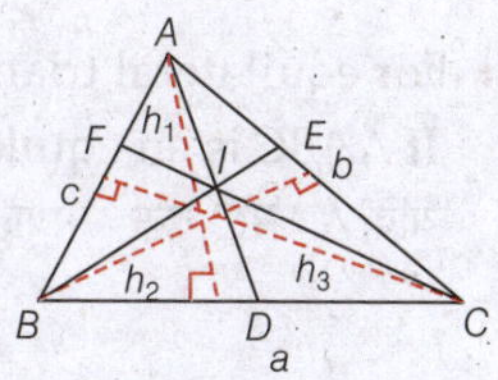

(i) $\angle BIC = 90° + \frac{\angle A}{2}$

$\angle AIC = 90° + \frac{\angle B}{2}$

$\angle AIB = 90° + \frac{\angle C}{2}$

(ii) Incentre divides each angle bisector in the ratio of length of sum of two adjacent sides and opposite side.
Here, AD, BE and CF are angle bisectors of $\angle A$, $\angle B$ and $\angle C$, respectively.
अन्तःकेन्द्र प्रत्येक कोण समद्विभाजक को दो आसन्न भुजाओं के योग की लम्बाई और विपरीत भुजा के अनुपात में विभाजित करता है।
यहाँ, AD, BE और CF, $\angle A$, $\angle B$ और $\angle C$ के कोण समद्विभाजक हैं।

$$AI : ID = b + c : a$$
$$BI : IE = a + c : b$$
$$CI : IF = a + b : c$$

(iii) If altitudes h_1, h_2, h_3 are given, then /यदि ऊँचाइयाँ h_1, h_2, h_3 दी गई हों, तब

$$\frac{1}{r} = \frac{1}{h_1} + \frac{1}{h_2} + \frac{1}{h_3}$$

- In right angle triangle/समकोण त्रिभुज में,
  If in ΔABC, $\angle B = 90°$ and I is the incentre, then
  यदि ΔABC में, $\angle B = 90°$ तथा I अन्तःकेन्द्र है, तब

  Inradius/अन्तःत्रिज्या, $r = \frac{P + B - H}{2}$

  Where/जहाँ, $P \rightarrow$ Perpendicular/लम्ब, $B \rightarrow$ Base/आधार,
  $H \rightarrow$ Hypotenuse/कर्ण

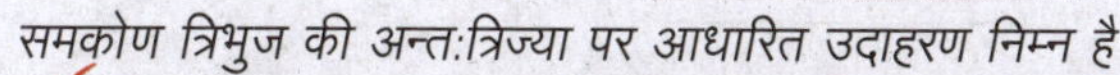

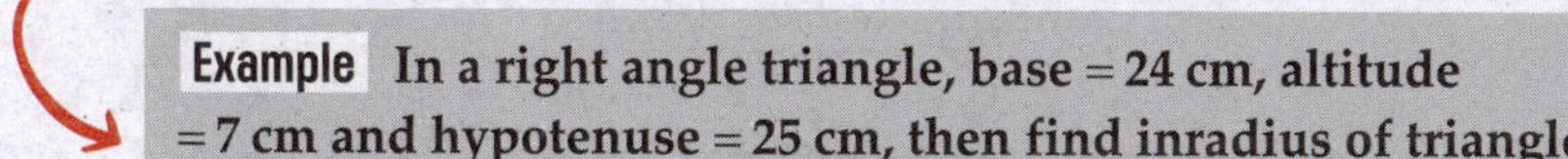

Example In a right angle triangle, base = 24 cm, altitude = 7 cm and hypotenuse = 25 cm, then find inradius of triangle.

एक समकोण त्रिभुज में आधार = 24 सेमी, ऊँचाई = 7 सेमी और कर्ण = 25 सेमी दिया है, तो त्रिभुज की अन्तःत्रिज्या ज्ञात कीजिए?

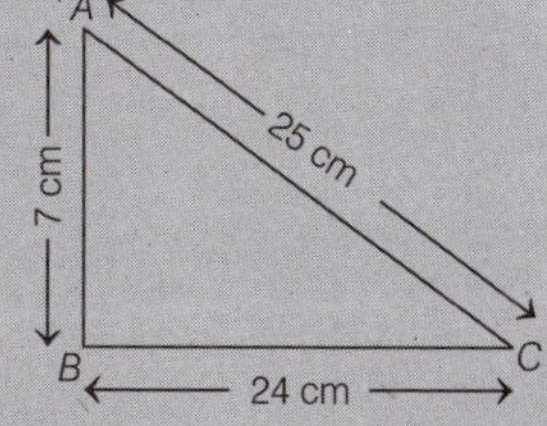

Sol. Inradius of right angled triangled

$$= \frac{P + B - H}{2}$$

$$= \frac{7 + 24 - 25}{2} = \frac{6}{2} = 3 \text{ cm}$$

- For equilateral triangle/समबाहु त्रिभुज के लिए,

  If ΔABC is an equilateral triangle and its side is 'a', then

  यदि ΔABC एक समबाहु त्रिभुज है और इसकी भुजा a है, तब

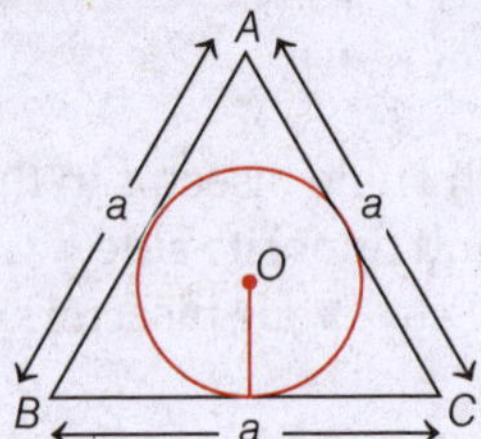

Inradius/अन्तःत्रिज्या, $r = \dfrac{a}{2\sqrt{3}}$

- If any incircle is in given form, then/यदि कोई अन्तः वृत्त दिए गए रूप में है, तब

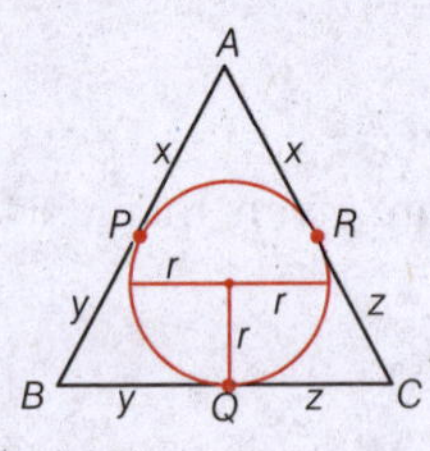

Inradius/अन्तःत्रिज्या, $r = \sqrt{\dfrac{xyz}{x+y+z}}$

Some Special Positions of Incircle अन्तः वृत्त की कुछ विशेष स्थितियाँ

(i) If $ABCD$ is a quadrilateral, then/यदि $ABCD$ एक चतुर्भुज है, तब

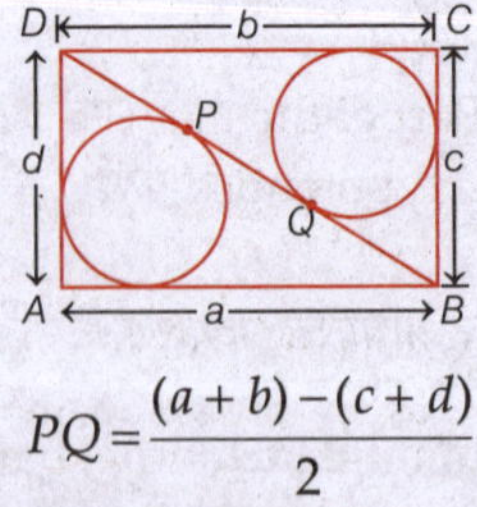

$$PQ = \frac{(a+b)-(c+d)}{2}$$

(ii) In the following figure/निम्न आकृति में,

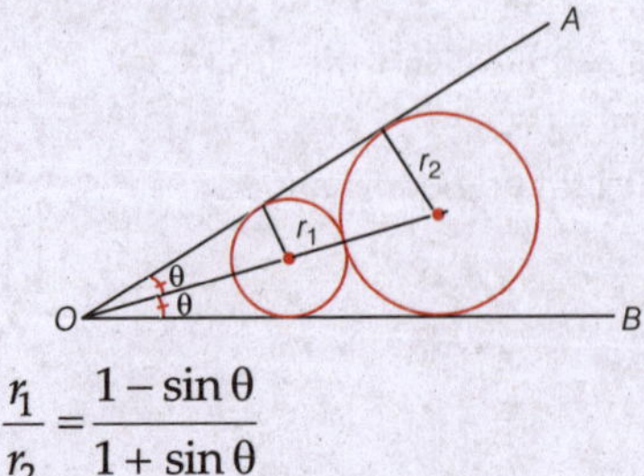

$$\frac{r_1}{r_2} = \frac{1-\sin\theta}{1+\sin\theta}$$

अन्तःवृत्त की विशेष स्थितियों पर आधारित उदाहरण निम्न है

Example **In the given figure, if $r_1 = 6$ cm and $\theta = 30°$, then find the value of r_2.**

दिए गए चित्र में, $r_1 = 6$ सेमी और $\theta = 30°$, तो r_2 का मान ज्ञात कीजिए।

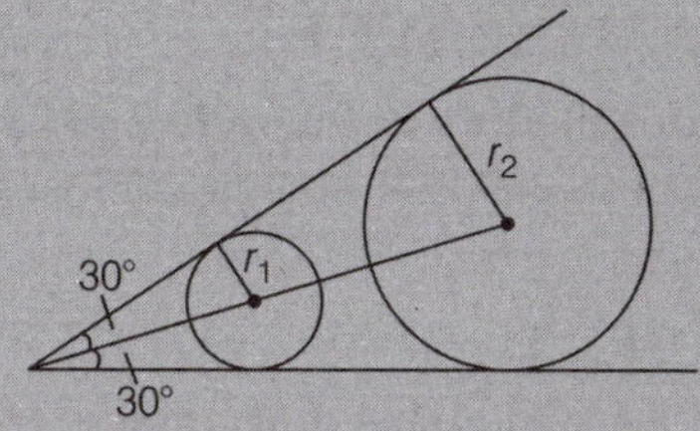

Sol. $\frac{r_1}{r_2} = \frac{1 - \sin\theta}{1 + \sin\theta}$ Here, $r_1 = 6$ cm and $\theta = 30°$

$$\frac{6}{r_2} = \frac{1 - \sin 30°}{1 + \sin 30°} \Rightarrow \frac{6}{r_2} = \frac{1 - \frac{1}{2}}{1 + \frac{1}{2}} = \frac{\frac{1}{2}}{\frac{3}{2}}$$

$$\frac{6}{r_2} = \frac{1}{3} \Rightarrow r_2 = 18 \text{ cm}$$

(iii) If I is the incentre in the following figure, then
यदि निम्न आकृति में I अन्तः केन्द्र है, तब

$$BI = \sqrt{ab} = \sqrt{2}r \quad \text{or/या} \quad r = \frac{\sqrt{ab}}{\sqrt{2}}$$

($r \rightarrow$ Inradius/अन्तः त्रिज्या)

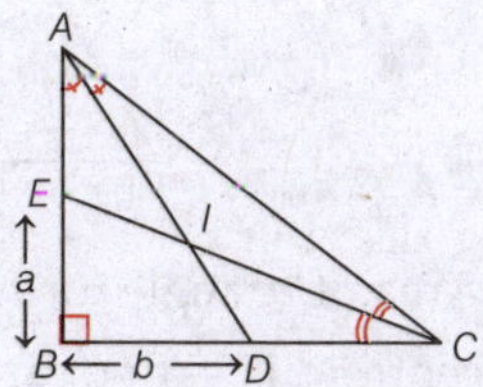

(iv) In any triangle ABC, DE is parallel to BC.
किसी ΔABC में, DE, BC के समान्तर है

$DE \parallel BC$, $DI = DB$, $IE = EC$

where I is incenter/जहाँ I अन्तःबिन्दु है।

ΔDBI and ΔCEI are isosceles triangle.

ΔBDI तथा ΔCEI समद्विबाहु त्रिभुज हैं।

$$DE = BD + EC$$

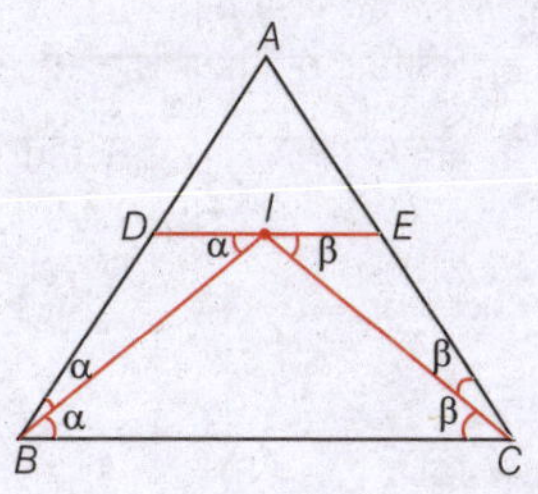

3. Centroid केन्द्रक

Median A median of a triangle is a line segment that connects a vertex to the midpoint of the opposite side bisecting that side. Every triangle has three median. Here, AD is the median.

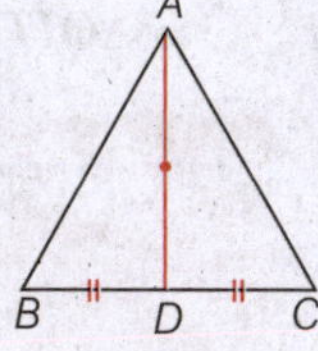

माध्यिका किसी त्रिभुज की माध्यिका एक शीर्ष के विपरीत भुजा के मध्य बिन्दु से जोड़ने वाली एक रेखाखण्ड होती है। जो उस भुजा को समद्विभाजित करती है। प्रत्येक त्रिभुज में तीन माध्यिकाएँ होती हैं। यहाँ, AD एक माध्यिका है।

The intersection point of all three medians of a triangle, is called centroid. It is denoted by G.
एक त्रिभुज की सभी तीनों माध्यिकाओं का प्रतिच्छेदन बिन्दु, केन्द्रक कहलाता है। इसे G से प्रदर्शित करते हैं।

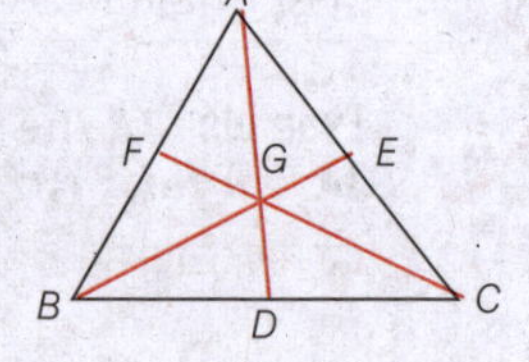

- Centroid always lies inside the triangle.
केन्द्रक सदैव त्रिभुज के अन्दर स्थित होता है।
- Centroid always divides median in the ratio 2 : 1.
केन्द्रक माध्यिका को सदैव 2 : 1 के अनुपात में विभाजित करता है।

i.e./अर्थात् $\dfrac{AG}{GD}=\dfrac{BG}{GE}=\dfrac{CG}{GF}=\dfrac{2}{1}$

Where, AD, BE and CF are medians of triangle.

जहाँ, AD, BE और CF त्रिभुज की माध्यिकाएँ हैं।
- If the area of $\Delta ABC = x$, AD, BE and CF are three medians and G is centroid, then
यदि ΔABC का क्षेत्रफल x हो, AD, BE और CF तीन माध्यिकाएँ और G केन्द्रक हो, तब

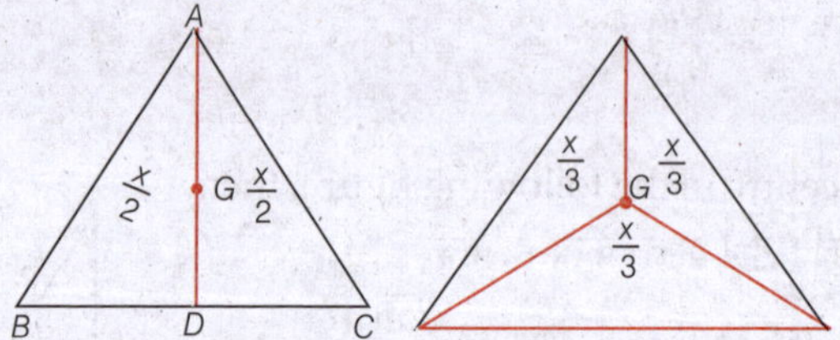

- Area of triangle formed by joining mid-points of two sides and centroid is $\dfrac{1}{12}$th of area of triangle.
दो भुजाओं के मध्य-बिन्दुओं और केन्द्रक को मिलाने से बने त्रिभुज का क्षेत्रफल, मूल त्रिभुज के क्षेत्रफल का $\dfrac{1}{12}$ भाग होता है।

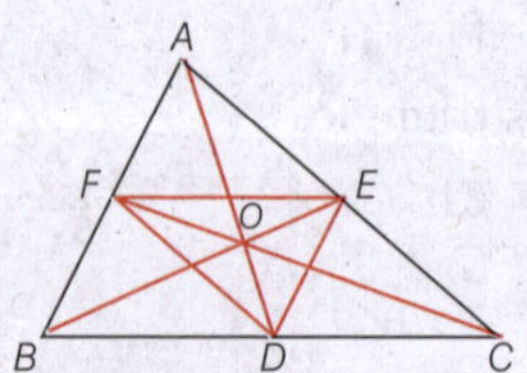

अतिसम्भावित! माध्यिकाओं से बनने वाले त्रिभुजों के क्षेत्रफल ज्ञात करने पर 1 प्रश्न SSC की परीक्षाओं में अवश्य पूछा जाता है

i.e. $\quad$ **ar ΔOFE = ar ΔOFD = arΔOED = ar = $\dfrac{1}{12}$ arΔABC**

अर्थात् **क्षेत्रफल** ΔOFE = **क्षेत्रफल** ΔOFD = **क्षेत्रफल** $\Delta OED = \dfrac{1}{12}$ **क्षेत्रफल** ΔABC

- How to find the area of the given triangle, if area of triangle made by using three medians as sides are given.

  दिए गए त्रिभुज का क्षेत्रफल कैसे ज्ञात करें, यदि तीन माध्यिकाओं को भुजाओं के रूप में उपयोग करके बनाए गए त्रिभुज का क्षेत्रफल दिया गया है।

- Area of $\Delta ABC = \frac{4}{3} \times$ Area of triangle made by using length of all 3 medians.

  ΔABC का क्षेत्रफल $= \frac{4}{3} \times$ सभी 3 माध्यिकाओं की लम्बाई का उपयोग करके बनाया गया त्रिभुज का क्षेत्रफल।

त्रिभुज की माध्यिकाओं के गुण या केन्द्रक पर आधारित उदाहरण निम्न है

Example **Length of 3 median of a triangle *ABC* is 15, 36 and 39, then find the area of triangle *ABC*.**

त्रिभुज *ABC* की 3 माध्यिकाओं की लम्बाई 15, 36 और 39 है, तो त्रिभुज *ABC* का क्षेत्रफल ज्ञात कीजिए।

Sol. Area of triangle made by 15, 36, 39 (triplet)

$$= \frac{1}{2} \times 15 \times 36 = 15 \times 18 = 270 \text{ cm}^2$$

Then area of $\Delta ABC = \frac{4}{3} \times 270 = 4 \times 90 = 360 \text{ cm}^2$

- The line segment joining the mid-points of two sides divides the line joining of vertex and the centroid in between the line in the ratio 3 : 1.

  दो भुजाओं के मध्य-बिन्दुओं को मिलाने वाला रेखाखण्ड, भुजाओं के मध्य के शीर्ष को केन्द्रक से मिलाने वाले रेखाखण्ड को 3 : 1 के अनुपात में विभाजित करता है।

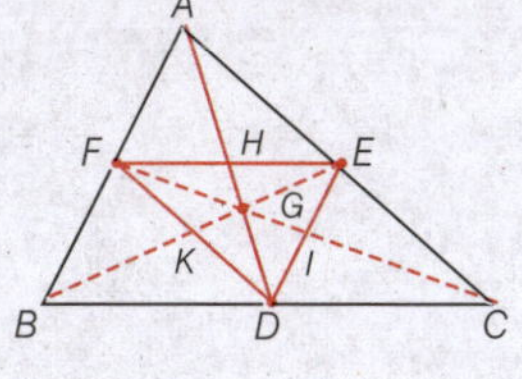

$$AH : GH = BK : GK = CI : GI = 3 : 1$$

Where, G is the centroid/जहाँ, G केन्द्रक है।

4. Orthocentre लम्बकेन्द्र

The point of intersections of the altitudes of a triangle is called the orthocentre of the triangle. It is denoted by H.

किसी त्रिभुज के शीर्ष लम्बों का प्रतिच्छेद बिन्दु, त्रिभुज का लम्बकेन्द्र कहलाता है। इसे H से प्रदर्शित करते हैं।

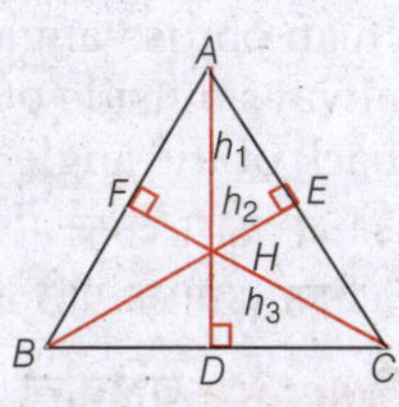

In $\Delta ABC/\Delta ABC$ में,

(i) $\angle BHC = 180° - \angle A$

(ii) $\angle CHA = 180° - \angle B$

(iii) $\angle AHB = 180° - \angle C$

(iv) $h_1 : h_2 : h_3 = \frac{1}{a} : \frac{1}{b} : \frac{1}{c}$

(v) $\Delta BFH \sim \Delta CEH$

$\therefore BH \times HE = CH \times HF$

or/या $BH \times HE = CH \times HF = AH \times HF$

(vi) $BD \times DC = AD \times DH$

$BE \times EH = CE \times EA$; $CF \times FH = AF \times FB$

त्रिभुज के लम्ब केन्द्र पर आधारित उदाहरण निम्न है

Example **In a triangle *ABC, AP, BQ* and *CR* are altitudes are drawn and they meet at point *H*. If angle $\angle A = 75°$, find $\angle BHC$.**

एक त्रिभुज *ABC* में, *AP, BQ* और *CR* शीर्षलम्ब खींचे गए हैं और वे बिन्दु *H* पर मिलते हैं। यदि $\angle A = 75°$, तो $\angle BHC$ ज्ञात करो।

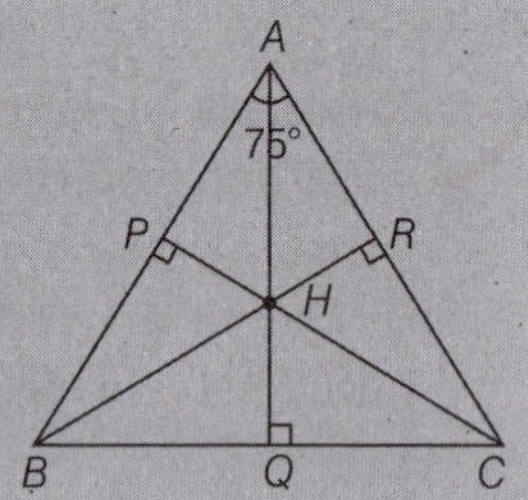

Sol. In ΔABC, $A = 75°$

Then, $\angle BHC = 180° - \angle A$

$\angle BHC = 180° - 75°$

$\angle BHC = 105°$

- In an acute-angled triangle, the orthocentre is always inside the triangle.
  किसी न्यूनकोण त्रिभुज में, लम्बकेन्द्र सदैव त्रिभुज के अन्दर होता है।

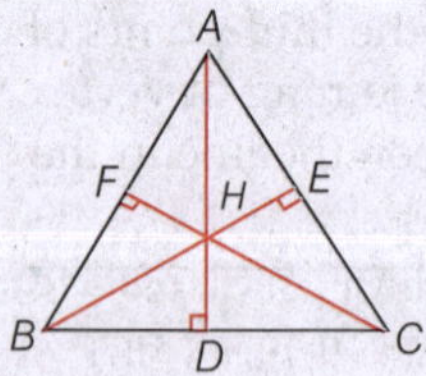

- In a right angled triangle, the orthocentre is on that vertex of the triangle at which triangle is right angled.
  किसी समकोण त्रिभुज में लम्ब केन्द्र समकोण बनाने वाले त्रिभुज के शीर्ष पर होता है।

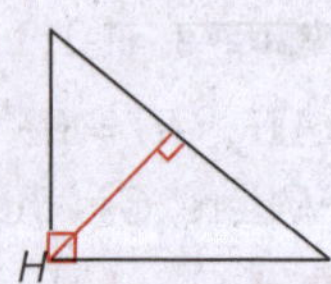

- In an obtuse-angled triangle, the orthocentre is always outside of the triangle and will be at the back of the angle which is obtuse.
  अधिक कोण त्रिभुज में लम्ब केन्द्र सदैव त्रिभुज के बाहर और अधिक कोण के पीछे की ओर होता है।

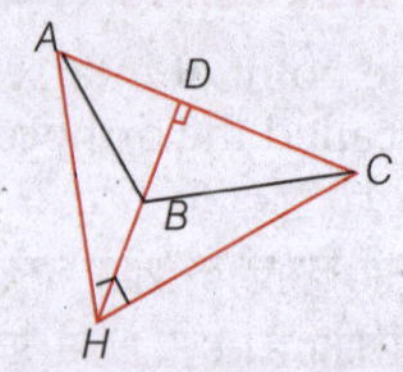

5. Excentre बहिकेन्द्र

The intersection point of internal angle bisector of one angle and bisectors of other two opposite exterior angles, is called excentre

एक कोण के आन्तरिक समद्विभाजक तथा अन्य दो बाह्य कोणों के समद्विभाजकों का प्रतिच्छेद बिन्दु, बहिकेन्द्र कहलाता है।

If ABC is a triangle, then/यदि ABC एक त्रिभुज है, तब

$$\angle BEC = 90^\circ - \frac{\angle A}{2}$$

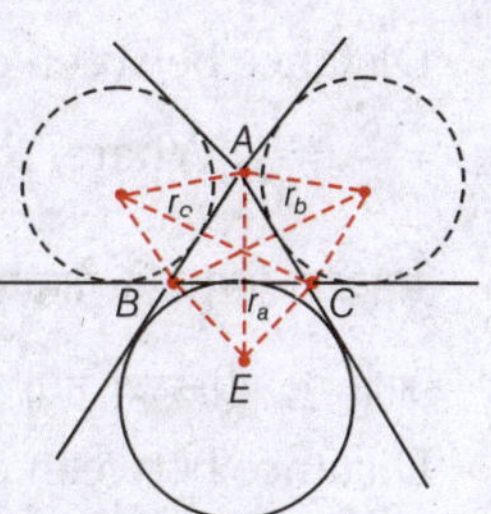

- Ex-radii/बहि-त्रिज्याएँ $r_a = \frac{\Delta}{s-a}$, $r_b = \frac{\Delta}{s-b}$, $r_c = \frac{\Delta}{s-c}$

Where, $\Delta \to$ Area of triangle, $\Delta = \sqrt{s(s-a)(s-b)(s-c)}$

and $s \to$ Semi-perimeter

जहाँ, $\Delta \to$ त्रिभुज का क्षेत्रफल तथा $s \to$ अर्द्धपरिमाप $\left(s = \frac{a+b+c}{2}\right)$

त्रिभुज के बहिकेन्द्र पर आधारित उदाहरण निम्न है

Example **In the given figure find r (inradius).**

दिए गए चित्र में r ज्ञात कीजिए।

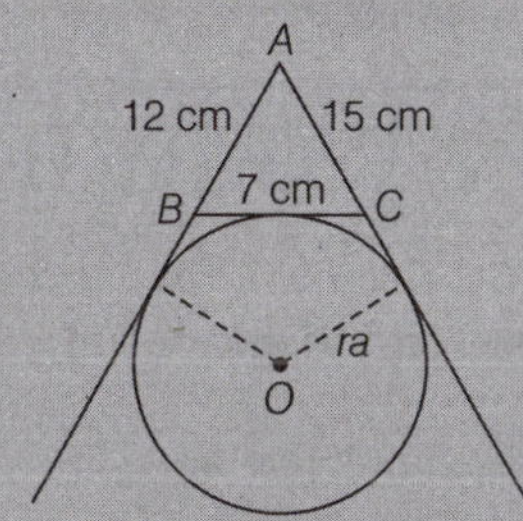

Sol. Semi-perimeter $= \frac{a+b+c}{2} = \frac{12+7+15}{2} = \frac{34}{2} = 17$ cm

Area of triangle, $\Delta = \sqrt{s(s-a)(s-b)(s-c)} = \sqrt{17 \times 5 \times 10 \times 2}$

$= \sqrt{17 \times 5 \times 5 \times 2 \times 2} = 10\sqrt{17}$ cm^2

Then, $r_a = \frac{\Delta}{s-a} = \frac{10\sqrt{17}}{17-7} = \frac{10\sqrt{17}}{10} = \sqrt{17}$ cm

Some Results Based on Centres of Triangle

त्रिभुज के केन्द्रों पर आधारित कुछ परिणाम

- Distance between orthocentre and incentre in right angle triangle, $= \sqrt{2}\, r$, where r is the inradius.

  समकोण त्रिभुज में लम्बकेन्द्र और अन्तःकेन्द्र के बीच की दूरी $= \sqrt{2}\, r$, जहाँ r अन्तःत्रिज्या है।

- Distance between orthocentre and circumcentre of right angled triangle = Median of hypotenuse $= \frac{H}{2}$

  समकोण त्रिभुज में लम्ब केन्द्र और परिकेन्द्र के बीच की दूरी

  $=$ **कर्ण की माध्यिका** $= \frac{H}{2}$

- Distance between centroid and circumcentre in right angled triangle $= \frac{R}{3} = \frac{H}{6}$. Where, R is circumradius and H is hypotenuse.

  समकोण त्रिभुज में केन्द्रक और परिकेन्द्र के बीच की दूरी $= \frac{R}{3} = \frac{H}{6}$

  जहाँ, R परित्रिज्या तथा H कर्ण है।

- Distance between orthocentre and centroid in right angled triangle $= \frac{H}{3}$, where H is hypotenuse.

  समकोण त्रिभुज में लम्ब केन्द्र और केन्द्रक के बीच की दूरी $= \frac{H}{3}$, जहाँ H कर्ण है।

- For any triangle distance between incentre and circumcentre,

$$d = \sqrt{R^2 - 2R \cdot r}$$

  Where, $R \rightarrow$ Circumradius, $r \rightarrow$ Inradius

  किसी भी प्रकार के त्रिभुज में अन्तःकेन्द्र और परिकेन्द्र के बीच की दूरी, $d = \sqrt{R^2 - 2R \cdot r}$

  जहाँ, $R \rightarrow$ परित्रिज्या, $r \rightarrow$ अन्तःत्रिज्या

त्रिभुज के केन्द्रकों के परिणाम पर आधारित उदाहरण निम्न है

Example **In a ΔABC circumradius and inradius are 11 cm and 3 cm, respectively. Find distance between circumcentre and incentre.**

ΔABC में, परित्रिज्या और अन्तःत्रिज्या क्रमशः 11 सेमी तथा 3 सेमी हैं। अन्तःकेन्द्र व परिकेन्द्र के बीच की दूरी ज्ञात कीजिए।

Sol. Circumradius of triangle $(R) = 11$ cm

Inradius of triangle $(r) = 3$ cm

Then, distance between circumcentre and incentre

$$d = \sqrt{R^2 - 2R \cdot r} \Rightarrow d = \sqrt{11^2 - 2 \times 11 \times 3}$$

$$d = \sqrt{121 - 66} \Rightarrow d = \sqrt{55} \text{ cm}$$

Thus, distance $= \sqrt{55}$ cm

- For any triangle/किसी त्रिभुज के लिए, $\left(\frac{R}{r}\right)_{min} = 2$ or/या $\frac{R}{r} \geq 2$

  Where/जहाँ, $R \rightarrow$ Circumradius/परित्रिज्या; $r \rightarrow$ Inradius/अन्तःत्रिज्या

- For equilateral triangle/समबाहु त्रिभुज के लिए,

$$\frac{\textbf{Area of Circumcircle}}{\textbf{Area of Incircle}} = \frac{4}{1}$$

$$\frac{\text{परिवृत्त का क्षेत्रफल}}{\text{अन्तः वृत्त का क्षेत्रफल}} = \frac{4}{1} \Rightarrow \frac{R}{r} = \frac{2}{1}$$

  Where/जहाँ, $R \rightarrow$ Circum radius/परित्रिज्या; $r \rightarrow$ In radius/अन्तःत्रिज्या

Theorems Based on Triangle
त्रिभुज पर आधारित प्रमेय

1. Pythagoras Theorem पाइथागोरस प्रमेय

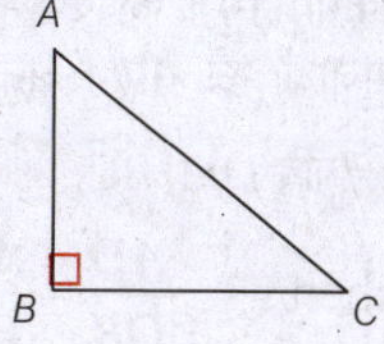

In a right angled triangle, the square of the hypotenuse is equal to the sum of the square of the other two sides.

एक समकोण त्रिभुज में, कर्ण का वर्ग शेष दो भुजाओं के वर्गों के योग के बराबर होता है।

If/यदि $\angle B = 90°$ Then/तब, $AB^2 + BC^2 = AC^2$

Pythagorean Triplets पाइथागोरियन त्रिक

A set of three integers a, b, c which satisfy Pythagoras theorem $(a^2 + b^2 = c^2)$ or are the sides of a right-angled triangle, is called Pythagorean triplets.

तीन पूर्णांकों a, b, c का समूह जो पाइथागोरस प्रमेय $(a^2 + b^2 = c^2)$ को सन्तुष्ट करे या किसी समकोण त्रिभुज की भुजाएँ हों, पाइथागोरियन त्रिक कहलाता है।

- The smallest Pythagorean triplet is (3, 4, 5) .
  सबसे छोटी पाइथागोरियन त्रिक (3, 4, 5) है।
- If (a, b, c) be the Pythagorean triplets, then (ak, bk, ck) or $\left(\frac{a}{k}, \frac{b}{k}, \frac{c}{k}\right)$ will also be Pythagorean triplets.
  यदि (a, b, c) पाइथागोरियन त्रिक हो, तब (ak, bk, ck) या $\left(\frac{a}{k}, \frac{b}{k}, \frac{c}{k}\right)$ भी पाइथागोरियन त्रिक होंगे।

  For Example (3, 4, 5), (6, 8, 10), (9, 12, 15), ... (7, 24, 25), (14, 48, 50), (8, 15, 17), (9, 40, 41), (11, 60, 61),
  $(1, 1, \sqrt{2})$, $(1, \sqrt{3}, 2)$, $(2n, n^2 - 1, n^2 + 1)$, $\left(\frac{n^2+1}{2}, \frac{n^2-1}{2}, n\right)$ and/और $(2ab, a^2 - b^2, a^2 + b^2)$.

पाइथागोरस प्रमेय पर आधारित उदाहरण निम्न है

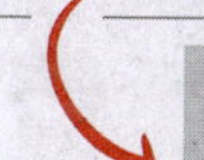

Example **In ΔABC, $AB = 6$ cm, $BC = 8$ cm, find hypotenuse of triangle.**

ΔABC में, $AB = 6$ सेमी, $BC = 8$ सेमी है त्रिभुज का कर्ण ज्ञात कीजिए।

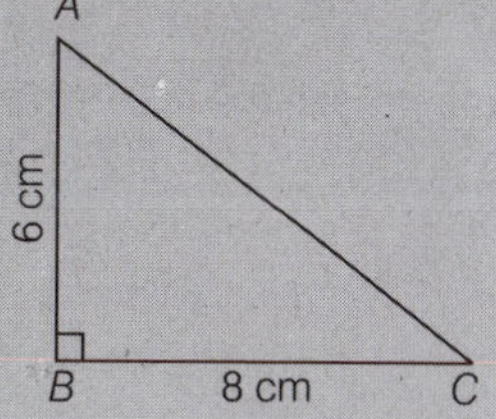

Sol. By Pythagoras theorem,

$$AB^2 + BC^2 = AC^2$$

$$\Rightarrow 6^2 + 8^2 = AC^2$$

$$100 = AC^2 \Rightarrow AC = 10 \text{ cm}$$

$\therefore$ Hypotenuse of triangle = 10 cm

2. Basic Proportionality Theorem (Thales Theorem)

आधारभूत आनुपातिका प्रमेय (थेल्स प्रमेय)

A line drawn parallel to one side of a triangle divides other two sides in the same ratio.

किसी त्रिभुज की एक भुजा के समान्तर खींची गई कोई रेखा अन्य दो भुजाओं को समान अनुपात में विभाजित करती है।

if/यदि $DE \parallel BC$,

then/तब $\dfrac{AD}{DB} = \dfrac{AE}{EC}$ here, $\dfrac{m}{n} = \dfrac{a}{b}$

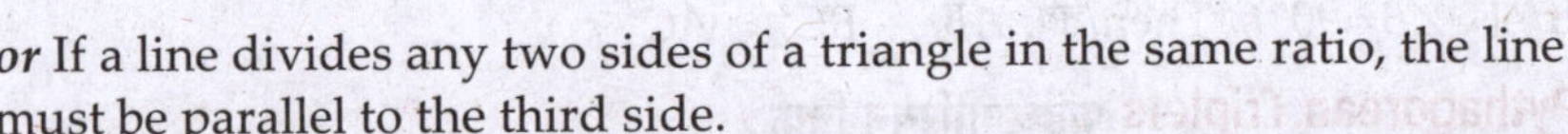

or If a line divides any two sides of a triangle in the same ratio, the line must be parallel to the third side.

या यदि कोई रेखा किसी त्रिभुज की किन्हीं दो भुजाओं को समान अनुपात में विभाजित करती है, तब वह रेखा तीसरी भुजा के समान्तर होनी चाहिए।

If/यदि $\dfrac{AD}{DB} = \dfrac{AE}{EC}$

then/तब $DE \parallel BC$

Some Results Based on Thales Theorem

थेल्स प्रमेय पर आधारित कुछ परिणाम

(i) $\dfrac{AD}{AB} = \dfrac{AE}{AC} = \dfrac{DE}{BC}$ (ii) $\Delta ADE \sim \Delta ABC$

(iii) $\dfrac{\text{ar}(\Delta ADE)}{\text{ar}(\Delta ABC)} = \left(\dfrac{AD}{AB}\right)^2 = \left(\dfrac{AE}{AC}\right)^2 = \left(\dfrac{DE}{BC}\right)^2$

थेल्स प्रमेय पर आधारित उदाहरण निम्न है

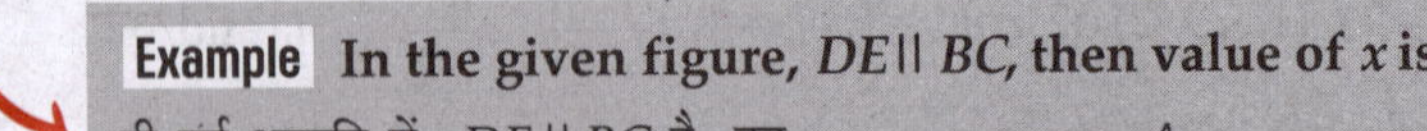

Example **In the given figure, $DE \parallel BC$, then value of x is**

दी गई आकृति में, $DE \parallel BC$ है, तब x का मान है

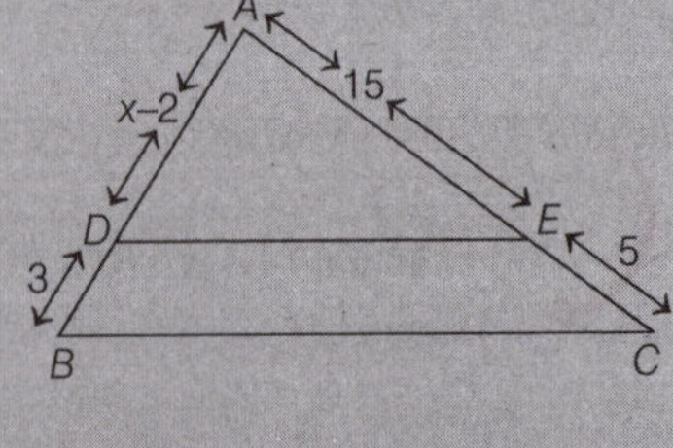

Sol. Now, from basic proportionality theorem (Thales' theorem)

$$\frac{x-2}{3} = \frac{15}{5}$$

$$x - 2 = 9$$

$$\Rightarrow \quad x = 11$$

3. Mid-Point Theorem मध्य-बिन्दु प्रमेय

The line segment joining the mid-points of any two sides of a triangle is parallel to the third side and is half of the third side.

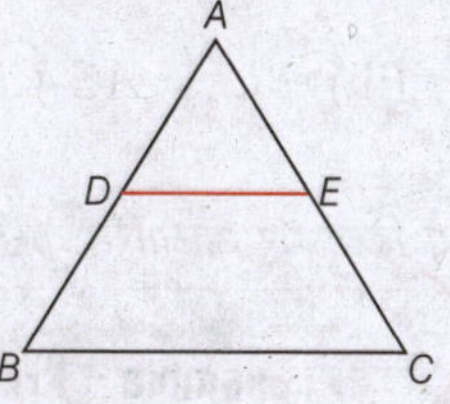

किसी त्रिभुज की किन्हीं दो भुजाओं के मध्य बिन्दुओं को मिलाने वाला रेखाखण्ड तीसरी भुजा के समान्तर और तीसरी भुजा का आधा होता है।

$$DE \parallel BC \text{ and/और } DE = \frac{BC}{2}$$

Some Results Based on Mid-Point Theorem
मध्य-बिन्दु प्रमेय पर आधारित कुछ परिणाम

(i) $\Delta ADE \sim \Delta ABC$ (ii) $\frac{\text{ar}(\Delta ADE)}{\text{ar}(\Delta ABC)} = \frac{1}{4}$ (iii) $\frac{\text{ar}(\Delta ADE)}{\text{ar}(\square BDEC)} = \frac{1}{3}$

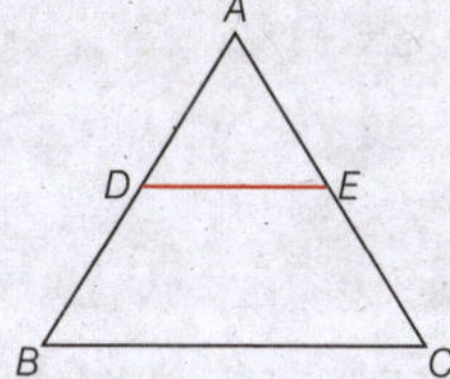

(ii) If DE is not parallel to BC and D and E are also not mid-points of AB and AC, then

यदि DE, BC के समान्तर नहीं है और D तथा E भी AB और AC के मध्य-बिन्दु नहीं हैं, तब

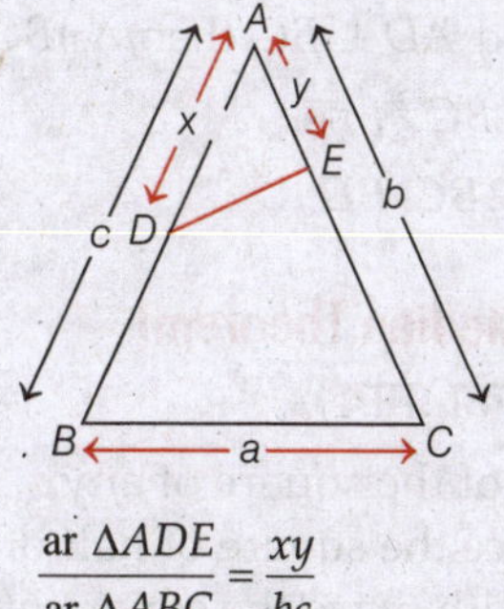

$$\frac{\text{ar } \Delta ADE}{\text{ar } \Delta ABC} = \frac{xy}{bc}$$

4. In a right angle triangle ABC, $\angle B = 90°$ and AC is hypotenuse. The perpendicular BD is dropped on hypotenuse AC from right angle vertex B, then

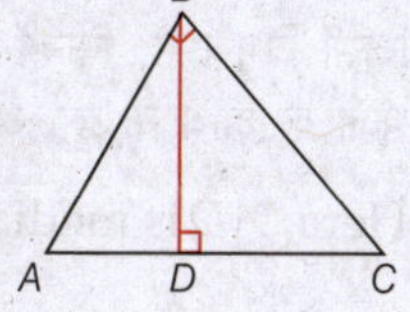

समकोण ΔABC में, $\angle B = 90°$ और AC कर्ण है। समकोण के शीर्ष B से BD, कर्ण AC पर लम्ब खींचा गया है, तब

$$\Delta ABC \sim \Delta ADB \sim \Delta BDC$$

(i) $BD = \dfrac{AB \times BC}{AC}$ (ii) $AD = \dfrac{AB^2}{AC}$

(iii) $BC^2 = AC \cdot CD$ (iv) $\dfrac{1}{BD^2} = \dfrac{1}{AB^2} + \dfrac{1}{BC^2}$

मध्य बिन्दु पर आधारित कुछ परिणाम (समकोण त्रिभुज) पर आधारित उदाहरण निम्न है

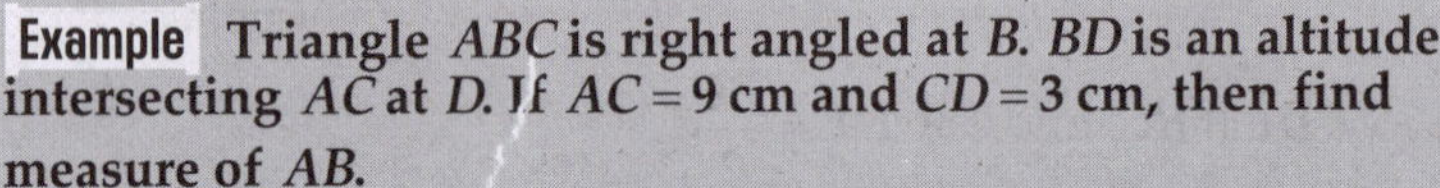

Example **Triangle ABC is right angled at B. BD is an altitude intersecting AC at D. If $AC = 9$ cm and $CD = 3$ cm, then find measure of AB.**

ΔABC, B पर समकोण है, BD एक शीर्षलम्ब है, जो AC को D पर प्रतिच्छेद करता है। यदि $AC = 9$ सेमी तथा $CD = 3$ सेमी है, तब AB की लम्बाई ज्ञात कीजिए। *SSC CGL 2021*

Sol. $AC = 9$ cm; $AD = 6$ cm; $CD = 3$ cm

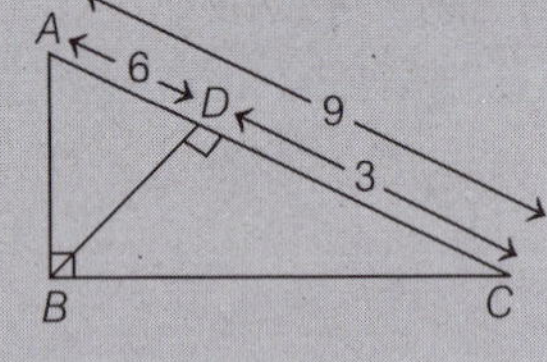

By formula,

$$AD = \frac{AB^2}{AC} \Rightarrow 6 = \frac{AB^2}{9}$$

$$\Rightarrow \quad AB^2 = 6 \times 9 = 54$$

$$\Rightarrow \quad AB = \sqrt{54} = 3\sqrt{6}$$

5. In ΔABC, if $\angle B > 90°$ and AD is perpendicular dropped on BC, then

ΔABC में, यदि $\angle B > 90°$ और AD, BC पर लम्ब है, तब

$$AC^2 = AB^2 + BC^2 + 2BC \cdot BD$$

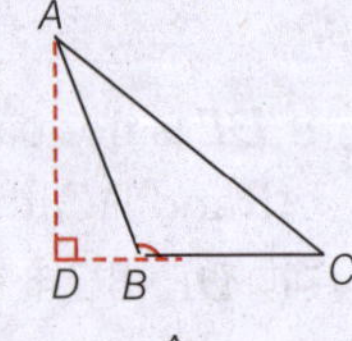

6. In ΔABC, if $\angle B < 90°$ and $AD \perp BC$, then/ΔABC में, यदि $\angle B < 90°$ और $AD \perp BC$ है, तब

$$AC^2 = AB^2 + BC^2 - 2BC \cdot BD$$

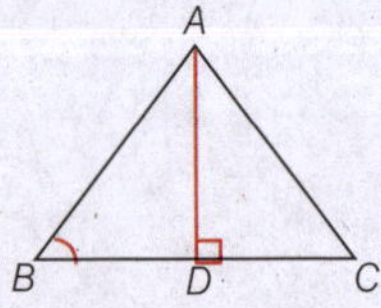

7. Apollonius Theorem (or Median Theorem)

अपोलोनियस प्रमेय (माध्यिका प्रमेय)

In any triangle, the sum of the square of any two sides is equal to twice the square of half of the third side together with twice the square of the median which bisect the third side.

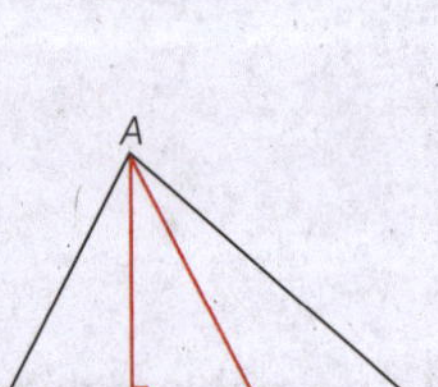

किसी त्रिभुज में, किन्हीं दो भुजाओं के वर्गों का योग, तीसरी भुजा के आधे के वर्ग और इसी भुजा की माध्यिका के वर्ग के योग का दोगुना होता है।

Here, AD is median/यहाँ, AD माध्यिका है।

$$\therefore \quad AB^2 + AC^2 = 2\left[AD^2 + \left(\frac{1}{2}BC\right)^2\right]$$

माध्यिका प्रमेय पर आधारित उदाहरण निम्न है

Example **In ΔABC $AB = 6$ cm, $AC = 8$ cm and $BC = 9$ cm, find the length of median AD.**

ΔABC में $AB = 6$ सेमी, $AC = 8$ सेमी और $BC = 9$ सेमी, माध्यिका AD की लम्बाई ज्ञात कीजिए।

Sol.

$$AB^2 + AC^2 = 2\left[AD^2 + \left(\frac{BC}{2}\right)^2\right]$$

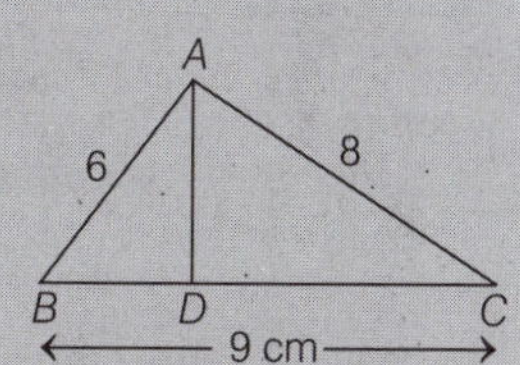

$$6^2 + 8^2 = 2\left[AD^2 + \left(\frac{9}{2}\right)^2\right]$$

$$100 = 2\left[AD^2 + \frac{81}{4}\right]$$

$$AD^2 = 50 - \frac{81}{4} = \frac{119}{4}$$

$$AD = \sqrt{\frac{119}{4}} = \frac{\sqrt{119}}{2} \text{ cm}$$

8. In a ΔABC, three times the sum of the squares of the sides of a triangle is equal to four times the sum of the squares of the medians of the triangle.

एक त्रिभुज ABC में, तीनों भुजाओं के वर्गों के योग का तीन गुना, तीनों माध्यिकाओं के वर्गों के योग के चार गुने के बराबर होता है।

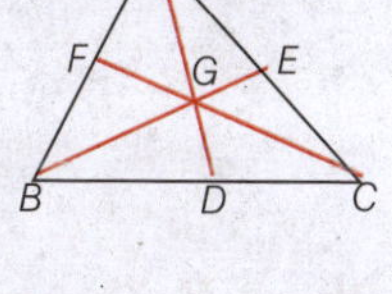

Here, AD, BE and CF are medians, then

यहाँ, AD, BE और CF माध्यिकाएँ हैं, तब

$$3(AB^2 + BC^2 + AC^2) = 4(AD^2 + BE^2 + CF^2)$$

9. The sum of any two sides of a triangle is greater than twice the median drawn to the third side and the sum of all three sides is always greater than sum of all three medians.

त्रिभुज की किन्हीं दो भुजाओं का योगफल तीसरी भुजा पर खींची गई माध्यिका के दोगुने से अधिक होता है और सभी तीनों भुजाओं का योगफल सदैव सभी तीनों माध्यिकाओं से अधिक होता है।

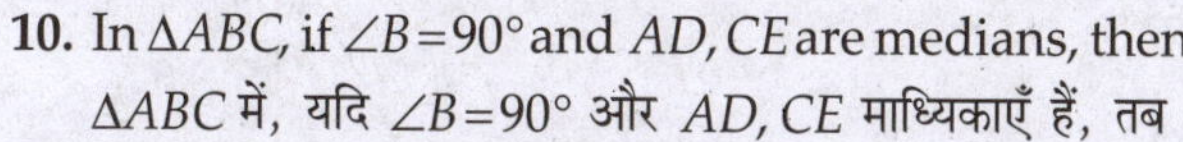
$AB + AC > 2AD$; $AB + BC > 2BE$; $AC + BC > 2CF$

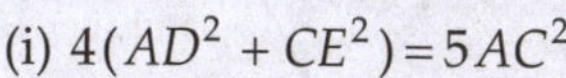
and/और $AB + BC + CA > AD + BE + CF$

10. In ΔABC, if $\angle B = 90°$ and AD, CE are medians, then

ΔABC में, यदि $\angle B = 90°$ और AD, CE माध्यिकाएँ हैं, तब

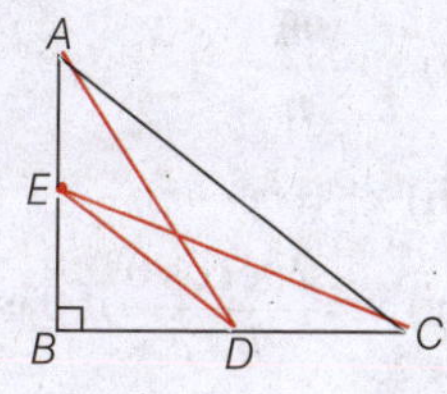

(i) $4(AD^2 + CE^2) = 5AC^2$

(ii) $AD^2 + CE^2 = 5DE^2$

(iii) $AD^2 + CE^2 = AC^2 + ED^2$

माध्यिकाओं के गुणों पर आधारित उदाहरण निम्न है

Example **Points M and N are on the sides PQ and QR respectively of a triangle PQR right angled at Q. If $PN = 9$ cm, $MR = 7$ cm and $MN = 3$ cm, then find the length of PR (in cm).**

बिन्दु M तथा N, त्रिभुज ΔPQR की क्रमशः भुजाएँ PQ तथा QR पर हैं तथा ΔPQR में $\angle Q$ समकोण है। यदि $PN = 9$ सेमी, $MR = 7$ सेमी तथा $MN = 3$ सेमी है, तब PR की लम्बाई (सेमी में) ज्ञात कीजिए। ***SSC CGL 2021***

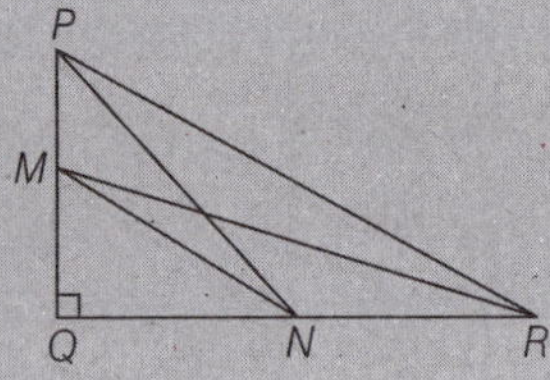

Sol. By formula,

$$(MN)^2 + (PR)^2 = (PN)^2 + (MR)^2$$

$$3^2 = (PR)^2 = 9^2 + 7^2$$

$$\Rightarrow \quad (PR)^2 = 81 + 49 - 9$$

$$\Rightarrow \quad PR^2 = 121 \Rightarrow PR = 11\,\text{cm}$$

11. Stewart's Theorem स्टीवर्ट की प्रमेय

In ΔABC / ΔABC में,

$$a^2 n + b^2 m = x^2 c + mnc$$

if/यदि $a = b$

Then/तब, $x^2 = a^2 - mn$

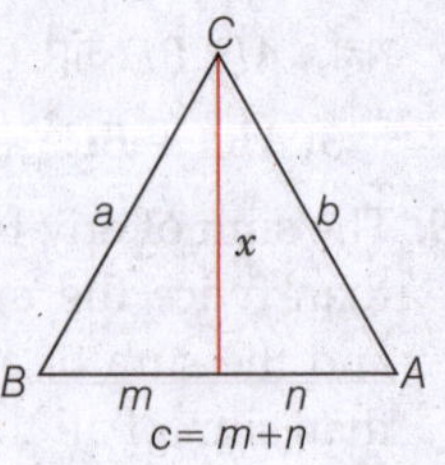

12. Interior Angle Bisector Theorem

आन्तरिक कोण द्विभाजक प्रमेय

In ΔABC, AD is bisector of angle A, then

ΔABC में, AD कोण A का समद्विभाजक है, तब

(i) $\dfrac{AB}{AC} = \dfrac{BD}{DC}$

(ii) $\dfrac{c}{b} = \dfrac{m}{n}$

(iii) $x^2 = bc - mn$

(iv) $x = \dfrac{2bc\cos\theta}{b+c}$ if θ is given./यदि θ दिया है।

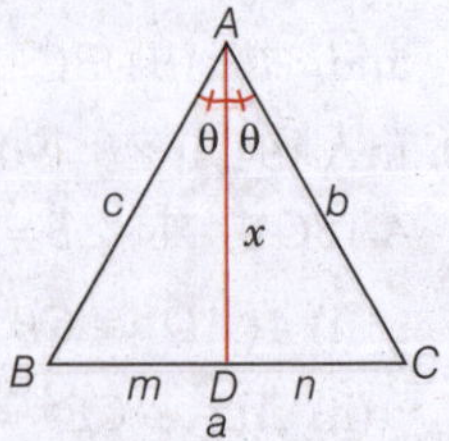

आन्तरिक कोण द्विभाजक प्रमेय पर आधारित उदाहरण निम्न है

Example **The bisector of $\angle QPR$ of $\angle PQR$ meets the side QR at S. If $PQ = 12$ cm, $PR = 15$ cm and $QR = 18$ cm, then length of SR.**

$\angle QPR$ और $\angle PQR$ का समद्विभाजक भुजा QR को S पर मिलता है। यदि $PQ = 12$ सेमी, $PR = 15$ सेमी और $QR = 18$ सेमी, तो SR की लम्बाई है

RRB NTPC CBT II 2022

Sol. Let $SR = x$ cm, then $QS = 18 - x$

By interior angle bisector theorem

$$\Rightarrow \frac{PQ}{PR} = \frac{QS}{SR}$$

$$\Rightarrow \frac{12}{15} = \frac{18-x}{x}$$

$$\Rightarrow \frac{4}{5} = \frac{18-x}{x}$$

$$\Rightarrow 4x = 90 - 5x$$

$$\Rightarrow 9x = 90$$

$$\Rightarrow x = 10 \text{ cm}$$

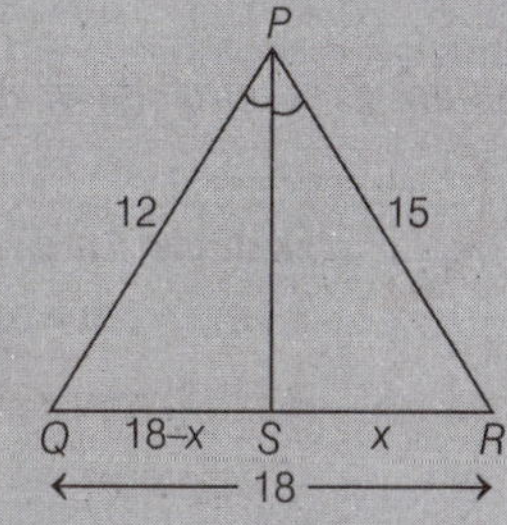

Hence, length of $SR = 10$ cm.

13. If we make an other right angle triangle in a right angle triangle, then big and small right angle triangle are always similar.

यदि हम एक समकोण त्रिभुज में एक अन्य समकोण त्रिभुज बनाते हैं, तो बड़ा और छोटा समकोण त्रिभुज सदैव समरूप होते हैं।

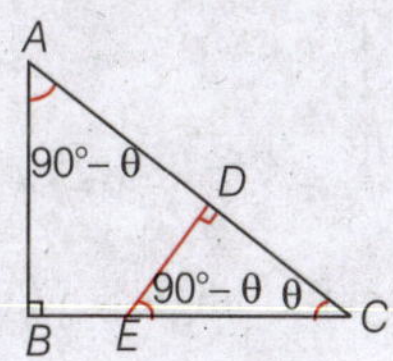

Here/यहाँ, $\Delta ABC \sim \Delta EDC$

14. Here/यहाँ, $AB \parallel CD$

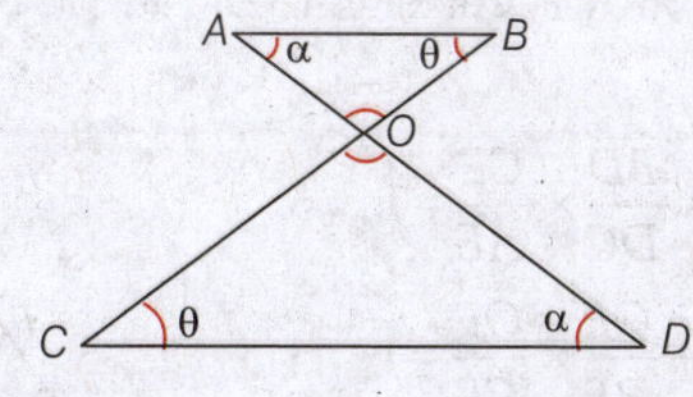

$\Delta AOB \sim \Delta DOC$

15. If P is a point inside an equilateral triangle ABC, then the sum of altitudes from the point P to the sides AB, BC and AC is equal to the median of the ΔABC.

यदि किसी समबाहु ΔABC के अन्दर P कोई बिन्दु है, तो P से भुजा AB, BC और AC पर डाले गए लम्बों का योगफल ΔABC की माध्यिका के बराबर होता है।

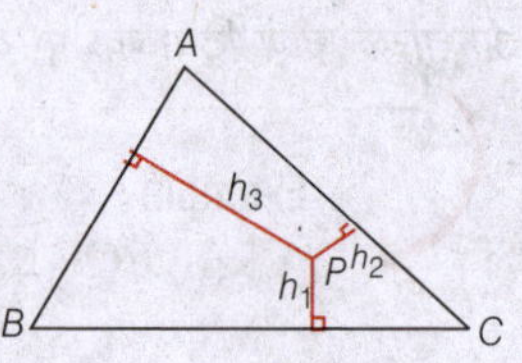

$$h_1 + h_2 + h_3 = \frac{\sqrt{3}\,a}{2} = \text{Height of triangle/त्रिभुज की ऊँचाई}$$

Where/जहाँ, $a \rightarrow$ Side of equilateral triangle/समबाहु त्रिभुज की भुजा

समबाहु त्रिभुज में शीर्षलम्बों के गुण पर आधारित उदाहरण निम्न है

Example **In given equilateral triangle, length of altitude are 2 cm, 3 cm and 4 cm, then find the side of equilateral triangle.**

दिए गए समबाहु त्रिभुज में शीर्षलम्बों की लम्बाई 2 सेमी, 3 सेमी तथा 4 सेमी है। समबाहु त्रिभुज की भुजा ज्ञात कीजिए।

Sol. Let side of triangle be a cm.

As we know,

$$h_1 + h_2 + h_3 = \frac{\sqrt{3}}{2} \times \text{side of equilateral triangle}$$

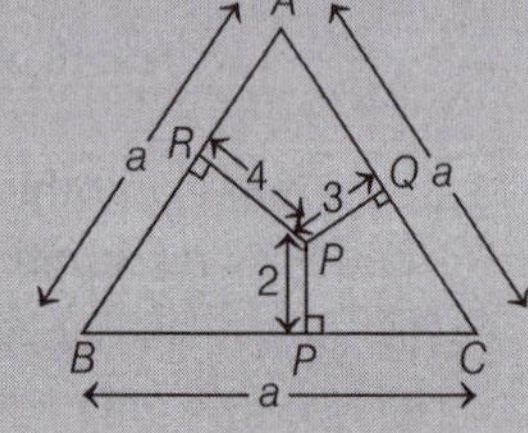

$$2 + 3 + 4 = \frac{\sqrt{3}}{2} \times a$$

$$\Rightarrow \quad 9 = \frac{\sqrt{3}}{2} a$$

$$\therefore \quad a = \frac{18}{\sqrt{3}} = \frac{18\sqrt{3}}{3} = 6\sqrt{3}$$

Thus, side of equilateral triangle $= 6\sqrt{3}$

16. Ceva's Theorem In ΔABC, AD, BE and CF are the Cevians which passes through a common point O.

ΔABC में AD, BE और CF ऐसी सेवियन रेखाएँ हैं, जो एक सामान्य बिन्दु O से होकर गुजरते हैं।

$$\frac{AF}{FB} \times \frac{BD}{DC} \times \frac{CE}{AE} = 1$$

$$\frac{OD}{AD} + \frac{OE}{BE} \times \frac{OF}{CF} = 1$$

$$\frac{AO}{OD} + \frac{BO}{OE} + \frac{CO}{OF} = 2$$

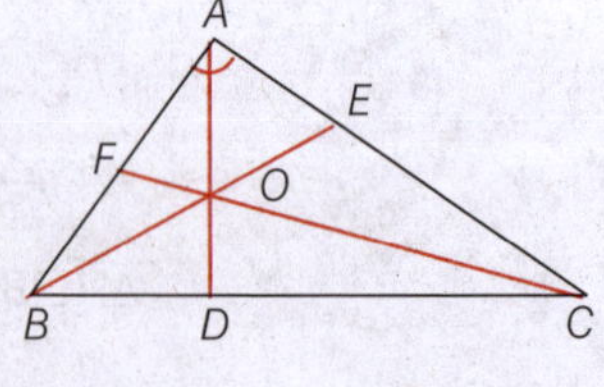

17. Ladder Theorem/लैडर प्रमेय

If/यदि $AB || XY || CD$,

Then/तब,

$$X \cdot Y = \frac{AB \cdot CD}{AB + CD}$$

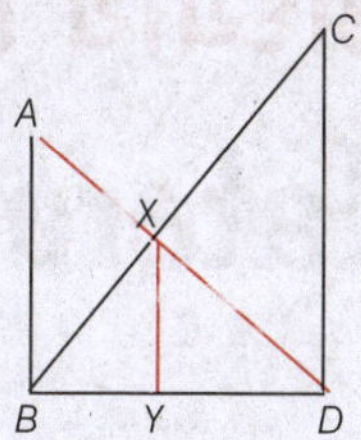

एक्सपर्ट टिप्स

लैडर प्रमेय पर आधारित 1 प्रश्न SSC की परीक्षाओं में अवश्य आता है। प्रमेय की सहायता से प्रश्न को हल करने में समय की बचत होती है।

लैडर प्रमेय पर आधारित उदाहरण निम्न है

Example **Three ladders AB, XY and CD are standing vertically parallel to each other and $AB || XY || CD$. If $AB = 8$ cm and $CD = 12$ cm, then find length of XY**

तीन सीढ़ियाँ AB, XY और CD भूमि पर सीधी खड़ी हैं ($AB || XY || CD$)। यदि $AB = 8$ सेमी और $CD = 12$ सेमी है, तो XY की लम्बाई ज्ञात करें।

Sol. By Ladder Theorem,

$$XY = \frac{AB \cdot CD}{AB + CD}$$

$$\Rightarrow \quad XY = \frac{8 \times 12}{8 + 12}$$

$$\Rightarrow \quad XY = \frac{8 \times 12}{20}$$

$$\Rightarrow \quad XY = 4.8 \text{ cm}$$

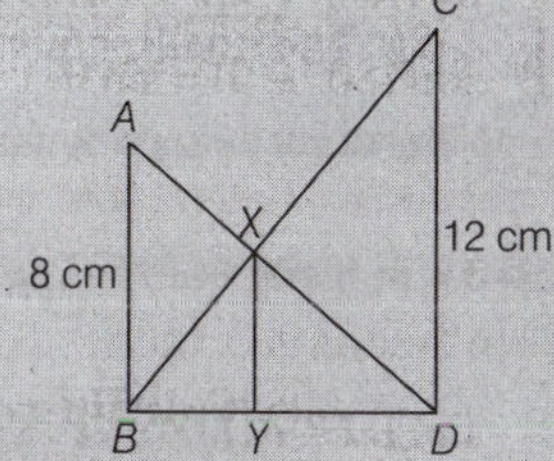

अध्याय 03

Quadrilateral चतुर्भुज

A figure enclosed by four line segments, is called a quadrilateral.
चार रेखाखण्डों द्वारा घिरी आकृति, **चतुर्भुज** कहलाती है।

Some Points Related to Quadrilateral

चतुर्भुज से सम्बन्धित कुछ बिन्दु

(i) $\angle A + \angle B + \angle C + \angle D = 360°$

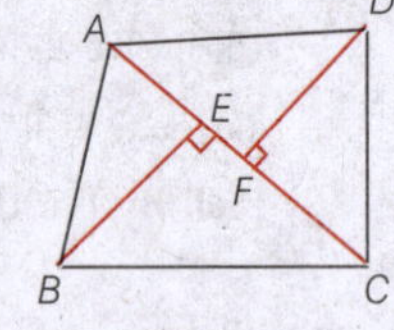

(ii) Area of quadrilateral

$= \frac{1}{2} \times$ Diagonal $\times$ Sum of perpendicular drop on it

चतुर्भुज का क्षेत्रफल

$= \frac{1}{2} \times$ विकर्ण $\times$ विकर्ण पर डाले गए लम्बों का योग

(iii) If diagonals of the quadrilateral intersect each other at 90°, then

यदि किसी चतुर्भुज के विकर्ण एक -दूसरे को 90° पर प्रतिच्छेद करें, तब

$$AB^2 + CD^2 = BC^2 + AD^2$$

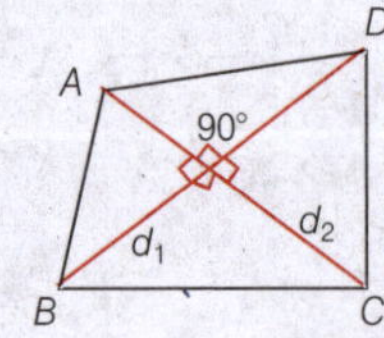

(iv) If A, B, C and D are the mid-points of sides SR, QR, PQ and PS respectively, then $ABCD$ is a parallelogram.

यदि A, B, C तथा D क्रमशः भुजाएँ SR, QR, PQ तथा PS का मध्य-बिन्दु है, तब $ABCD$ एक समान्तर चतुर्भुज है।

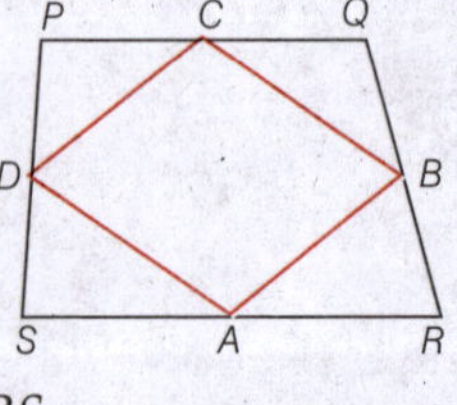

Area of $\square ABCD = \frac{1}{2} \times$ Area of quadrilateral $PQRS$

चतुर्भुज $ABCD$ का क्षेत्रफल $= \frac{1}{2} \times$ चतुर्भुज $PQRS$ का क्षेत्रफल

(v) $ABCD$ is a quadrilateral and BO and CO are angle bisector of $\angle B$ and $\angle C$, then

$ABCD$ एक चतुर्भुज है और BO तथा CO, $\angle B$ और $\angle C$ के समद्विभाजक हैं, तब

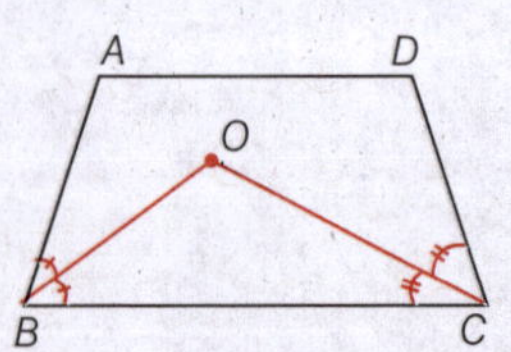

$$\angle BOC = \frac{(\angle A + \angle D)}{2}$$

(vi) If a circle touches all four sides of a quadrilateral, then
यदि एक वृत्त, चतुर्भुज की चारों भुजाओं को स्पर्श करे, तब

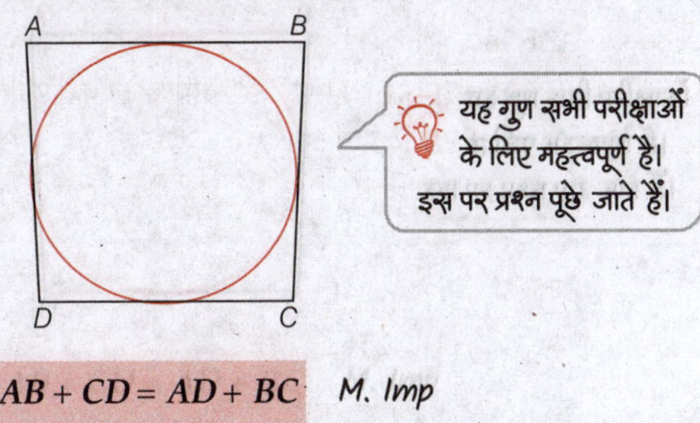

$AB + CD = AD + BC$ M. Imp

चतुर्भुज पर आधारित उदाहरण निम्नलिखित हैं

Example **In given figure $ABCD$ is a quadrilateral $AC = 12$ cm, $DP = 4$ cm and $FB = 5$ cm. Find area of quadrilateral.**

दी गई आकृति में $ABCD$ एक चतुर्भुज है जिसमें $AC = 12$ सेमी, $DP = 4$ सेमी तथा $FB = 5$ सेमी है। चतुर्भुज का क्षेत्रफल ज्ञात कीजिए।

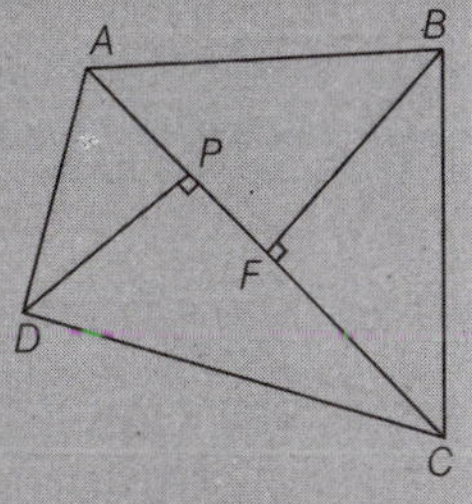

Sol. Area of quadrilateral

$$= \frac{1}{2} \times \text{Diagonal} \times \text{Sum of perpendicular drop on it}$$

$$= \frac{1}{2} \times 12 \times (4 + 5) = \frac{1}{2} \times 12 \times 9$$

$$= 6 \times 9 = 54 \text{ cm}^2$$

Example **In given figure $AB = 10$ cm, $CD = 12$ cm, $AD = 8$ cm, find BC.**

दिए गए चित्र में, $AB = 10$ सेमी, $CD = 12$ सेमी, $AD = 8$ सेमी है, तब BC ज्ञात कीजिए।

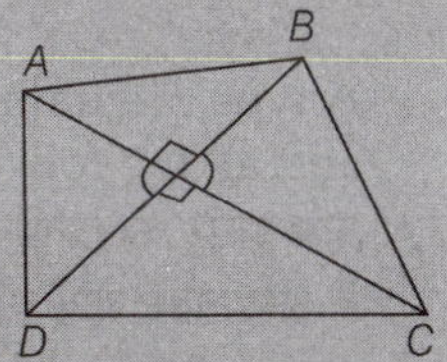

Sol. $\because$ Diagonals intersect each other at 90°.

$\therefore \quad AB^2 + CD^2 = AD^2 + BC^2$

$\Rightarrow \quad (10)^2 + 12^2 = 8^2 + BC^2$

$\Rightarrow \quad 100 + 144 = 64 + BC^2$

$\Rightarrow \quad BC^2 = 180$

$\Rightarrow \quad BC = 6\sqrt{5}$ cm

Example **In a quadrilateral *ABCD*, the bisector of $\angle C$ and $\angle D$ meet at point *E*. If $\angle CED = 57°$ and $\angle A = 47°$, then measure of $\angle B$ is**

एक चतुर्भुज *ABCD* में, $\angle C$ और $\angle D$ का समद्विभाजक बिन्दु *E* पर मिलता है। यदि $\angle CED = 57°$ तथा $\angle A = 47°$ है, तब $\angle B$ की माप है *SSC CGL 2022*

Sol. $\angle CED = \frac{1}{2} \times (\angle A + \angle B)$

$\Rightarrow \quad 57° = \frac{1}{2} \times (47° + \angle B)$

$\Rightarrow \quad 114° = 47° + \angle B$

$\Rightarrow \quad \angle B = 114° - 47° = 67°$

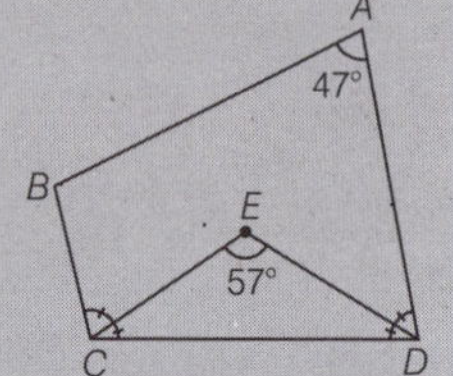

Example **A circle touches all four sides of a quadrilateral *ABCD*. If $AB = 18$ cm, $BC = 21$ cm and $AD = 15$ cm, then length of *CD* is**

एक वृत्त, चतुर्भुज *ABCD* की चारों भुजाओं को स्पर्श करता है। यदि $AB = 18$ सेमी, $BC = 21$ सेमी और $AD = 15$ सेमी है, तब *CD* की लम्बाई है

SSC CGL (TIER II) 2023

Sol. As, we know,

$AB + CD = AD + BC$

$\Rightarrow \quad 18 + CD = 15 + 21$

$\Rightarrow \quad 18 + CD = 36$

$\Rightarrow \quad CD = 18$ cm

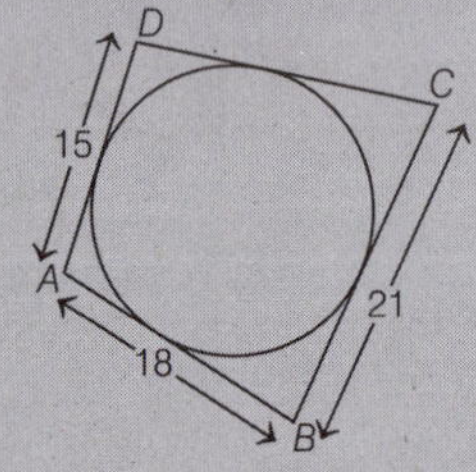

- A type of quadrilateral whose opposite sides are equal and parallel and all the four vertex angles are equal to 90°.
  एक प्रकार का चतुर्भुज जिसकी विपरीत (आमने-सामने) की भुजाएँ बराबर और समान्तर होती हैं और चारों शीर्ष कोण 90° के बराबर होते हैं।

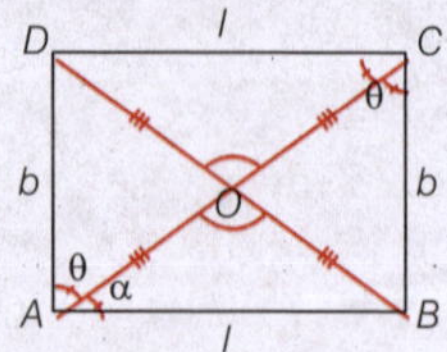

Some Points Related to Rectangle आयत से सम्बन्धित कुछ बिन्दु

(i) If $AB = CD = l$ and $BC = AD = b$, then
यदि $AB = CD = l$ और $BC = AD = b$, तब

Area of $ABCD$ = Length × Breadth = $l \times b$
$ABCD$ का क्षेत्रफल = लम्बाई × चौड़ाई = $l \times b$
Perimeter of $ABCD$ = 2 (Length + Breadth) = $2(l + b)$
$ABCD$ का परिमाप = 2 (लम्बाई + चौड़ाई) = $2(l + b)$
Diagonal/विकर्ण, $AC = BD = \sqrt{l^2 + b^2}$

(ii) The diagonal does not bisect vertex angle.
विकर्ण, शीर्ष कोण को समद्विभाजित नहीं करता है।

(iii) $\Delta ABC \cong \Delta CDA$ and/और $\Delta DAB \cong \Delta BCD$

(iv) Diagonals are equal and bisect each other but not at 90°.
विकर्ण बराबर तथा एक-दूसरे को समद्विभाजित करते हैं, परन्तु 90° पर नहीं।

$$AC = BD$$
$$AO = OC = BO = DO = \frac{AC}{2} = \frac{BD}{2}$$

(v) $\Delta DOC \cong \Delta BOA$ and/और $\Delta AOD \cong \Delta COB$
BO is median of ΔABC./ BO, ΔABC की माध्यिका है।

Then/तब, ar/क्षेत्रफल ΔAOB = ar/क्षेत्रफल ΔBOC
Similarly, ar/क्षेत्रफल ΔCOD = ar/क्षेत्रफल ΔAOD

(vi) In all the rectangles of perimeter, the square will have the maximum area.
परिमाप वाले सभी आयतों में से वर्ग का क्षेत्रफल अधिकतम होता है।

(vii) Bisector of the four angles enclose a square.
चार कोणों के समद्विभाजक एक वर्ग बनाते हैं।

(viii) When the rectangle is inscribed in a circle, it will have the maximum area when it is a square.
वृत्त के अन्दर बने आयत का क्षेत्रफल अधिकतम होगा, जब यह वर्ग होता है।

(ix) $ABCD$ is a rectangle or a square. P is a point inside it and W, X, Y and Z are areas, then
$ABCD$ एक आयत या वर्ग है।
P इसके अन्दर एक बिन्दु इस प्रकार है, कि W, X, Y और Z क्षेत्रफल हैं, तब

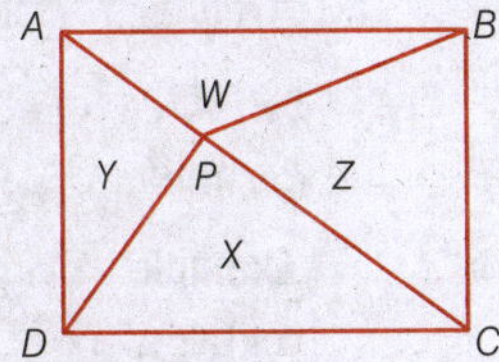

$$W + X = Y + Z$$

(xi) $ABCD$ is a rectangle and P is any point on side CD, then
$ABCD$ एक आयत है तथा P कोई बिन्दु है जो कि CD पर है, तब

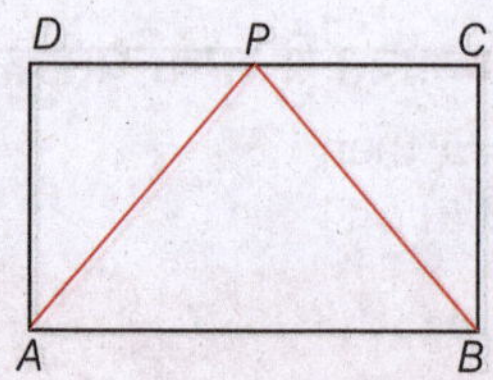

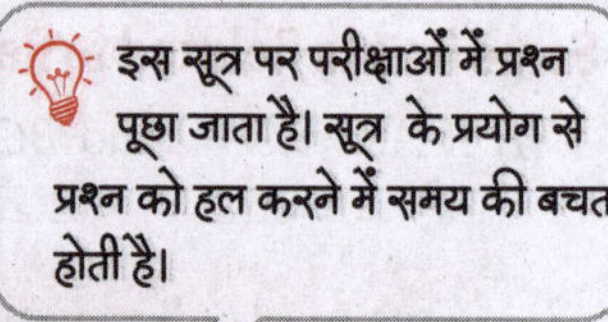

Area of $\Delta APB = \frac{1}{2} \times$ Area of rectangle $ABCD$

त्रिभुज APB का क्षेत्रफल $= \frac{1}{2} \times$ आयत $ABCD$ का क्षेत्रफल

British Flag Theorem/ब्रिटिश ध्वज प्रमेय (For square/Rectangle, वर्ग/आयत के लिए)

P is any point inside the rectangle $ABCD$./ आयत $ABCD$ के अन्दर कोई बिन्दु P है।

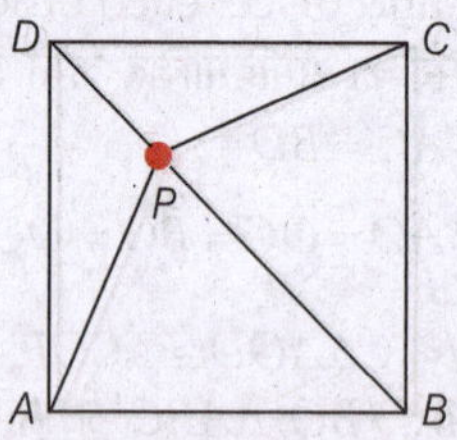

Then/तब,

$$AP^2 + PC^2 = PB^2 + PD^2$$

आयत पर आधारित उदाहरण निम्नलिखित हैं

Example **In figure $ABCD$ is a rectangle in which $AB = 11$ cm, $BC = 5$ cm. Find area and perimeter of rectangle.**

दी गई आकृति में, $ABCD$ एक आयत है जिसमें $AB = 11$ सेमी, $BC = 5$ सेमी। आयत का क्षेत्रफल और परिमाप ज्ञात कीजिए।

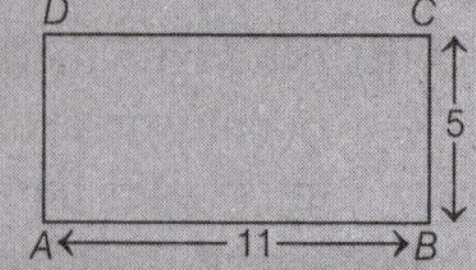

Sol. Area of rectangle $= l \times b = 11 \times 5 = 55$ cm^2

Perimeter of rectangle $= 2(l + b) = 2(11 + 5) = 2 \times 16 = 32$ cm

Example **In given figure $ABCD$ is a rectangle of length 15 cm and breadth 10 cm. Find area of ΔABP.**

दी गई आकृति में, $ABCD$ एक आयत है जिसकी लम्बाई 15 सेमी तथा चौड़ाई 10 सेमी है। ΔABP का क्षेत्रफल ज्ञात कीजिए।

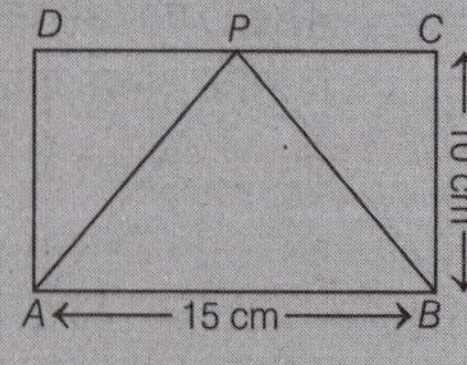

Sol. Area of $\Delta APB = \frac{1}{2} \times$ Area of rectangle $ABCD$

$$= \frac{1}{2} \times L \times B = \frac{1}{2} \times 15 \times 10 = 75 \text{ cm}^2$$

Example **P is any point inside the rectangle $ABCD$. If $PA = 27$ cm, $PB = 21$ cm and $PC = 6$ cm, then length of PD.**

P, चतुर्भुज $ABCD$ के अन्दर कोई बिन्दु है। यदि $PA = 27$ सेमी, $PB = 21$ सेमी तथा $PC = 6$ सेमी है, तब PD की लम्बाई ज्ञात कीजिए। *RRC GROUP D 2022*

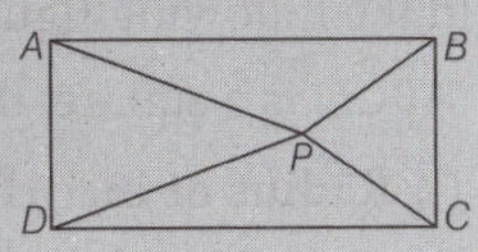

Sol. $AP^2 + PC^2 = PB^2 + PD^2$

$\Rightarrow \quad 27^2 + 6^2 = (21)^2 + PD^2$

$\Rightarrow PD^2 = 729 + 36 - 441$

$\Rightarrow PD^2 = 324 \Rightarrow PD = 18$ cm

Square वर्ग

- A quadrilateral whose all four sides are equal and all angles are 90°.
  एक चतुर्भुज जिसकी चारों भुजाएँ बराबर हों और सभी कोण 90° हों।

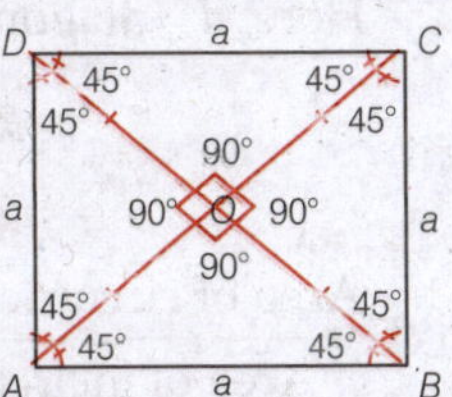

Some Points Related to Square
वर्ग से सम्बन्धित कुछ बिन्दु

(i) Diagonals are equal and bisect each other at right angles.
विकर्ण बराबर और एक-दूसरे को समकोण पर समद्विभाजित करते हैं।

$$AC = BC, AO = BO = CO = DO = \frac{AC}{2} = \frac{BD}{2}$$

(ii) Diagonal bisects the vertex angle. /विकर्ण शीर्ष कोण को समद्विभाजित करता है।

(iii) Area/क्षेत्रफल $= a^2 = \frac{d^2}{2}$; Diagonal/विकर्ण $= \sqrt{2}a$; Perimeter/परिमाप $= 4a$

(Where/जहाँ, a = side/भुजा, d = Diagonal/विकर्ण)

(iv) $ABCD$ is a square and P and Q are two points on DC and AD respectively, then
$ABCD$ एक वर्ग है और P तथा Q क्रमशः CD और AD पर दो बिन्दु हैं, तब

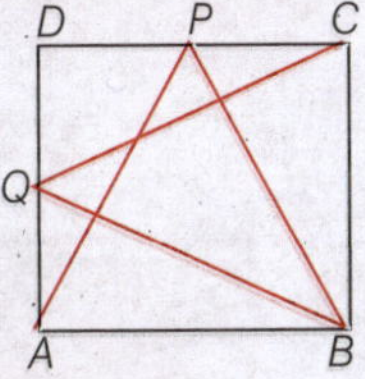

Area of ΔAPB = Area of $\Delta BQC = \frac{1}{2} \times$ Area of square $ABCD$

ΔAPB का क्षेत्रफल $= \Delta BQC$ का क्षेत्रफल $= \frac{1}{2} \times$ वर्ग $ABCD$ का क्षेत्रफल

(v) $\Delta DOC \cong \Delta AOB \cong \Delta AOD \cong \Delta BOC$

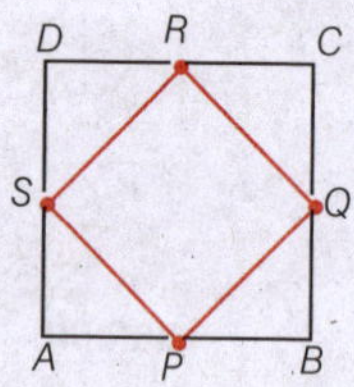

- Straight lines joining the mid-points of adjacent sides of any square forms a square.
  वर्ग की संगत भुजाओं के मध्य-बिन्दुओं को मिलाने से बनने वाली आकृति वर्ग होती है।

$$\text{Area of } PQRS = \frac{1}{2} \text{ area of } ABCD$$

$$PQRS \text{ का क्षेत्रफल} = \frac{1}{2} \times ABCD \text{ का क्षेत्रफल}$$

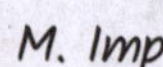

- In the following figure/निम्न आकृति में,
  $ABCD$ is a square, then/ $ABCD$ एक वर्ग है, तब
  (i) **Radius of incircle/अन्तःवृत्त की त्रिज्या,** $r = \frac{a}{2}$
  (ii) **Radius of circumcircle/परिवृत्त की त्रिज्या,**
  $R = \frac{d}{2} = \frac{a}{\sqrt{2}}$
  Here, d = diagonal of square/यहाँ, d = वर्ग का विकर्ण

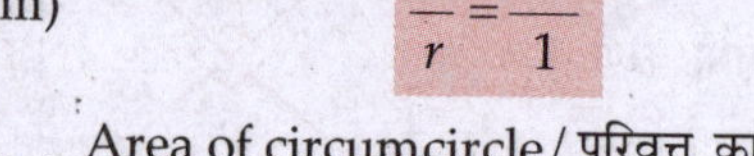

(iii) $\frac{R}{r} = \frac{\sqrt{2}}{1}$

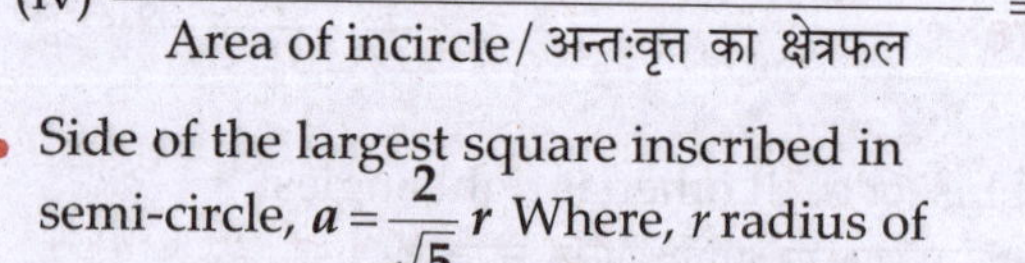

(iv) $\frac{\text{Area of circumcircle/ परिवृत्त का क्षेत्रफल}}{\text{Area of incircle/ अन्तःवृत्त का क्षेत्रफल}} = \frac{2}{1}$

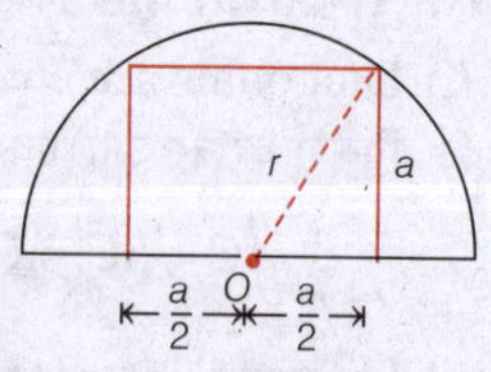

- Side of the largest square inscribed in semi-circle, $a = \frac{2}{\sqrt{5}} r$ Where, r radius of semi-circle.
  अर्द्धवृत्त में बने सबसे बड़े वर्ग की भुजा, $a = \frac{2}{\sqrt{5}} r$
  जहाँ, r अर्द्धवृत्त की त्रिज्या है।
- P, Q, R, S are the mid-points of respective sides of square $ABCD$, then
  P, Q, R, S वर्ग $ABCD$ की संगत भुजाओं के मध्य-बिन्दु हैं, तब

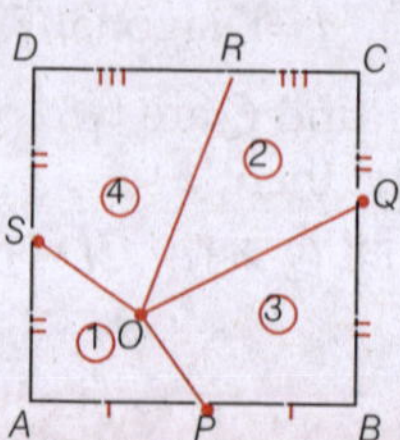

$$\text{Area } (1 + 2) = \text{Area } (3 + 4) = \frac{1}{2} \text{ Area } ABCD.$$

$$\text{क्षेत्रफल } (1 + 2) = \text{क्षेत्रफल } (3 + 4) = \frac{1}{2} \text{ क्षेत्रफल } ABCD$$

- Side of the largest square inscribed in a triangle
  एक त्रिभुज में अंकित एक सबसे बड़े वर्ग की भुजा

(i) In the following figure, *PQRS* is a square and *ABC* is a triangle.
दी गई आकृति में, *PQRS* एक वर्ग है तथा *ABC* एक त्रिभुज है।

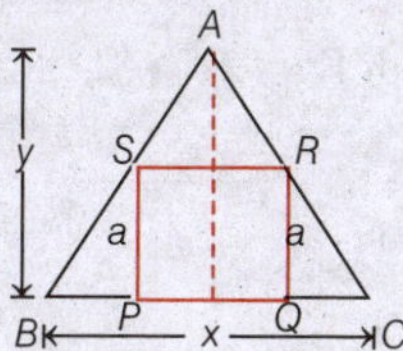

Then/तब, Side of square/वर्ग की भुजा, $a = \dfrac{xy}{x+y}$

(ii) In the following figure, side of square/निम्न आकृति में, वर्ग की भुजा, $a = \dfrac{xy}{x+y}$

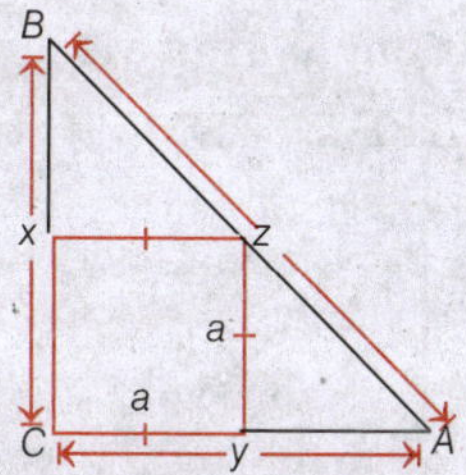

(iii) In the following figure, side of square
निम्न आकृति में, वर्ग की भुजा, $a = \dfrac{xyz}{x^2 + y^2 + xy}$

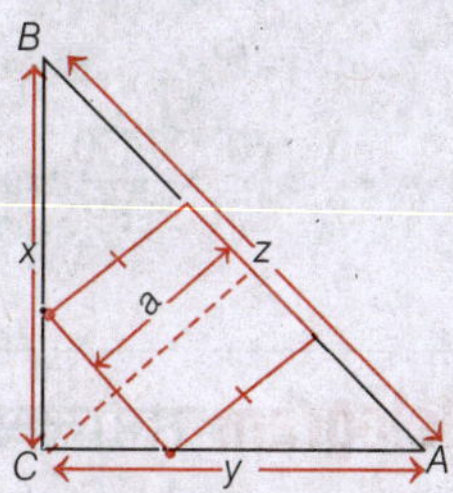

(iv) In the following figure, side of square/ निम्न आकृति में, वर्ग की भुजा, $a = \sqrt{xy}$

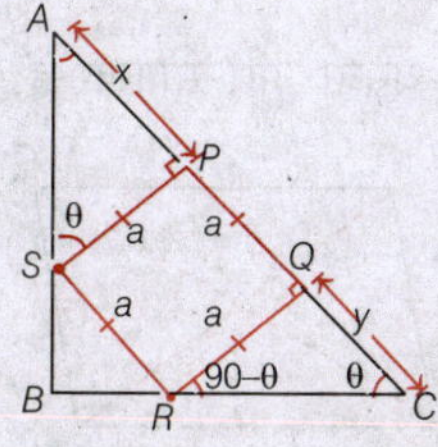

वर्ग पर आधारित उदाहरण निम्नलिखित हैं

Example ***ABCD* is a square of side 9 cm. Find its area and perimeter.**

ABCD एक वर्ग है जिसकी भुजा 9 सेमी है। इसका क्षेत्रफल और परिमाप ज्ञात कीजिए।

Sol. Area of square $= (\text{Side})^2 = (9)^2 = 81\ \text{cm}^2$

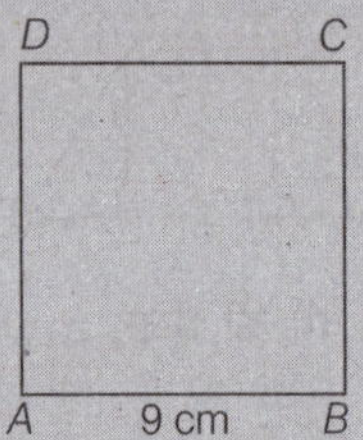

Perimeter of $\square ABCD = 4 \times \text{Side} = 4 \times 9 = 36$ cm

Example **In given triangle, find area of square *PDBE*.**

दिए गए त्रिभुज में, वर्ग *PDBE* का क्षेत्रफल ज्ञात कीजिए।

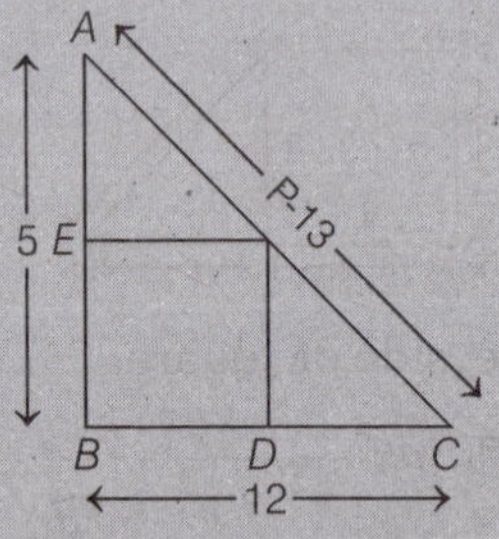

Sol. Side of square $= \dfrac{12 \times 5}{12 + 5} = \dfrac{60}{17}$ cm

Area of square $= (\text{Side})^2 = \dfrac{60}{17} \times \dfrac{60}{17} = \dfrac{3600}{289}\ \text{cm}^2$

Parallelogram समान्तर चतुर्भुज

- A quadrilateral whose opposite sides are equal and parallel, is called parallelogram.
  एक चतुर्भुज जिसकी सम्मुख भुजाएँ बराबर और समान्तर हैं, समान्तर चतुर्भुज कहलाता है।

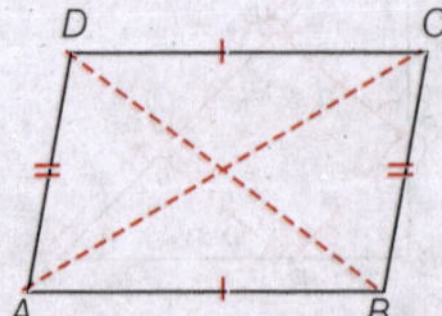

Some Points Related to Parallelogram
समान्तर चतुर्भुज से सम्बन्धित कुछ बिन्दु

(i) $AB \parallel DC$ and/और $AD \parallel BC$

(ii) $\angle A = \angle C, \angle B = \angle D$ *i.e.* Opposite angles are equal.
$\angle A = \angle C, \angle B = \angle D$ अर्थात् सम्मुख कोण बराबर होते हैं।

(iii) Sum of two adjacent angle is 180°.
दो आसन्न कोणों का योग 180° होता है।

(iv) Diagonal/विकर्ण, $AC \neq BD$

(v) Diagonals bisect each other, but not at 90°.
विकर्ण एक-दूसरे को समद्विभाजित करते हैं, परन्तु 90° पर नहीं।

(vi) Bisectors of the four angles enclose a rectangle.
चारों कोणों के समद्विभाजक आयत बनाते हैं।

- Each diagonal divides the parallelogram in two triangles of equal area.
प्रत्येक विकर्ण समान्तर चतुर्भुज को दो बराबर क्षेत्रफल के त्रिभुजों में विभाजित करता है।

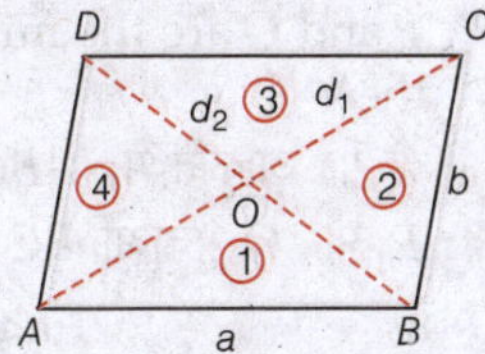

Here/यहाँ,

(i) Area of ΔABC = Area of ΔADC
ΔABC का क्षेत्रफल = ΔADC का क्षेत्रफल

(ii) Area of all four triangles is equal.
सभी 4 त्रिभुजों का क्षेत्रफल समान है।
i.e. ar/क्षेत्रफल ΔAOB = ar/क्षेत्रफल ΔBOC = ar/क्षेत्रफल ΔCOD

$$= \text{ar/क्षेत्रफल } \Delta AOD = \frac{1}{4} \text{ ar/क्षेत्रफल } \square ABCD$$

Area of Parallelogram / समान्तर चतुर्भुज का क्षेत्रफल

In the following figure, $ABCD$ is a parallelogram, then
निम्न आकृति में, $ABCD$ एक समान्तर चतुर्भुज है, तब

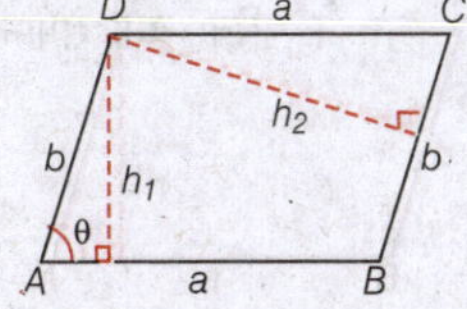

$$\text{Area/क्षेत्रफल} = ab \sin\theta = ah_1 = bh_2$$

- In the given figure, $ABCD$ is a parallelogram and P is any point on diagonals AC, then
दी गई आकृति में, $ABCD$ एक समान्तर चतुर्भुज है तथा विकर्ण AC पर P कोई बिन्दु है,
तब, Area $(x + y)$ = Area $(z + w)$

$$= \frac{1}{2} \text{ Area of parallelogram } ABCD = \frac{1}{2} ah_1 = \frac{1}{2} bh_2$$

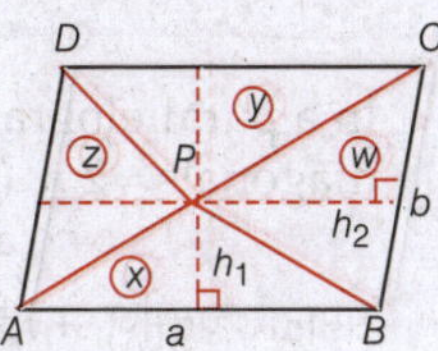

क्षेत्रफल $(x+y)$ = क्षेत्रफल $(z+w) = \frac{1}{2}$ समान्तर चतुर्भुज $ABCD$ का क्षेत्रफल

$$= \frac{1}{2} ah_1 = \frac{1}{2} bh_2$$

Where, h_1 and h_2 are height of parallelogram $ABCD$

जहाँ, h_1 और h_2 समान्तर चतुर्भुज $ABCD$ की ऊँचाई हैं।

- In the given figure, $ABCD$ is a parallelogram, then

  दी गई आकृति में, $ABCD$ एक समान्तर चतुर्भुज है, तब

  (i) Area/क्षेत्रफल $\Delta APB = \frac{1}{2}$ Area/क्षेत्रफल $\square ABCD$

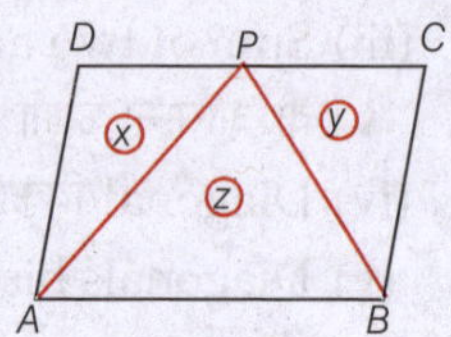

  (ii) Area/क्षेत्रफल $(x+y)$ = Area/क्षेत्रफल $(z) = \frac{1}{2}$ Area/क्षेत्रफल $\square ABCD$

- In the given figure, $ABCD$ is a parallelogram and P and Q are the mid-points of side BC and DC, then

  दी गई आकृति में, $ABCD$ एक समान्तर चतुर्भुज है तथा P और Q भुजाओं BC और DC के मध्य-बिन्दु हैं, तब

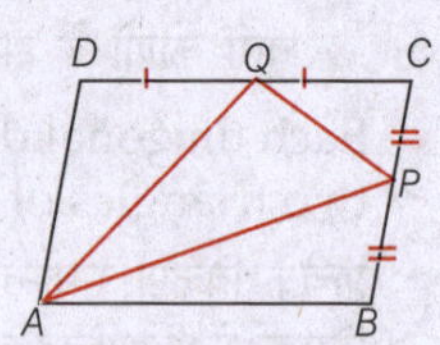

$$\frac{\textbf{Area/ क्षेत्रफल } \Delta APQ}{\textbf{Area/ क्षेत्रफल } \square ABCD} = \frac{3}{8}$$

- A parallelogram inscribed in a circle is always a rectangle.

  वृत्त के अन्दर बना हुआ समान्तर चतुर्भुज, सदैव आयत होता है।

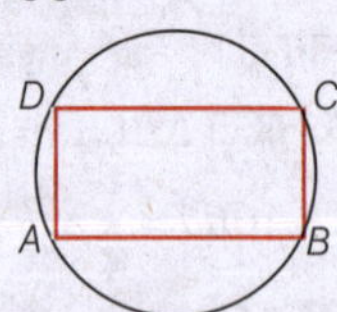

- A parallelogram circumscribed around a circle is always a rhombus

  वृत्त के परितः बना समान्तर चतुर्भुज, सदैव समचतुर्भुज होता है।

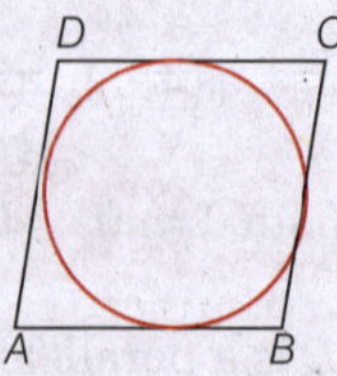

- In a parallelogram the sum of the square of the diagonals = 2 × (Sum of the squares of the two adjacent sides)

  समान्तर चतुर्भुज में, दोनों विकर्णों के वर्गों का योगफल = 2 × (दो संगत भुजाओं के वर्गों का योगफल)

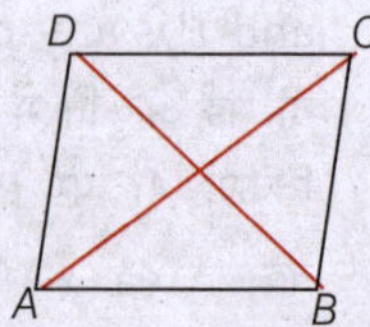

i.e., $AC^2 + BD^2 = 2(AB^2 + BC^2)$ *M. Imp*

समान्तर चतुर्भुज पर आधारित उदाहरण निम्नलिखित हैं

Example **In a parallelogram, one of the parallel side is 16 cm and the other side is 12 cm. If the perpendicular distance between two parallel side of dimension 16 cm is 24 cm, then the perpendicular distance between its other two parallel sides.**

एक समान्तर चतुर्भुज में, समान्तर भुजाओं में एक 16 सेमी तथा दूसरी भुजा 12 सेमी है। यदि दो समान्तर भुजाओं के बीच की लम्बवत् दूरी परिमाण 16 सेमी की ओर 24 सेमी है, तब दूसरी ओर लम्बवत् दूरी ज्ञात कीजिए।

SSC SELECTION POST 2023

Sol. As, we know

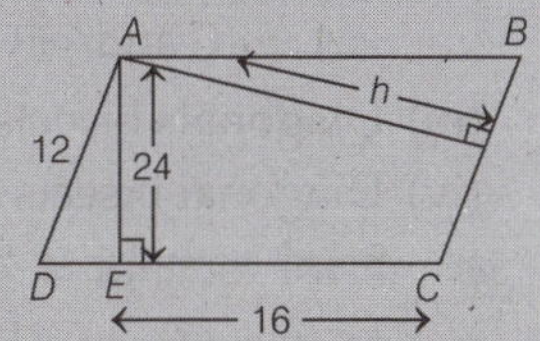

$$ah_1 = bh_2$$

$\Rightarrow$ $$16 \times 24 = 12 \times h$$

$\therefore$ $$h = 32 \text{ cm}$$

Example ***ABCD* is a parallelogram having area 144 cm^2 and *P* and *Q* are the mid-points of side *BC* and *DC*. Find area of ΔAPQ.**

ABCD एक समान्तर चतुर्भुज है जिसका क्षेत्रफल 144 सेमी^{2} और *P* तथा *Q* भुजा *BC* और *DC* के मध्य-बिन्दु हैं। ΔAPQ का क्षेत्रफल ज्ञात कीजिए।

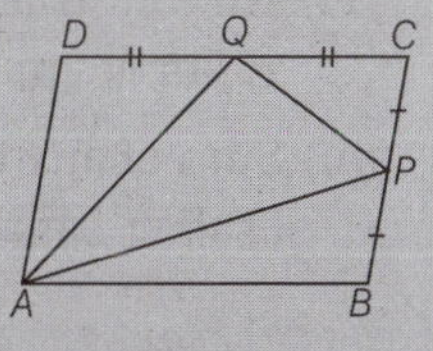

Sol. Area of $\Delta APQ = \frac{3}{8} \times$ Area of parallelogram

$$ABCD = \frac{3}{8} \times 144 = 3 \times 18 = 54 \text{ cm}^2$$

Example ***ABCD* is a parallelogram, where $AB = 12$ cm, $BC = 8$ cm and $AC = 13$ cm. Find measure of *BD*.**

ABCD एक समान्तर चतुर्भुज है, जहाँ $AB = 12$ सेमी, $BC = 8$ सेमी तथा $AC = 13$ सेमी। *BD* की माप ज्ञात कीजिए।

Sol. As, we know

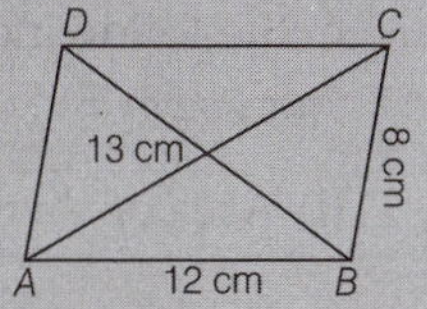

$$AC^2 + BD^2 = 2(AB^2 + BC^2)$$

$\Rightarrow$ $$13^2 + BD^2 = 2(12^2 + 8^2)$$

$\Rightarrow$ $$169 + BD^2 = 2(144 + 64)$$

$\Rightarrow$ $$169 + BD^2 = 2 \times 208$$

$\Rightarrow$ $$BD^2 = 416 - 169 = 247 \Rightarrow BD = \sqrt{247} \text{ cm}$$

Rhombus समचतुर्भुज

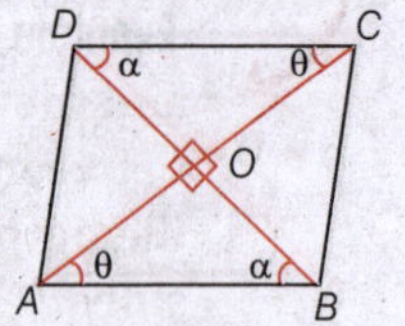

- A parallelogram in which all the sides are equal is called a **rhombus**.
  एक समान्तर चतुर्भुज जिसकी सभी भुजाएँ समान हों, **समचतुर्भुज** कहलाता है।

Some Points Related to Rhombus
समचतुर्भुज से सम्बन्धित कुछ बिन्दु

(i) Opposite sides are parallel/सम्मुख भुजाएँ समान्तर होती हैं।

(ii) Opposite angles are equal /सम्मुख कोण बराबर होते हैं।
$\angle A = \angle C$ and/और $\angle B = \angle D$

(iii) Diagonals bisects vertex angle./ विकर्ण शीर्ष कोण को समद्विभाजित करते हैं।

(iv) Diagonal bisects rhombus into two equal areas.
विकर्ण समचतुर्भुज को दो बराबर क्षेत्रफलों में विभाजित करता है।

(v) Diagonals are not equal and bisect each other at 90°.
विकर्ण बराबर नहीं होते हैं और एक-दूसरे को 90° पर समद्विभाजित करते हैं।
i.e., $AC \neq BD$

(vi) All four triangles made by two diagonals are congruent.
दो विकर्णों द्वारा बनाए गए सभी चार त्रिभुज सर्वांगसम होते हैं।

(vii) Sum of any two adjacent angles is 180°.
किन्हीं दो आसन्न कोणों का योगफल 180° होता है।
i.e., $\angle A + \angle B = 180°$

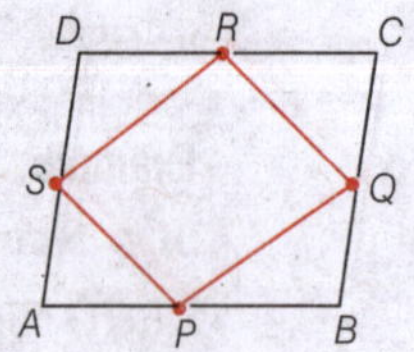

(viii) Figure formed by joining the mid-points of the adjacent sides of a rhombus is a rectangle.
Therefore, *PQRS* is a rectangle.
समचतुर्भुज की भुजाओं के मध्य-बिन्दुओं को मिलाने से बनी आकृति आयत होती है।
इस प्रकार, *PQRS* एक आयत है।

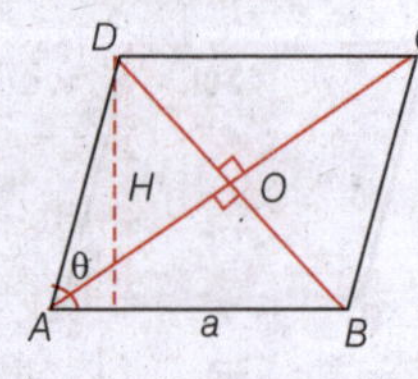

- In the following figure, *ABCD* is a rhombus, then
  निम्न आकृति में, *ABCD* एक समचतुर्भुज है, तब

(i) Area of rhombus $= \frac{1}{2} \times$ **Product of diagonals**
$$= \frac{1}{2} \times d_1 \times d_2$$
समचतुर्भुज का क्षेत्रफल $= \frac{1}{2} \times$ (**विकर्णों का गुणनफल**)
$$= \frac{1}{2} \times d_1 \times d_2$$

> **अतिसम्भावित!**
> इस सूत्र पर सभी परीक्षाओं SSC, Railway आदि में 1 प्रश्न अवश्य पूछा जाता है।

(ii) Area of *ABCD*/ *ABCD* का क्षेत्रफल $= a^2 \sin\theta$

(iii) $a=\sqrt{\left(\frac{d_1}{2}\right)^2+\left(\frac{d_2}{2}\right)^2} \Rightarrow d_1^2+d_2^2=4a^2$

(iv) Perimeter of rhombus/समचतुर्भुज का परिमाप $=4a$

(v) Height of rhombus/ समचतुर्भुज की ऊँचाई $(H)=\frac{d_1 d_2}{\sqrt{d_1^2+d_2^2}}$

समचतुर्भुज पर आधारित उदाहरण निम्नलिखित हैं

Example **If $ABCD$ is a rhombus such that $\angle ACB = 40°$, then $\angle ADB$ is**

यदि $ABCD$ एक समचतुर्भुज है जिसमें $\angle ACB = 40°$, तब $\angle ADB$ का मान है

RRC GROUP D 2022

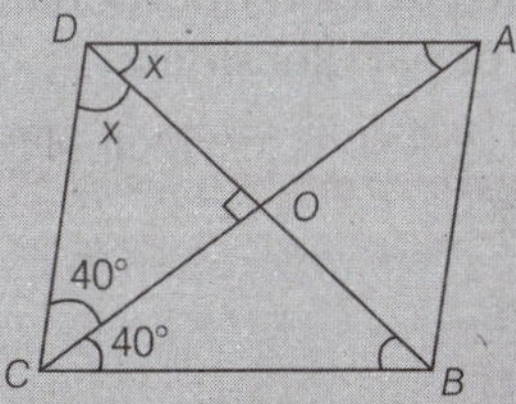

Sol. $\angle ACB = \angle ACD$ and $\angle ADB = \angle BDC$

($\because$ Diagonals bisects vertex angle)

Now, In ΔDOC

$$\angle x + 90° + 40° = 180° \Rightarrow \angle x = 180° - 90° - 40°$$

$$\Rightarrow \angle x = 180° - 130° \Rightarrow \angle x = 50°$$

$$\angle ADB = x = 50°$$

Example **The length of the diagonals of a rhombus are 16 cm and 12 cm. The length of the side of rhombus.**

एक समचतुर्भुज के विकर्णों की लम्बाई 16 सेमी तथा 12 सेमी है। समचतुर्भुज की भुजा की लम्बाई ज्ञात कीजिए।

RRC GROUP D 2022

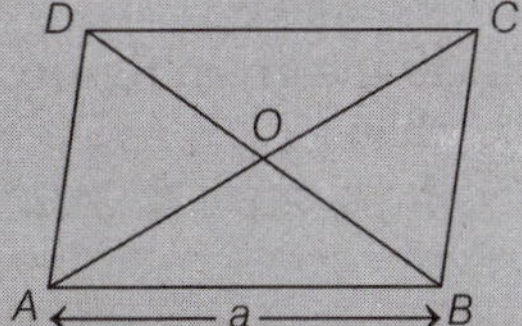

Sol. As, we know

$$\Rightarrow d_1^2 + d_2^2 = 4a^2$$

$$\Rightarrow (16)^2 + (12)^2 = 4a^2$$

$$\Rightarrow 256 + 144 = 4a^2$$

$$\Rightarrow 400 = 4a^2$$

$$\Rightarrow a^2 = \frac{400}{4} = 100$$

$$\therefore a = \sqrt{100} = 10 \text{ cm}$$

Trapezium समलम्ब

- A quadrilateral with only two sides parallel, but unequal in length is called a trapezium.
एक चतुर्भुज जिसकी केवल दो भुजाएँ समान्तर हों, परन्तु लम्बाई में बराबर न हो, समलम्ब कहलाता है।

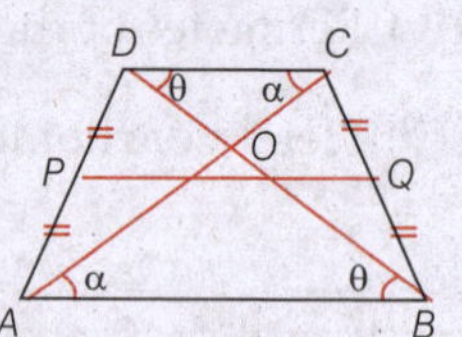

Some Related Points कुछ सम्बन्धित बिन्दु

(i) $AB \| CD$ (ii) $\Delta AOB \sim \Delta COD$

(iii) $PQ = \frac{AB + CD}{2}$, where P and Q are the mid points of AD and BC, respectively./जहाँ, P और Q क्रमशः AD और BC के मध्य-बिन्दु हैं।

(iv) Diagonals intersect each other proportionally in the ratio of lengths of parallel sides.
विकर्ण एक-दूसरे को समान्तर भुजाओं की लम्बाई के अनुपात में समानुपाती रूप से काटते हैं।

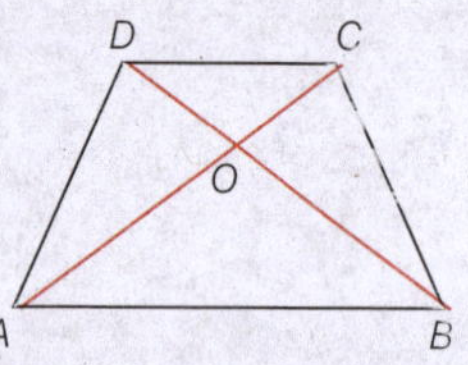

$$\frac{AO}{CO} = \frac{BO}{DO} = \frac{AB}{CD}$$

(v) $AC^2 + BD^2 = AD^2 + BC^2 + 2(AB \times CD)$

(vi) In the given figure, $ABCD$ is a trapezium, then
दी गई आकृति में, $ABCD$ एक समलम्ब चतुर्भुज है, तब

$$\frac{\text{Area/ क्षेत्रफल } \square DCQP}{\text{Area/ क्षेत्रफल } \square PQBA} = \frac{m}{n}$$

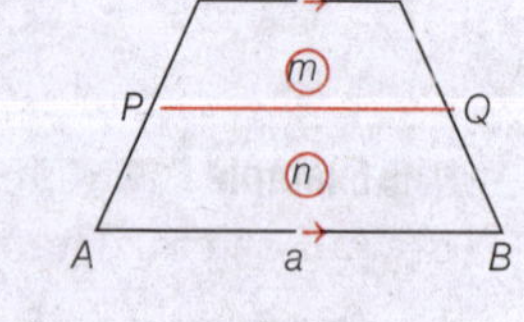

$$PQ = \sqrt{\frac{ma^2 + nb^2}{m + n}}$$

If/यदि $m : n = 1 : 1$, then/तब $PQ = \sqrt{\frac{a^2 + b^2}{2}}$

(vii) In the given figure, $ABCD$ is a trapezium, then
दी गई आकृति में, $ABCD$ एक समलम्ब चतुर्भुज है, तब

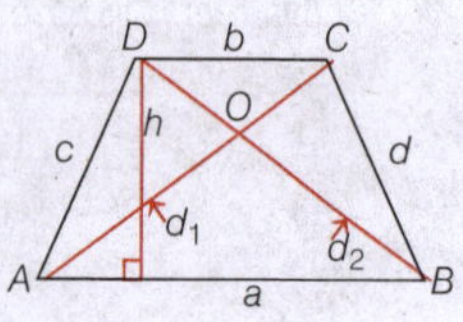

$$d_1^2 + d_2^2 = c^2 + d^2 + 2ab$$

Area of trapezium $ABCD = \frac{1}{2}$ (Sum of parallel sides) × Distance between parallel sides

समलम्ब $ABCD$ का क्षेत्रफल $= \frac{1}{2}$ (समान्तर भुजाओं का योग) × समान्तर भुजाओं के बीच की दूरी

Area of trapezium/समलम्ब $ABCD$ का क्षेत्रफल $= \frac{1}{2} \times (a + b) \times h$ *M. Imp*

(viii) The line segment joining the mid-points of the diagonals of trapezium is parallel to each of the parallel sides and is equal to half the difference of these sides.

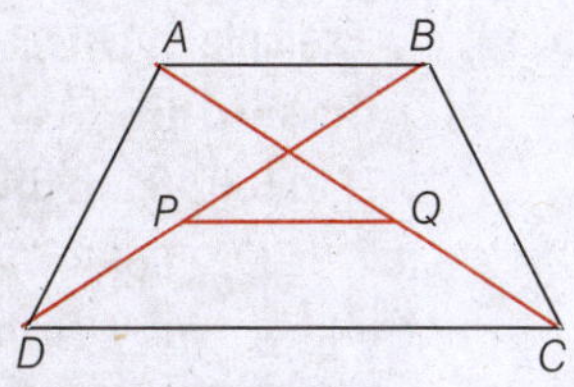

समलम्ब चतुर्भुज के विकर्ण के मध्य-बिन्दुओं को मिलाने वाला रेखाखण्ड प्रत्येक समान्तर भुजा के समानान्तर होता है तथा इन भुजाओं के अन्तर के आधे के बराबर होता है।

If P and Q are mid-point of both diagonals respectively.

यदि P और Q क्रमशः दोनों विकर्णों के मध्य-बिन्दु हैं।

(i) $PQ \parallel AB$ or DC

(ii) $PQ = \frac{1}{2}(DC - AB)$

(ix) A trapezium $ABCD$ is divided into two parts by line EF, which divide the non-parallel sides in the ratio of $m : n$, then the length of line EF is

एक समलम्ब चतुर्भुज $ABCD$ रेखा EF द्वारा दो भागों में विभाजित होता है जो असमान्तर भुजाओं को $m : n$ के अनुपात में विभाजित करता है, तब रेखा EF की लम्बाई है

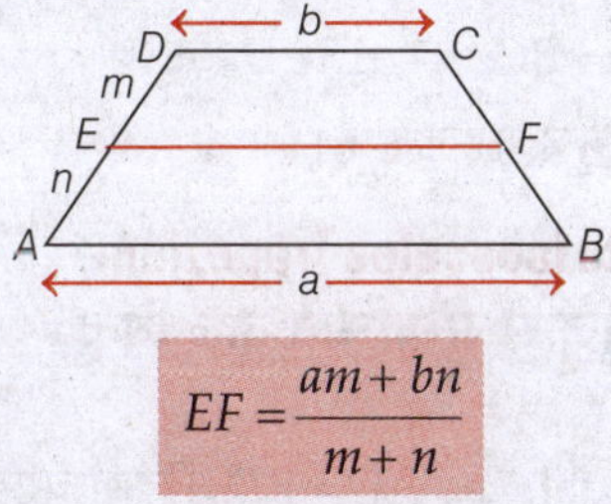

$$EF = \frac{am + bn}{m + n}$$

समलम्ब पर आधारित उदाहरण निम्नलिखित हैं

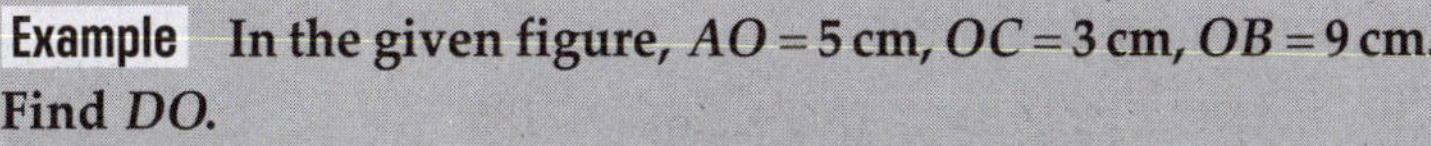

Example **In the given figure, $AO = 5$ cm, $OC = 3$ cm, $OB = 9$ cm. Find DO.**

दी गई आकृति में, $AO = 5$ सेमी, $OC = 3$ सेमी, $OB = 9$ सेमी है, DO ज्ञात कीजिए।

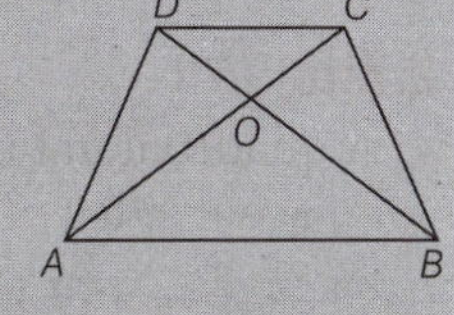

Sol. As, we know

$$\frac{AO}{OC} = \frac{BO}{OD} \Rightarrow \frac{5}{3} = \frac{9}{OD}$$

$$\therefore \quad OD = \frac{27}{5} = 5.4 \text{ cm}$$

Example **$ABCD$ is a trapezium in which $AB \parallel DC$, E and F are the mid-points of the diagonal AC and BD, respectively. If $AB = 18$ cm and $CD = 6$ cm, then find EF.**

$ABCD$ एक समलम्ब चतुर्भुज है जिसमें $AB \parallel DC$ तथा E और F क्रमशः विकर्ण AC और BD के मध्य-बिन्दु हैं। यदि $AB = 18$ सेमी तथा $CD = 6$ सेमी हैं, तब EF ज्ञात कीजिए। *SSC CGL (TIER II) 2022*

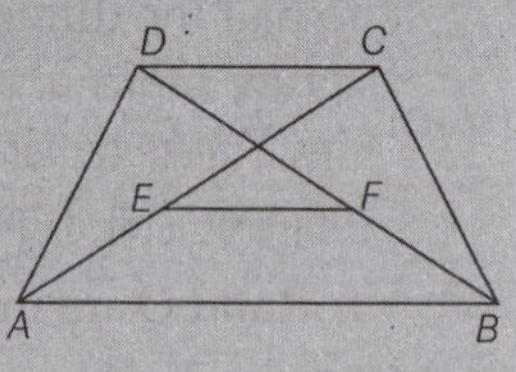

Sol. $EF = \frac{1}{2} \times (AB - CD) = \frac{1}{2} \times (18 - 6)$

$EF = 6$ cm

Some Special Figures कुछ विशेष आकृतियाँ

1. Isosceles Trapezium समद्विबाहु समलम्ब चतुर्भुज

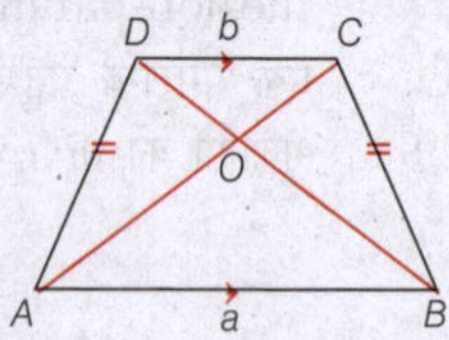

- If $AB \parallel CD$ and $AD = BC$, then $ABCD$ is called an **isosceles trapezium**.
  यदि $AB \parallel CD$ और $AD = BC$, तब $ABCD$ एक **समद्विबाहु समलम्ब चतुर्भुज** कहलाता है।

Some Points Related to Isosceles Trapezium
समद्विबाहु समलम्ब चतुर्भुज से सम्बन्धित कुछ बिन्दु

(i) $\angle DAB = \angle CBA$

(ii) Diagonals are equal, *i.e.*, $AC = BD$./विकर्ण बराबर होते हैं, अर्थात् $AC = BD$.

(iii) If a trapezium is inscribed in a circle, it has to be an isosceles trapezium.
यदि समलम्ब चतुर्भुज को एक वृत्त के अन्दर बनाया जाता है, तब यह एक समद्विबाहु समलम्ब चतुर्भुज होता है।

(iv) $\angle ADC + \angle DAB = \angle ABC + \angle BCD = 180°$

2. Kite पतंग

- A quadrilateral is called a **kite**, if it has two pairs of equal and adjacent sides.
  एक चतुर्भुज, **पतंग** कहलाती है, यदि इसमें समान और आसन्न भुजाओं के दो युग्म होते हैं।

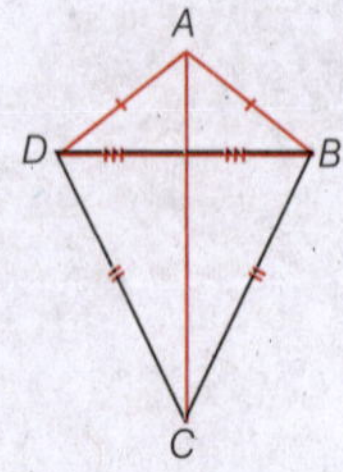

Some Points Related to Kite पतंग से सम्बन्धित कुछ बिन्दु

(i) Two pairs of adjacent sides are equal.
संगत भुजाओं के दो युग्म बराबर होते हैं।

(ii) The diagonals intersect at right angles.
विकर्ण समकोण पर प्रतिच्छेद करते हैं।

(iii) The longer diagonal bisects the shorter diagonal.
लम्बा विकर्ण छोटे विकर्ण को समद्विभाजित करता है।

(iv) Area $= \frac{1}{2} \times$ Product of diagonals $= \frac{1}{2} \times AC \times BD$

क्षेत्रफल $= \frac{1}{2} \times$ विकर्णों का गुणनफल $= \frac{1}{2} \times AC \times BD$ *M. Imp*

पतंग पर आधारित उदाहरण को ध्यानपूर्वक समझें

Example ***ABCD*** **is a kite having diagonals are 15 cm and 24 cm. Find area of kite.**

ABCD एक पतंग है जिसके विकर्ण 15 सेमी तथा 24 सेमी हैं। पतंग का क्षेत्रफल ज्ञात कीजिए।

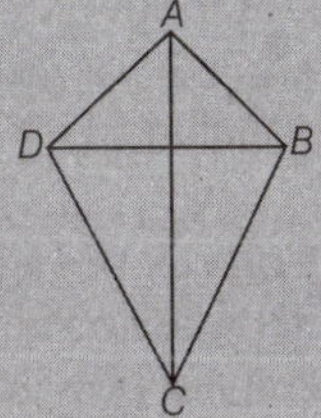

Sol. Area of kite $= \frac{1}{2} \times 15 \times 24$

$= 15 \times 12 = 180 \text{ cm}^2$

अध्याय 04

Polygons बहुभुज

A figure enclosed by three or more sides is called a **polygon**.
तीन या तीन से अधिक भुजाओं से घिरी आकृति को **बहुभुज** कहते हैं।

Types of Polygon बहुभुज के प्रकार

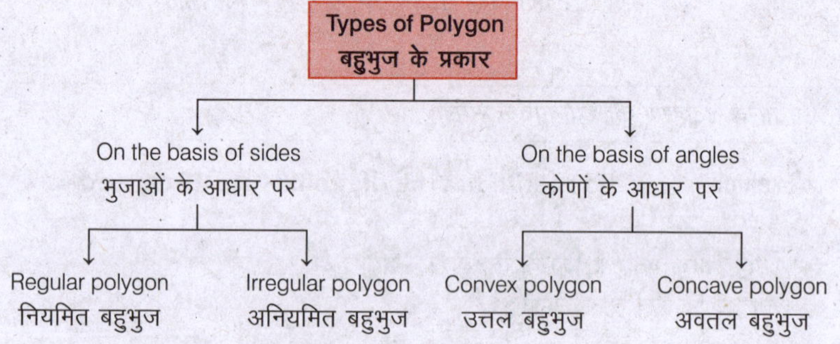

- **Convex polygon :** A polygon in which none of it's interior angles is more than 180°, is called a convex polygon.
उत्तल बहुभुज : एक बहुभुज, जिसमें कोई भी आन्तरिक कोण 180° से अधिक न हो, उत्तल बहुभुज कहलाता है।

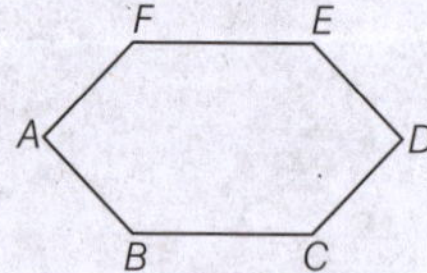

- **Concave polygon :** A polygon in which atleast one angle is more than 180°, is called a concave polygon.
अवतल बहुभुज : एक बहुभुज, जिसमें कम-से-कम एक कोण 180° से अधिक हो, अवतल बहुभुज कहलाता है।
Here/यहाँ, $\angle DEF > 180°$

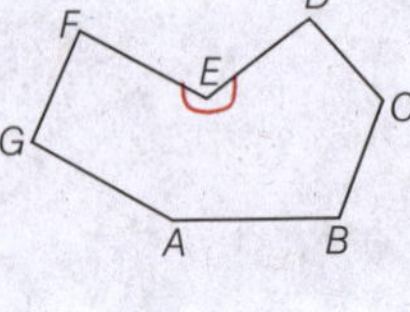

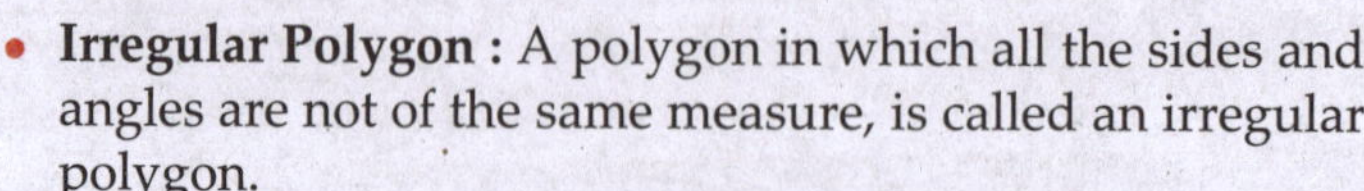

- **Irregular Polygon :** A polygon in which all the sides and angles are not of the same measure, is called an irregular polygon.
अनियमित बहुभुज : एक बहुभुज, जिसमें सभी भुजाएँ और कोण समान माप के नहीं होते, अनियमित बहुभुज कहलाता है।

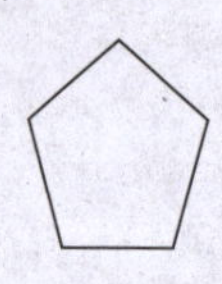

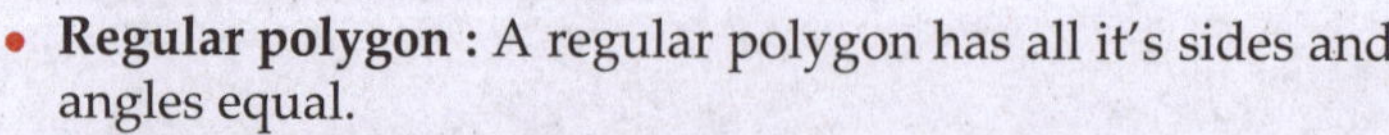

- **Regular polygon :** A regular polygon has all it's sides and angles equal.
नियमित बहुभुज या समबहुभुज : एक बहुभुज, जिसकी सभी भुजाएँ और कोण बराबर हों।

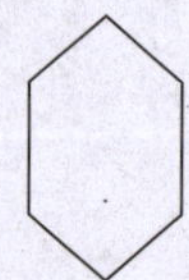

Some Important Formulae Related to Regular Polygon
नियमित बहुभुज से सम्बन्धित कुछ महत्त्वपूर्ण सूत्र

(i) Sum of all interior angles of a polygon of n sides

n भुजाओं वाले बहुभुज के सभी अन्तः कोणों का योगफल $= (n-2) \times 180°$ M. Imp.

(ii) Each interior angle of a regular polygon of n sides

n भुजाओं वाले समबहुभुज का प्रत्येक अन्तः कोण $= \dfrac{(n-2) \times 180°}{n}$ M. Imp.

(iii) Sum of all exterior angles of a polygon of n sides

n भुजाओं वाले बहुभुज के सभी बाह्य कोणों का योगफल $= 360°$

(iv) Each exterior angle of a regular polygon of n sides

n भुजाओं वाले समबहुभुज का प्रत्येक बाह्य कोण $= \dfrac{360°}{n}$ M. Imp.

(v) Interior angle + Exterior angle $= 180°$

अन्तः कोण + बाह्य कोण $= 180°$

(vi) Number of diagonals in a polygon of n sides

n भुजाओं वाले बहुभुज में विकर्णों की संख्या $= \dfrac{n(n-3)}{2}$ M. Imp.

(vii) Area of a regular polygon of n sides $= \dfrac{na^2}{4} \cot \dfrac{180°}{n}$,

where, a is the length of each side.

n भुजाओं वाले समबहुभुज का क्षेत्रफल

$= \dfrac{na^2}{4} \cot \dfrac{180°}{n}$, जहाँ, a प्रत्येक भुजा की लम्बाई है।

(viii) Perimeter of a regular polygon of n sides $= n \times a$; where, a is the length of each side.

n भुजाओं वाले नियमित बहुभुज का परिमाप $= n \times a$, जहाँ, a प्रत्येक भुजा की लम्बाई है।

> **अतिसम्भावित!**
> ये सूत्र सभी परीक्षाओं की दृष्टि से M. Imp है। इन पर आधारित आधारित 1 प्रश्न पूछे जाने की सम्भावना रहती है।

(ix) Number of sides in a regular polygon

समबहुभुज में भुजाओं की संख्या $= \dfrac{360°}{\text{Exterior angle/बाह्य कोण}}$ M. Imp.

(x) Ratio of the measure of interior angle and exterior angle of a regular polygon of n sides $= \left(\dfrac{n}{2}-1\right):1$ or $(n-2):2$

n भुजाओं वाले समबहुभुज के अन्तः और बाह्य कोणों की माप का अनुपात

$= \left(\dfrac{n}{2}-1\right):1$ या $(n-2:2)$

बहुभुज पर पूछे जाने वाले प्रश्नों पर आधारित उदाहरण निम्नलिखित हैं

Example **Find the number of diagonals of a regular polygon sum of whose interior angle is 2700°.**

एक नियमित बहुभुज के विकर्णों की संख्या ज्ञात कीजिए, जिसके आन्तरिक कोणों का योग 2700° है। *SSC CGL 2023*

Sol. Sum of interior angle $= 2700°$

$\Rightarrow \quad (n-2) \times 180° = 2700°$

$\Rightarrow \quad (n-2) = \frac{2700}{180} \Rightarrow n-2 = 15$

$\Rightarrow \quad n = 17$

$\therefore$ Number of diagonals $= \frac{n(n-3)}{2} = \frac{17 \times 14}{2} = 119$

Example **Find the number of sides of a regular polygon, if the measure of each of its exterior angle is 36°.**

एक नियमित बहुभुज की भुजाओं की संख्या ज्ञात कीजिए, यदि प्रत्येक बाह्य कोण की माप 36° है। *RRB NTPC 2022*

Sol. Each exterior angle $= \frac{360°}{n}$

$\therefore \quad n = \frac{360°}{36°} = 10$

Example **In a polygon, the interior and exterior angles are in the ratio 3 : 2. The number of sides of the polygon is**

किसी बहुभुज में अन्तः और बाह्य कोणों का अनुपात 3 : 2 है। बहुभुज की भुजाओं की संख्या है

Sol. Using property (X), $\left(\frac{n}{2}-1\right) : 1 = 3 : 2$

$\Rightarrow \quad \frac{n}{2} = \frac{3}{2} + 1 \Rightarrow \frac{n}{2} = \frac{5}{2} \Rightarrow n = 5$

Regular Hexagon समषट्भुज

A regular hexagon is a closed shape of polygon having six equal sides and six equal angles.

एक नियमित षट्भुज (समषट्भुज) एक बन्द आकार का बहुभुज है, जिसमें छः समान भुजाएँ और छः समान कोण होते हैं।

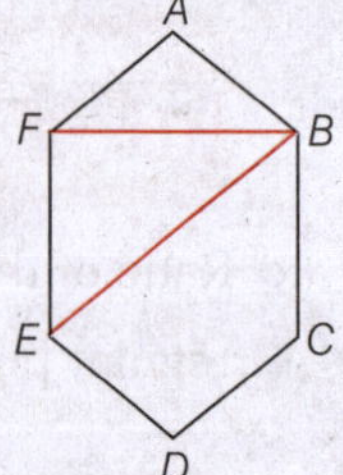

Here/यहाँ, BE = Larger diagonal/बड़ा विकर्ण

BF = Smaller diagonal/छोटा विकर्ण

(i) Perimeter of regular hexagon $= 6 \times$ Each side's length
नियमित षट्भुज का परिमाप $= 6 \times$ प्रत्येक भुजा की लम्बाई

(ii) Number of diagonals in regular hexagon
समषट्भुज में विकर्णों की संख्या $= 9$

(iii) Area of regular hexagon/नियमित षट्भुज का क्षेत्रफल

(a) When side is given/जब भुजा दी गई हो,

$$\text{Area/क्षेत्रफल} = \frac{3\sqrt{3}}{2}\,(\text{Side})^2/(\text{भुजा})^2$$

(b) When larger diagonal is given/जब बड़ा विकर्ण दिया गया हो,

$$\text{Area/क्षेत्रफल} = \frac{3\sqrt{3}}{8}\,(D)^2,\ D \rightarrow \text{Larger diagonal/बड़ा विकर्ण}$$

(c) When smaller diagonal is given/जब छोटा विकर्ण दिया गया हो,

$$\text{Area/क्षेत्रफल} = \frac{\sqrt{3}}{2}\,(d)^2,\ d \rightarrow \text{Smaller diagonal/छोटा विकर्ण}$$

(iv) Circumradius of regular hexagon is equal to length of side of hexagon.
नियमित षट्भुज की परित्रिज्या, षट्भुज की भुजा की लम्बाई के बराबर होती है।

(v) Inradius of regular hexagon/समषट्भुज की अन्तःत्रिज्या $= \frac{\sqrt{3}}{2}$ **Side/भुजा**

(vi) Inradius : circumradius $= \sqrt{3} : 2$
अन्तःत्रिज्या : परित्रिज्या $= \sqrt{3} : 2$

षट्भुज पर आधारित उदाहरण निम्नलिखित है

Example **If the side of regular hexagon is 7 cm, then what is the area?**
यदि नियमित षट्भुज की भुजा की लम्बाई 7 सेमी है, तो क्षेत्रफल क्या है?

Sol. Area of regular hexagon $= \frac{3\sqrt{3}}{2}(a)^2 = \frac{3\sqrt{3}}{2} \times (7)^2 = 73.5\sqrt{3}\ \text{cm}^2$

Regular Octagon नियमित अष्टभुज

A regular octagon has eight equal sides and eight equal angles.
एक नियमित अष्टभुज में आठ समान भुजाएँ व आठ समान कोण होते हैं।

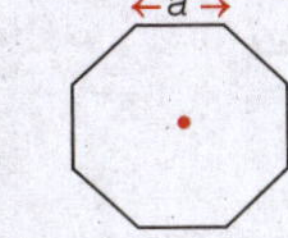

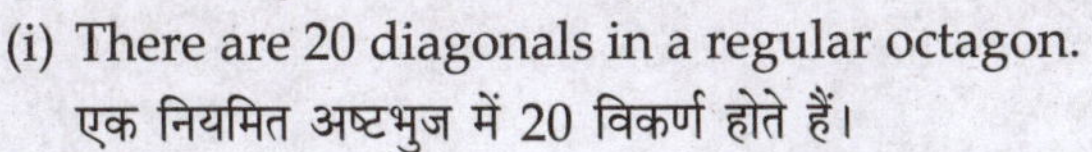

(i) There are 20 diagonals in a regular octagon.
एक नियमित अष्टभुज में 20 विकर्ण होते हैं।

(ii) The measure of each interior angle of regular octagon is 135°.
नियमित अष्टभुज के प्रत्येक आन्तरिक कोण की माप 135° होती है।

(iii) Sum of all interior angles is 1080°.
सभी आन्तरिक कोणों का योगफल 1080° होता है।

(iv) Perimeter of regular octagon/नियमित अष्टभुज का परिमाप $= 8 \times$ side/भुजा

(v) Area of regular octagon/समअष्ट्भुज का क्षेत्रफल

$$= \mathbf{2 \times (Side)^2\ (1 + 2\sqrt{2}) = 2a^2\,(1 + 2\sqrt{2})}$$

Where/जहाँ, a = Side/भुजा,

(vi) Inradius/अन्तःत्रिज्या $(r) = \dfrac{a}{2\sqrt{2} - 2}$

(vii) Circumradius/बाह्य त्रिज्या $(R) = \dfrac{a}{\sqrt{2 - \sqrt{2}}}$

अष्टभुज पर आधारित उदाहरण निम्नलिखित है

Example **The interior angles of 8-sided polygon are in the ratio 1 : 3 : 4 : 6 : 7 : 11 : 13 : 15. Find the measure of the largest interior angle in this polygon.**

एक 8-भुजा वाले बहुभुज के आन्तरिक कोणों का अनुपात 1 : 3 : 4 : 6 : 7 : 11 : 13 : 15 है। इस बहुभुज में सबसे बड़े आन्तरिक कोण की माप ज्ञांत कीजिए।

RRC GROUP D 2022

Sol. Sum of all Interior angle of polygon

$$= (n - 2) \times 180° = (8 - 2) \times 180° = 1080°$$

According to the question,

$$x + 3x + 4x + 6x + 7x + 11x + 13x + 15x = 1080°$$

$$\Rightarrow \quad 60x = 1080°$$

$$\therefore \quad x = \frac{1080°}{60} = 18°$$

Largest angle of polygon = $15x = 15 \times 18° = 270°$

अध्याय 05

Circle वृत्त

Circle A round plane figure whose boundary consists of points equidistant from a fixed point is called a circle. The fixed point is called centre of circle (O).

वृत्त एक निश्चित बिन्दु से समान दूरी पर स्थित बिन्दुओं का बिन्दुपथ, वृत्त कहलाता है। यह निश्चित बिन्दु (O) वृत्त का केन्द्र कहलाता है।

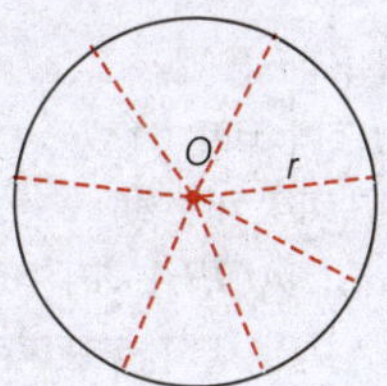

Related Definitions सम्बन्धित परिभाषाएँ

- **Centre** The centre of a circle is the point that is equidistant from all points on the circumference.

  केन्द्र वृत्त का केन्द्र वह बिन्दु है, जो वृत्त की परिधि पर सभी बिन्दुओं से समान दूरी पर है।
- **Radius** The distance between the centre and circumference of the circle is called radius of circle (r).

  त्रिज्या वृत्त के केन्द्र और परिधि के बीच की दूरी, वृत्त की त्रिज्या (r) कहलाती है।
- **Chord** The line which touches the circumference of a circle at two points is called a chord of circle.

  जीवा वह रेखा जो वृत्त की परिधि को दो बिन्दुओं पर स्पर्श करती है, वृत्त की जीवा कहलाती है।
- **Diameter** A chord which passes through the centre of the circle is called the diameter of the circle. It is twice of the radius.

  व्यास वृत्त के केन्द्र से गुजरने वाली जीवा को वृत्त का व्यास कहा जाता है। यह वृत्त की त्रिज्या का दोगुना होता है।
- **Arc** A continuous piece/part of a circle is called an arc of the circle.

  चाप किसी वृत्त का एक सतत् भाग/टुकड़ा, वृत्त का चाप कहलाता है।
- **Circumference** The length of boundary of circle is called the circumference of a circle.

  परिधि वृत्त के चारों ओर सीमा की लम्बाई को वृत्त की परिधि कहा जाता है।

- **Segment of Circle** A region bounded by a chord and corresponding arc lying between the end points of chord is called segment of circle.
  वृत्तखण्ड जीवा और जीवा के अन्तिम बिन्दुओं के बीच स्थित संगत चाप से घिरे क्षेत्र को वृत्तखण्ड कहा जाता है।

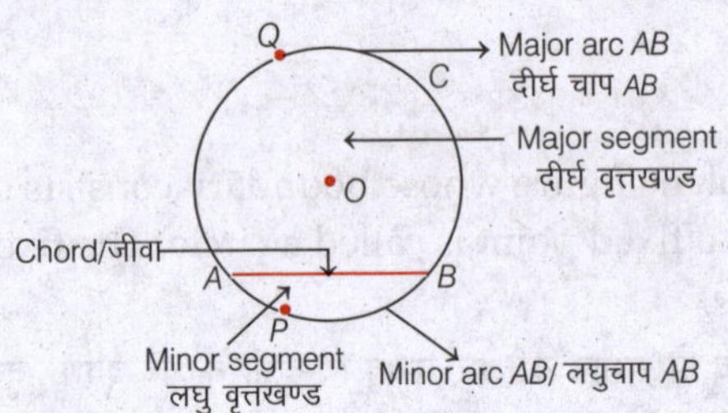

Here/यहाँ,
AB is a chord/ *AB* एक जीवा है।
APB is the minor *arc*/ *APB* लघु चाप है।
AQB is the major arc/ *AQB* दीर्घ चाप है।
AQBA is the major segment/ *AQBA* दीर्घ वृत्तखण्ड है।
APBA is minor segment/ *APBA* लघु वृत्तखण्ड है।

- **Central Angle** An angle formed at the centre of the circle by two points on the circumference of the circle is called central angle.
  केन्द्रीय कोण परिधि के दो बिन्दुओं द्वारा वृत्त के केन्द्र पर बना कोण केन्द्रीय कोण कहलाता है।
  Here, $\angle AOB$ is the central angle/यहाँ, $\angle AOB$ एक केन्द्रीय कोण है।

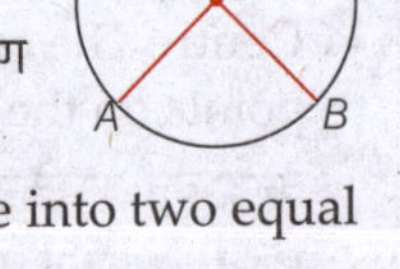

- **Semicircle** The diameter of the circle divides the circle into two equal parts. Each part is known as semicircle.
  अर्द्धवृत्त वृत्त का व्यास, वृत्त को दो समान भागों में विभाजित करता है। प्रत्येक भाग को अर्द्धवृत्त कहा जाता है।

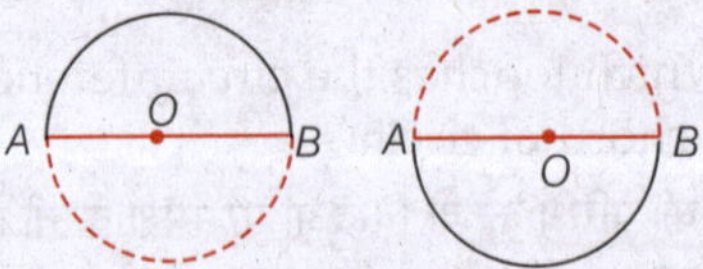

- **Concentric Circles** Circles having the same centre are called concentric circles.
  In the given figure r_1 and r_2 is the radius of two different circle with same centre *O*.
  संकेन्द्रीय वृत्त एक ही केन्द्र वाले वृत्तों को संकेन्द्रीय वृत्त कहा जाता है।
  दिए गए चित्र में r_1 तथा r_2 दो अलग-अलग वृत्तों की त्रिज्याएँ, जिसका समान केन्द्र *O* है।

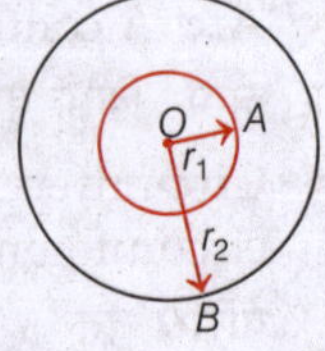

- **Congruent Circles** Circles with equal radii are called congruent circles.
सर्वांगसम वृत्त समान त्रिज्या वाले वृत्तों को सर्वांगसम वृत्त कहा जाता है।

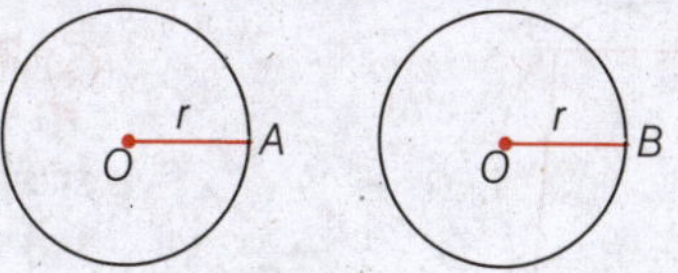

Here/यहाँ, $OA = OB = r$

Properties of Circle वृत्त की विशेषताएँ

- The diameter of a circle is the longest chord of a circle.
वृत्त का व्यास, वृत्त की सबसे लम्बी जीवा होती है।
- The chords that are equidistant from the centre are equal in length.
केन्द्र से समान दूरी पर स्थित जीवाएँ एकसमान लम्बाई की होती हैं।
- Equal chords of a circle subtend equal angles at the centre.
वृत्त की समान जीवाएँ वृत्त के केन्द्र पर समान कोण अन्तरित करती हैं।
- The radius drawn perpendicular to the chord, bisects the chord in equal parts.
जीवा पर लम्बवत् खींची गई त्रिज्या, जीवा को समद्विभाजित करती है।
- If two chords of a circle bisect each other they must be diameters.
यदि किसी वृत्त की दो जीवाएँ एक-दूसरे को समद्विभाजित करती हैं, तो वे सदैव व्यास होंगी।

Cyclic Quadrilateral चक्रीय चतुर्भुज

A quadrilateral whose all vertices lie on a circle is called a cyclic quadrilateral.
Here, $ABCD$ is a cyclic quadrilateral.

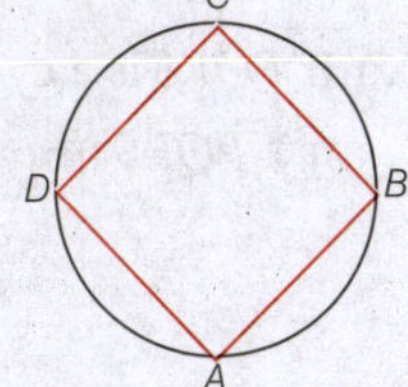

एक चतुर्भुज जिसके सभी शीर्ष एक वृत्त पर स्थित हों, चक्रीय चतुर्भुज कहलाता है।
यहाँ, $ABCD$ एक चक्रीय चतुर्भुज है।

Properties of Cyclic Quadrilateral चक्रीय चतुर्भुज के गुण

(i) The sum of opposite angles of a cyclic quadrilateral is 180° or opposite angles of a cyclic quadrilateral are supplementary.
चक्रीय चतुर्भुज के सम्मुख कोणों का योग 180° होता है या चक्रीय चतुर्भुज के सम्मुख कोण सम्पूरक होते हैं।

$\angle A + \angle C = 180°$ and/और $\angle B + \angle D = 180°$

(ii) The exterior angle formed by producing a side of a cyclic quadrilateral is equal to the opposite interior angle.

चक्रीय चतुर्भुज की एक भुजा बढ़ा देने पर बनने वाला बाह्य कोण, सम्मुख अन्तःकोण के बराबर होता है।

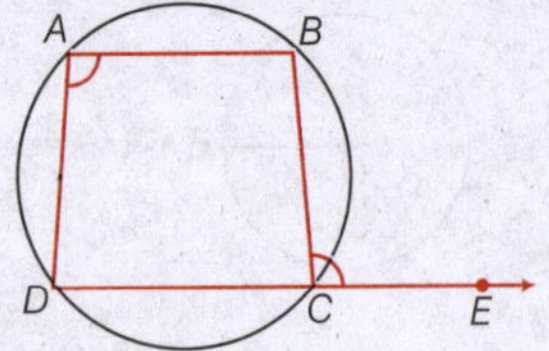

अतिसम्भावित!
चक्रीय चतुर्भुज के गुणों पर आधारित 1 प्रश्न परीक्षा में पूछे जाने की अधिक सम्भावना रहती है।

Here/यहाँ, $\angle BAD = \angle BCE$

(iii) An isosceles trapezium is always cyclic.
एक समद्विबाहु समलम्ब सदैव चक्रीय होता है।

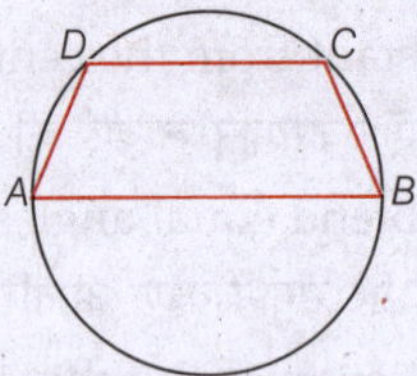

(iv) If two opposite sides of cyclic quadrilateral are equal, then the other two sides are parallel.
यदि किसी चक्रीय चतुर्भुज की दो विपरीत भुजाएँ समान हैं, तो अन्य दो भुजाएँ समानान्तर होती हैं।

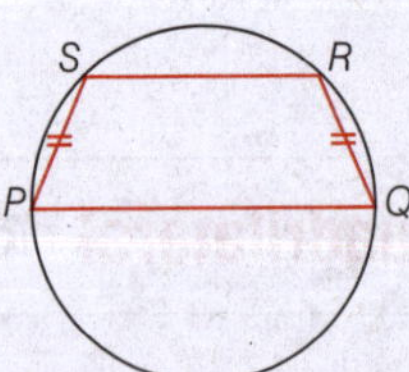

(v) If $PQRS$ is a cyclic quadrilateral and PR and QS are diagonals, then
यदि $PQRS$ एक चक्रीय चतुर्भुज है तथा PR तथा QS उनके विकर्ण हैं, तब

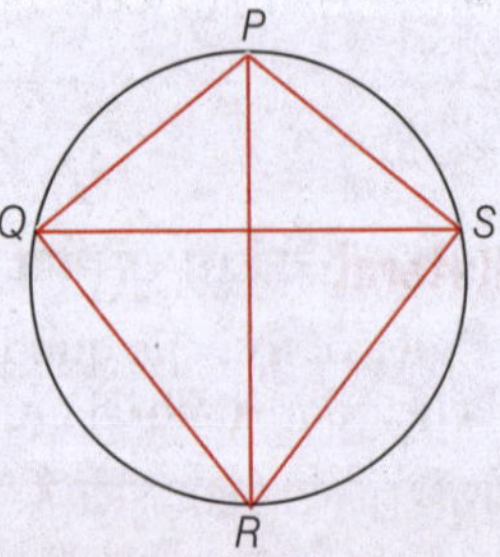

$$PR \times QS = (PQ \times RS) + (QR \times PS)$$

चक्रीय चतुर्भुज के गुणों पर आधारित उदाहरण निम्नलिखित है

Example $ABCD$ **is a cyclic quadrilateral and** BC **is a diameter of the related circle on which** A **and** D **also lie. If** $\angle BCA = 19°$ **and** $\angle CAD = 32°$**, then what is the measure of** $\angle ACD$**?**

$ABCD$ एक चक्रीय चतुर्भुज है और BC सम्बन्धित वृत्त का व्यास है, जिस पर A और D भी स्थित हैं। यदि $\angle BCA = 19°$ तथा $\angle CAD = 32°$ है, तो $\angle ACD$ की माप क्या है? **SSC CGL 2023**

Sol. $\angle BAC = 90°$ (Angle made in semicircle)

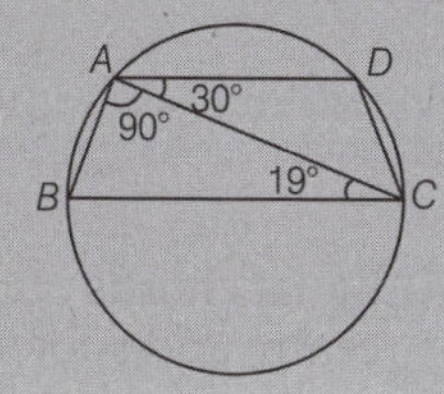

$$\angle A + \angle C = 180°$$

$$\Rightarrow 90° + 32° + 19° + \angle ACD = 180°$$

$$\Rightarrow \angle ACD = 180° - 141°$$

$$\therefore \angle ACD = 39°$$

Tangent स्पर्श रेखा

A line that touches the circle at only one point is known as tangent of a circle.

एक रेखा जो वृत्त को केवल एक बिन्दु पर स्पर्श करती है, वृत्त की स्पर्श रेखा कहलाती है।

Here, RPQ is a tangent/RPQ एक स्पर्श रेखा है।

Note $\angle OPQ = 90°$ or/या $OP \perp PQ$

Secant Line छेदक रेखा

A straight line that intersects a circle at two points is called a secant line of circle.

एक सीधी रेखा जो एक वृत्त को दो बिन्दुओं पर प्रतिच्छेद करती है। वृत्त की छेदक रेखा कहलाती है।

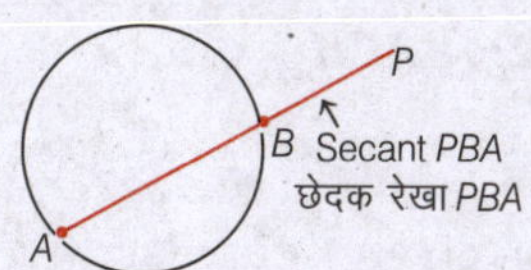

Theorems Based on Circle वृत्त पर आधारित प्रमेय

(i) A perpendicular drawn from the centre of a circle to a chord bisects the chord.

एक वृत्त के केन्द्र से जीवा पर डाला गया लम्ब जीवा को समद्विभाजित करता है।

In the given figure, $OP \perp AB$, then $AP = PB$.

दी गई आकृति में, $OP \perp AB$ है, तब $AP = PB$

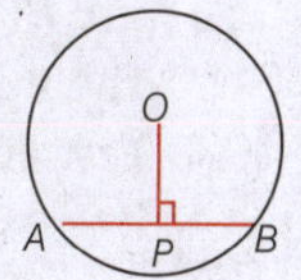

(ii) If two chords intersects each other at 90°.
यदि दो जीवाएँ एक-दूसरे को 90° पर प्रतिच्छेद करती हैं।
i.e./अर्थात् $AB \perp CD$ then/तब, $r = \frac{\sqrt{a^2 + b^2 + c^2 + d^2}}{2}$ or/ या $r = \frac{\sqrt{x^2 + y^2}}{2}$

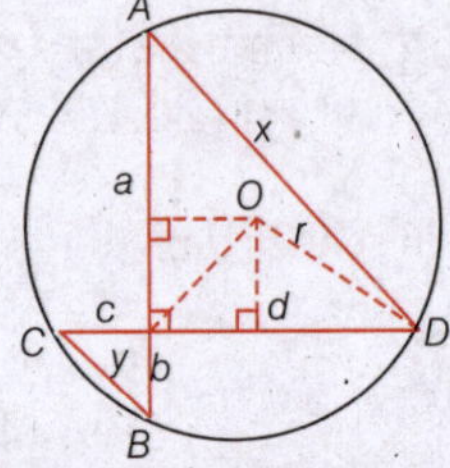

परीक्षा ब्रह्मास्त्र
ये प्रमेय प्रत्येक परीक्षा SSC, Railway, CDS की दृष्टि से M. Imp हैं। इन पर 1-2 प्रश्न अवश्य पूछे जाते हैं।

(iii) Equal chords of a circle are equidistant from the centre and subtends equal angles at the centre.
एक वृत्त की समान जीवाएँ केन्द्र से समदूरस्थ होती हैं तथा केन्द्र पर समान कोण अन्तरित करती है।
If/यदि $AB = CD$, then/तब $OP = OQ$
Where/जहाँ, $OP \perp AB$ and/और $OQ \perp CD$
and/और $\angle AOB = \angle COD$

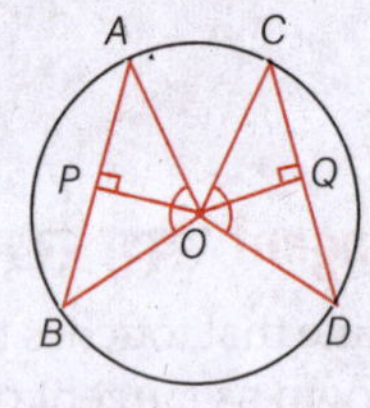

(iv) Angle subtended by an arc at the centre is twice the angle subtended by the same arc at any point on the circumference.
किसी चाप द्वारा केन्द्र पर बनाया गया कोण उसी चाप द्वारा परिधि के किसी बन्दु पर बनाए गए कोण का दोगुना होता है।

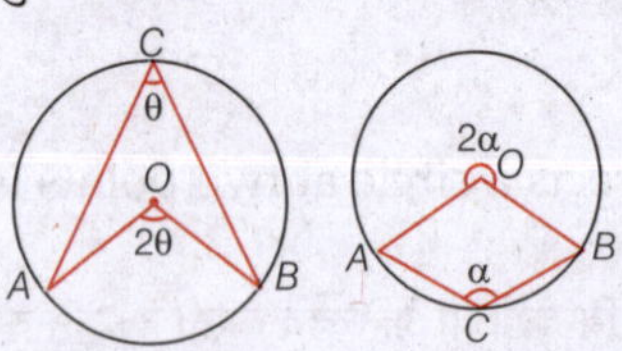

$\angle AOB = 2 \times \angle ACB$

Example **If O is the centre of the circle, then the value of x in the given figure is**
यदि O वृत्त का केन्द्र है, तो दी गई आकृति में x का मान है

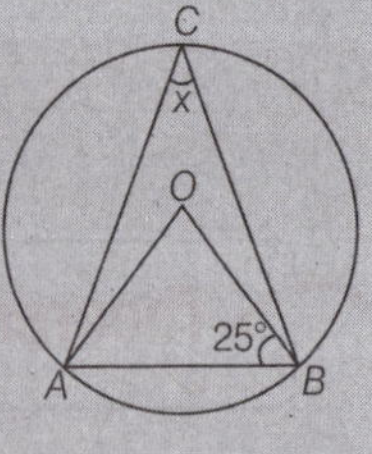

Sol. In ΔAOB,
$$\angle AOB = 180° - (25° + 25°) = 130°$$
$$\therefore \quad \angle ACB = x = \frac{\angle AOB}{2} = \frac{130°}{2} = 65°$$

Example **O is the centre of a circle and A is a point on a major arc BC of the circle. $\angle BOC$ and $\angle BAC$ are the angles made by the minor arc BC on the centre and circumference, respectively. If $\angle ABO = 40°$ and $\angle ACO = 30°$, then find $\angle BOC$.**

O एक वृत्त का केन्द्र है और A वृत्त के बड़े चाप BC पर एक बिन्दु है। $\angle BOC$ और $\angle BAC$ लघुचाप BC द्वारा क्रमशः केन्द्र और परिधि पर बनाए गए कोण हैं। यदि $\angle ABO = 40°$ और $\angle ACO = 30°$ हैं, तो $\angle BOC$ ज्ञात कीजिए।

SSC (10+2) 2023

Sol. In given circle, $OA = OB = OC = r$ (radii of the circle)

$$\angle ABO = 40° \text{ and } \angle ACO = 30°$$

In ΔAOB, $OA = OB$

$\therefore \quad \angle OAB = \angle ABO = 40°$

Now, in ΔAOC

$OA = OC \quad \therefore \angle OAC = \angle ACO = 30°$

Now, $\angle BAC = \angle BAO + \angle OAC = 40° + 30° = 70°$

$\because \quad \angle BAC = \frac{1}{2} \angle BOC$ [$\because$ central angle theorem]

$\therefore \quad \angle BOC = 2 \times 70° = 140°$

Example **In a ΔXYZ, C is circumcentre of this triangle. $\angle Y = 50°$ and $\angle Z = 70°$. What is the value of $\angle YCZ$?**

ΔXYZ में, C त्रिभुज का परिकेन्द्र है। $\angle Y = 50°$ तथा $\angle Z = 70°$ है। $\angle YCZ$ का मान क्या है?

SSC (10+2) 2023

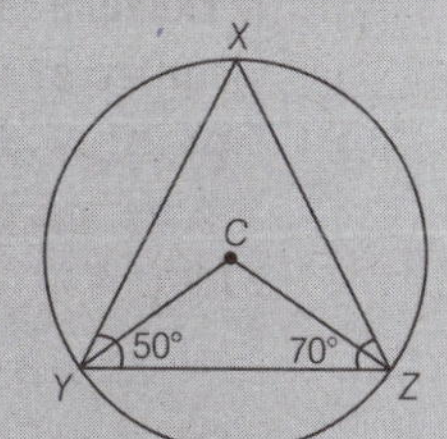

Sol. In ΔXYZ, $\angle X + \angle Y + \angle Z = 180°$

[angle sum property of triangle]

$\Rightarrow \angle X + 50° + 70° = 180°$

$\Rightarrow \angle X = 180° - 120° \Rightarrow \angle X = 60°$

Now, $\quad \angle YCZ = 2\angle X$

$\therefore \quad \angle YCZ = 2 \times 60° = 120°$

(v) Angles in the same segment of a circle are equal.
वृत्त के एक ही वृत्तखण्ड में बने कोण बराबर होते हैं।

$\angle ACB = \angle ADB$ *M. Imp*
and/और $\angle AEB = \angle AFB$

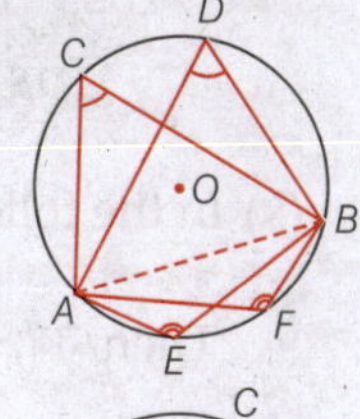

(vi) Angle made in semi-circle is right angle.
अर्द्धवृत्त में बना कोण समकोण होता है।

$$\angle ACB = 90°$$

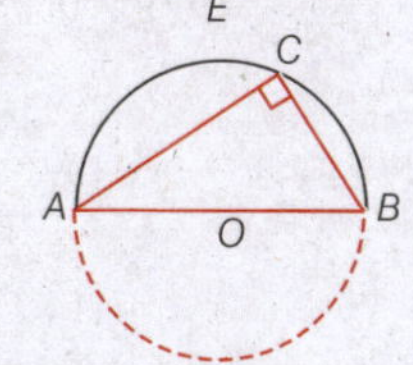

(vii) In the given figure
दी गई आकृति में,
$\angle ACB + \angle OAB = 90°$
or/या $\angle ACB + \angle OBA = 90°$
(where/जहाँ, O = centre/केन्द्र)
$\therefore \quad x + \theta = 90°$

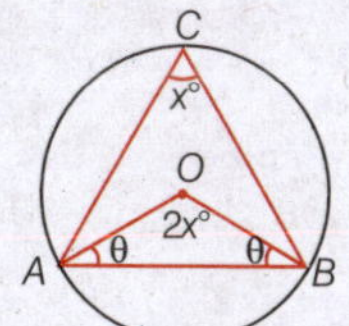

(viii) The angles in the major segment and the minor segment are supplementary.
दीर्घ वृत्तखण्ड और लघु वृत्तखण्ड में बने कोण सम्पूरक होते हैं।

$\angle ACB + \angle ADB = 180°$ *M. Imp*

Where/जहाँ, AB = chord/जीवा

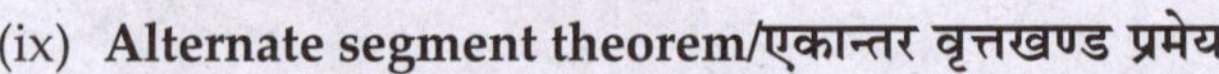

(ix) **Alternate segment theorem/एकान्तर वृत्तखण्ड प्रमेय**

Angle made by a chord and tangent is equal to the angle made by the chord in other segment of the circle.
एक जीवा और स्पर्श रेखा द्वारा बनाया गया कोण जीवा द्वारा वृत्त के दूसरे खण्ड में बनाए गए कोण के बराबर होता है।

$\angle BAP = \angle BCA = \theta$ and/और $\angle BAQ = \angle BCA = \alpha$

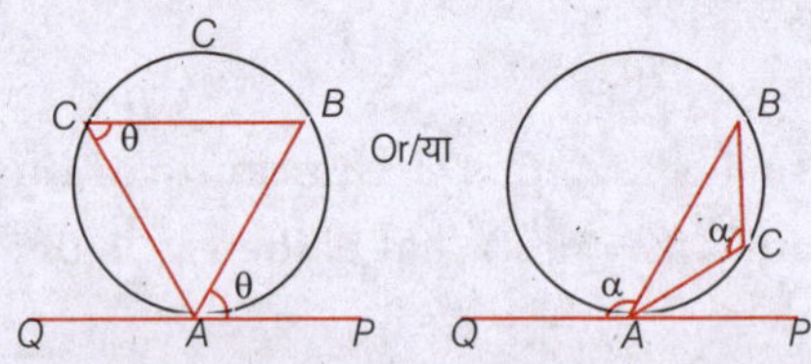

Example **In the given figure, O is the centre of the circle and AB is tangent at point A. If $\angle DAB = 40°$, then $\angle DBA = ?$**

दी गई आकृति में, O वृत्त का केन्द्र है और AB, बिन्दु A पर स्पर्श रेखा है, यदि $\angle DAB = 40°$ है, तब $\angle DBA = ?$

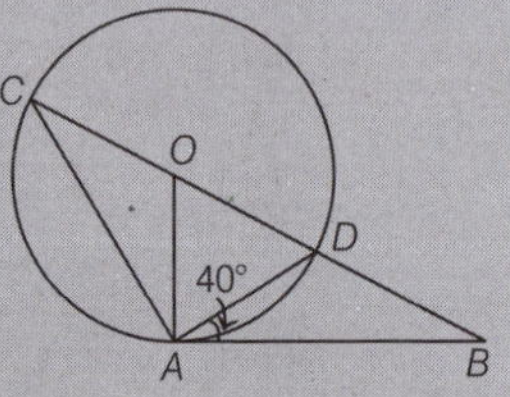

Sol. Here, $\angle BAD = \angle ACD = 40°$ (alternate segment theorem)

Now, $\angle CAB = 40° + 90° = 130°$

$\therefore \angle DBA = 180° - (130° + 40°) = 180° - 170° = 10°$

(x) In the following figure/निम्न आकृति में, $\angle APB = \theta$,

Then/तब, $\angle ADB = 90° + \dfrac{\theta}{2}$, $\angle ACB = 90° - \dfrac{\theta}{2}$ *M. Imp*

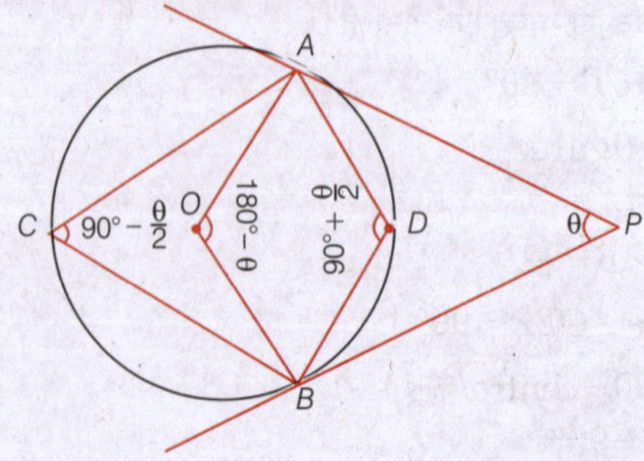

Example **Points A and B are on a circle with centre O. PA and PB are tangents to the circle from an external point P. If PA and PB are inclined to each other at 42°, then find the measure of $\angle OAB$.**

बिन्दु A और B केन्द्र O वाले वृत्त पर दो बिन्दु हैं। PA और PB वृत्त के बाहरी बिन्दु P से स्पर्श रेखाएँ हैं। यदि PA और PB एक-दूसरे से 42° पर झुके हैं, तो $\angle OAB$ की माप ज्ञात कीजिए। *SSC CGL (TIER I) 2022*

Sol. Here, $\angle OAP = \angle OBP = 90°$

From quadrilateral $OAPB$,

$$\angle OAP + \angle APB + \angle OBP + \angle AOB = 360°$$

$$\Rightarrow \quad 90° + 42° + 90° + \angle AOB = 360°$$

$$\angle AOB = 360° - 222° = 138°$$

From ΔOAB, $OA = OB$ (radii)

So, $\angle OAB = \angle OBA$

Also, $\angle OAB + \angle OBA + \angle AOB = 180°$

$$\Rightarrow \quad 2\angle OAB + 138° = 180°$$

$$\therefore \quad \angle OAB = \frac{180° - 138°}{2} = 21°$$

(xi) In the following figure/निम्न आकृति में,

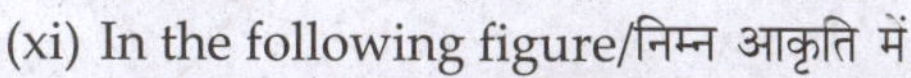

$$\angle APB = \alpha$$

Then/तब, $\angle COD = 180° - 2\alpha$

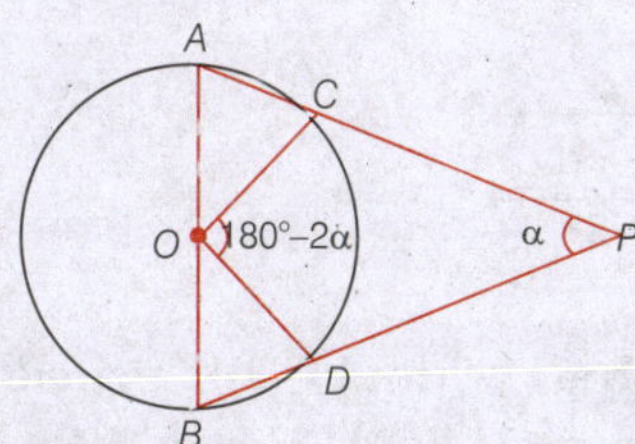

(xii) Angle made in quarter circle (quadrant) is 135°.
चौथाई वृत्त (चतुर्थांश) में बना कोण 135° होता है।

$$\angle AOB = 90°$$

Then/तब, $\angle ACB = 135°$

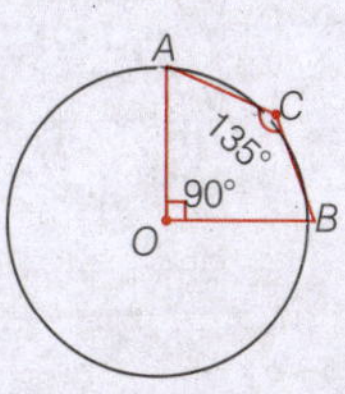

(xiii) AB and CD are two chords which intersect each other internally at point P, then
AB और CD दो जीवाएँ आन्तरिक रूप से बिन्दु P पर काटती हैं, तब

$$\angle APD = \angle BPC = \frac{\angle AOD + \angle BOC}{2}$$

$$\angle APC = 180° - \angle APD$$

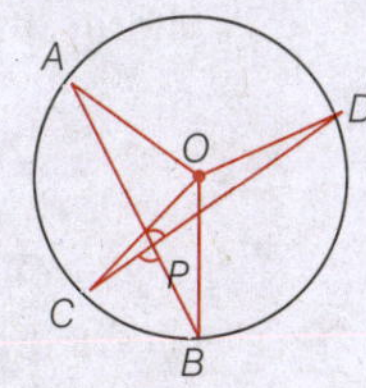

(xiv) AB and CD are two chords which intersect each other externally at point P, then
AB और CD दो जीवाएँ एक-दूसरे को वृत्त के बाहर बिन्दु P पर काटती हैं, तब

$$\angle APC = \angle BPD = \frac{\theta - \alpha}{2}$$

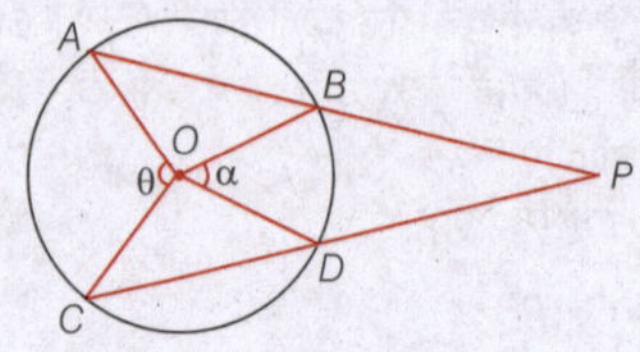

(xv) If a circle is inscribed in a quadrilateral or a quadrilateral is circumscribing a circle. Then,
यदि एक वृत्त एक चतुर्भुज में बना हुआ है या एक चतुर्भुज एक वृत्त के परिगत है, तब

$$AB + CD = BC + AD$$

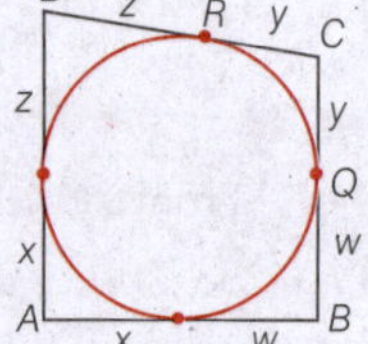

(xvi) Two chords AB and CD of a circle intersect each other at point P internally.
एक वृत्त की दो जीवाएँ AB और CD एक-दूसरे को बिन्दु P पर आन्तरिक प्रतिच्छेद करती हैं।

Then/तब, $PA \times PB = PC \times PD$

and/और $$r = \sqrt{\frac{AP^2 + PD^2 + CP^2 + BP^2}{2}}$$

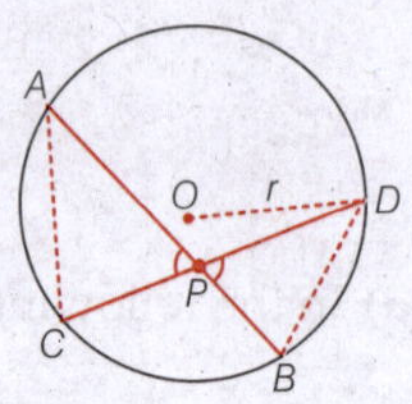

If/यदि $\angle APD = 90°$,

Then/तब $r = \sqrt{\frac{BC^2 + AD^2}{2}}$

Example **Two chords AB and CD of a circle intersect at point F inside the circle. If $AF = 4$ cm, $BF = 6$ cm and $CF = 3$ cm, then find the length of CD.**

वृत्त की दो जीवाएँ AB और CD वृत्त के आन्तरिक बिन्दु F पर प्रतिच्छेद करती हैं। यदि $AF = 4$ सेमी, $BF = 6$ सेमी और $CF = 3$ सेमी हैं, तब CD की लम्बाई ज्ञात कीजिए। ***RRB NTPC 2022***

Sol. $AF \times BF = CF \times DF$

[∵ chords of a circle intersect inside a circle]

$\Rightarrow \quad 4 \times 6 = 3 \times DF$

$\Rightarrow \quad \frac{24}{3} = DF$

$\Rightarrow \quad DF = 8$ cm

$\therefore \quad CD = DF + FC = 8 + 3 = 11$ cm

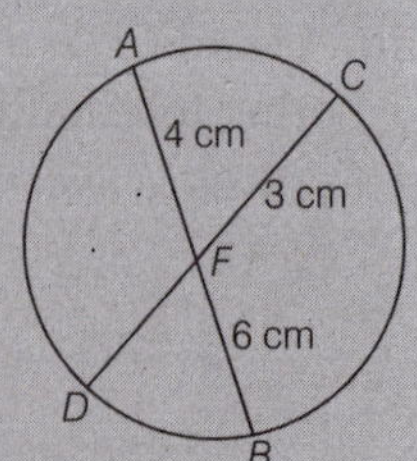

(xvii) Two chords AB and CD of a circle intersect each other at point P externally.
एक वृत्त की दो जीवाएँ AB और CD एक-दूसरे को बिन्दु P पर बाह्यतः प्रतिच्छेद करती हैं।

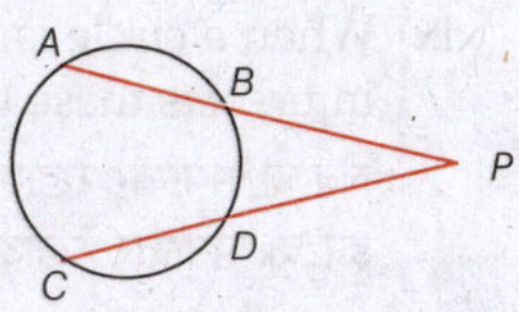

Then/तब, $PB \times PA = PD \times PC$

Example **In the given figure, diameter AB and chord CD of a circle meet at P. PT is a tangent to the circle at T. If $CD = 7.8$ cm, $PD = 5$ cm, $PB = 4$ cm, then find AB.**

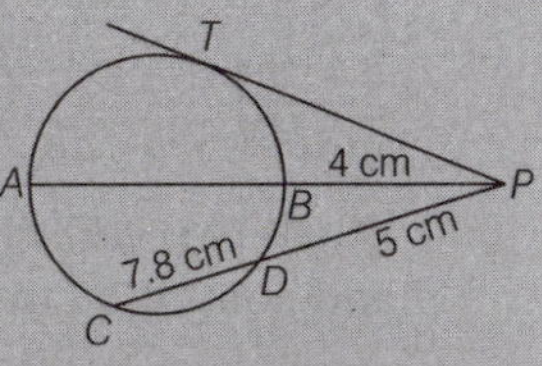

दी गई आकृति में, वृत्त का व्यास AB और जीवा CD बिन्दु P पर मिलते हैं। PT वृत्त के बिन्दु T पर स्पर्श रेखा है। यदि $CD = 7.8$ सेमी, $PD = 5$ सेमी, $PB = 4$ सेमी, तब AB ज्ञात कीजिए। ***DP HEAD CONSTABLE 2023***

Sol. By intersecting chord theorem,

$\because \quad PA \times PB = PC \times PD \Rightarrow PA \times 4 = (7.8 + 5) \times 5$

$\Rightarrow \quad PA = \frac{12.8 \times 5}{4} = 16 \text{ cm}$

$\therefore \quad AB = PA - PB = 16 - 4 = 12 \text{ cm}$

(xviii) If a chord and a tangent of a circle intersect each other at a point P externally, then
यदि एक वृत्त की एक जीवा और एक स्पर्श रेखा एक-दूसरे को बिन्दु P पर बाह्यतः प्रतिच्छेद करती हैं, तब

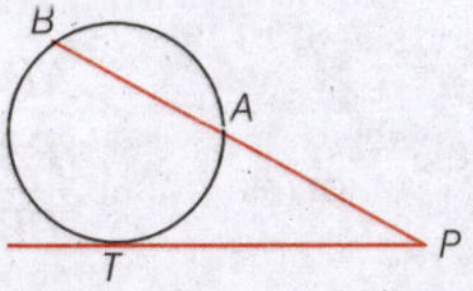

$$PA \times PB = PT^2$$

Example **In the following figure, AD is tangent of the circle and ABC is secant. If $AB = 4$ cm and $BC = 5$ cm, then the length of tangent AD is**

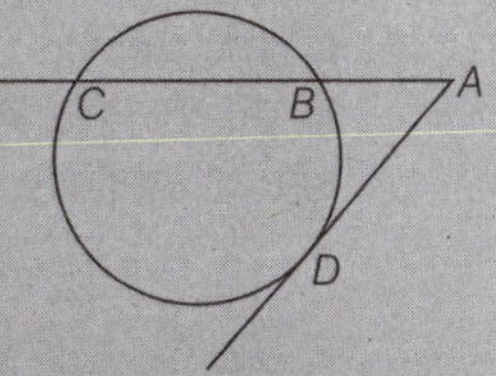

दी गई आकृति में, AD वृत्त की स्पर्श रेखा है तथा ABC एक छेदक रेखा है। यदि $AB = 4$ सेमी तथा $BC = 5$ सेमी है, तब स्पर्श रेखा AD की लम्बाई है ***UP TGT 2021, UP TET 2018***

Sol. We know, $AD^2 = AC \times AB$

$\Rightarrow \quad AD = \sqrt{(5 + 4) \times 4}$

$= \sqrt{9 \times 4} = 6 \text{ cm}$

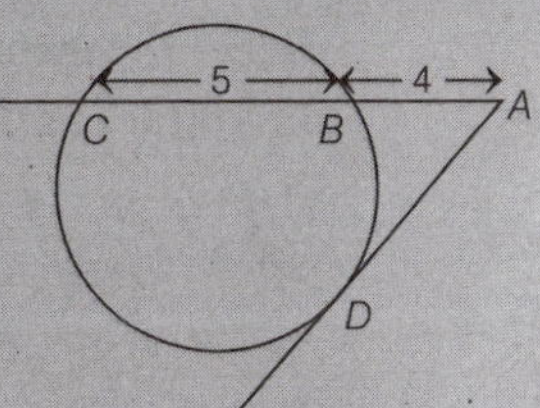

(xix) When a circle is drawn between two parallel lines and a tangent intersects these two parallel lines.

जब दो समान्तर रेखाओं के बीच एक वृत्त खींचा जाता है और इन दो समान्तर रेखाओं को एक स्पर्श रेखा काटती है।

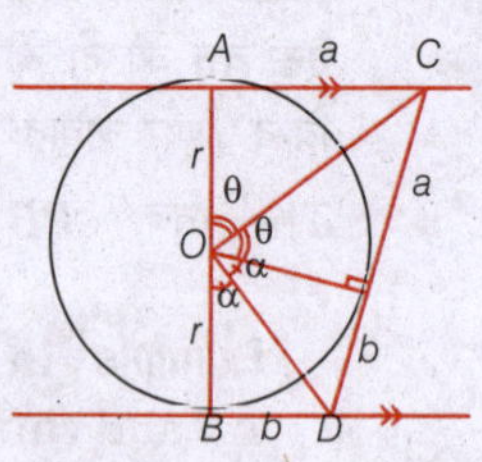

$$\angle COD = 90^\circ$$

$$2\theta + 2\alpha = 180^\circ$$

$$\theta + \alpha = 90^\circ$$

and/तथा $r = \sqrt{ab}$

(xx) When two circles touch each other internally and common tangent is given, then

जब दो वृत्त एक-दूसरे को आन्तरिक रूप से स्पर्श करते हैं और उभयनिष्ठ स्पर्श रेखा दी जाती है, तब

$\Delta PAC \sim \Delta PBD$

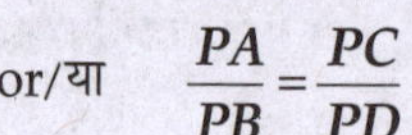
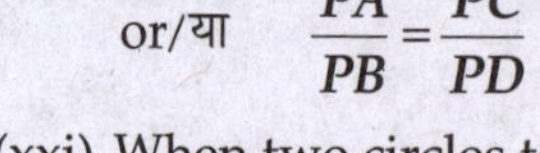

or/या $\dfrac{PA}{PB} = \dfrac{PC}{PD}$

(xxi) When two circles touch each other externally and common tangent is given, then

जब दो वृत्त एक-दूसरे को बाह्य रूप से स्पर्श करते हैं और उभयनिष्ठ स्पर्श रेखा दी जाती है, तब

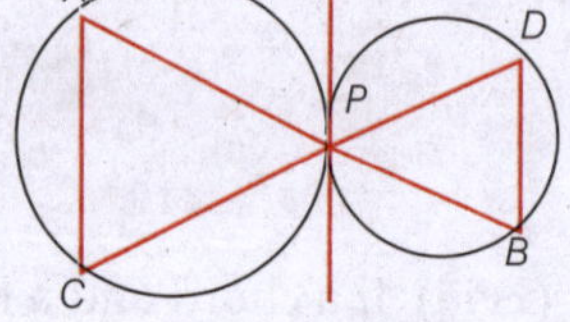

$$\Delta PAC \sim \Delta PBD$$

or/या $$\frac{PA}{PB} = \frac{PC}{PD}$$

(xxii) Two circles intersect each other at two distinct points and two common tangents are given, then

दो वृत्त एक-दूसरे को दो भिन्न बिन्दुओं पर प्रतिच्छेद करते हैं और दो उभयनिष्ठ स्पर्श रेखाएँ दी गई हैं, तब

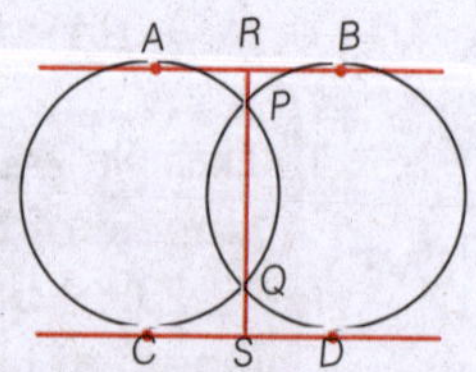

$AB = CD$ and/तथा $RP = QS$,

$RS^2 = AB^2 + PQ^2$

Example **Two circles touch each other externally at P. AB is a direct common tangent to the two circles, A and B are points of contact and $\angle PAB = 40^\circ$. The measure of $\angle ABP$ is**

दो वृत्त P पर एक-दूसरे को बाह्य रूप से स्पर्श करते हैं। AB एक उभयनिष्ठ स्पर्श रेखा है, जिसके A और B सम्पर्क बिन्दु हैं तथा $\angle PAB = 40^\circ$ है। $\angle ABP$ की माप है

SSC CGL (TIER I) 2022

Sol. Given, $\angle PAB = 40°$

Tangent at P be PC.

In ΔACP, $\quad AC = PC \qquad$ [tangents]

$\therefore \qquad \angle PAC = \angle APC = 40°$

[angles opposite to equal sides of a triangle are equal]

By angle sum property,

$$\angle ACP = 180° - 2 \times 40° = 100°$$

$$\therefore \qquad \angle PCB = 180° - 100° = 80°$$

Similarly, in ΔBCP,

$$PC = BC \qquad \text{[tangents]}$$

$$\therefore \qquad \angle CBP = \angle BPC$$

Now, $\qquad \angle ABP = \angle CBP = \dfrac{180° - 80°}{2}$

[by angle sum property, in ΔPBC]

$$= \frac{100°}{2} = 50°$$

(xxiii) In the given figure,

दी गई आकृति में,

$AB \times AC + AE \times DE = AE^2$

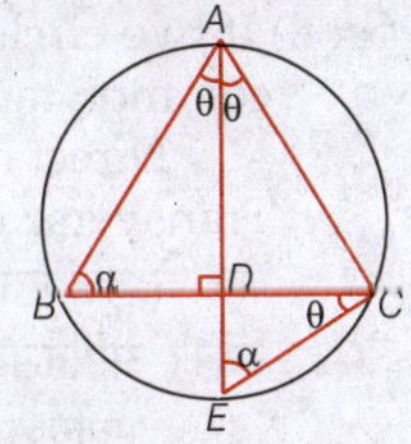

$\Delta ABD \quad \Delta AEC$

$$\Rightarrow \qquad \frac{AD}{AC} = \frac{AB}{AE}$$

$$\Rightarrow \quad \boldsymbol{AB \times AC = AD \times AE}$$

Example ***PQR* is an equilateral triangle inscribed in a circle. *S* is any point on the arc *QR*. Measure of $\frac{1}{2} \angle PSQ$ is**

PQR एक समबाहु त्रिभुज है, जो वृत्त के अन्दर है। चाप QR पर S कोई एक बिन्दु है। $\frac{1}{2} \angle PSQ$ का मान है

SSC CPO 2023

Sol. Here, $\angle QRP = 60°$ (equilateral triangle)

$\angle QSP = \angle QRP = 60°$ (angles in same segment)

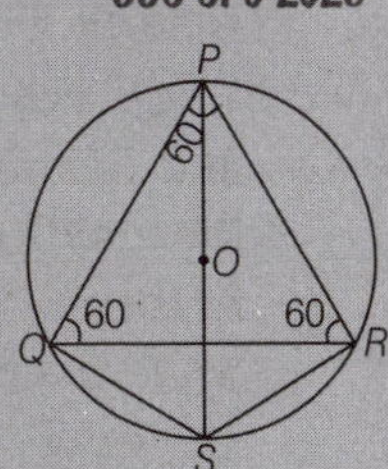

$$\therefore \qquad \frac{1}{2} \angle PSQ = \frac{60°}{2}$$

$$= 30°$$

(xxiv) When two circles touch each other externally. Then distance between their centres is equal to the sum of their radii, then maximum common tangents $= 3$

जब दो वृत्त एक-दूसरे को बाह्य स्पर्श करते हैं, तब उनके केन्द्रों की बीच की दूरी उनकी त्रिज्याओं के योग के बराबर होती है, तब

अधिकतम उभयनिष्ठ स्पर्श रेखाएँ $= 3$

and/तथा $AB = CD = EF$

(xxv) When two circles touch each other internally, then distance between their centres is equal to the difference of their radii.

जब दो वृत्त एक-दूसरे को अन्तःस्पर्श करते हैं, तब उनके केन्द्रों के बीच की दूरी उनकी त्रिज्याओं के अन्तर के बराबर होती है।

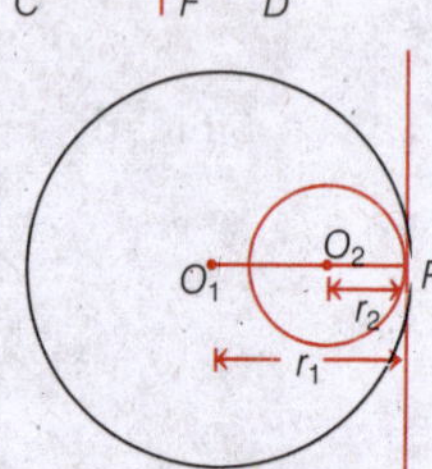

O_1O_2P is a straight line/O_1O_2P एक सीधी रेखा है

$$O_1O_2 = r_1 - r_2$$

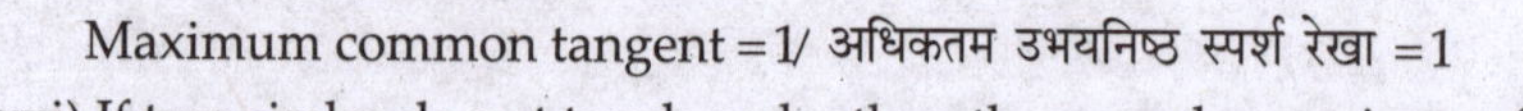

Maximum common tangent $= 1$/ अधिकतम उभयनिष्ठ स्पर्श रेखा $= 1$

(xxvi) If two circles do not touch each other, there can be maximum four common tangents.

Direct common tangents $= 2$

Transverse common tangents $= 2$

जब दो वृत्त एक-दूसरे को स्पर्श नहीं करते हैं, तब उनकी अधिकतम चार स्पर्श रेखाएँ हो सकती हैं।

उभयनिष्ठ अनुस्पर्श रेखाएँ $= 2$

उभयनिष्ठ तिर्यक स्पर्श रेखाएँ $= 2$

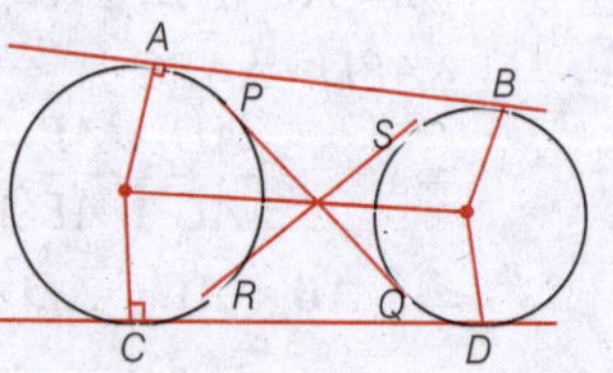

Direct common tangents are AB and CD.

Transverse common tangents are PQ and RS.

Maximum common tangents $= 4$

उभयनिष्ठ अनुस्पर्श रेखाएँ AB और CD हैं।

उभयनिष्ठ तिर्यक स्पर्श रेखाएँ PQ और RS हैं।

अधिकतम उभयनिष्ठ स्पर्श रेखाओं की संख्या 4 है।

(xxvii) PQR is a triangle with inradius r. AB, CD and EF are 3 tangents to the incircle of PQR the inradii of PAB, QCD, REF are r_1, r_2 and r_3, then

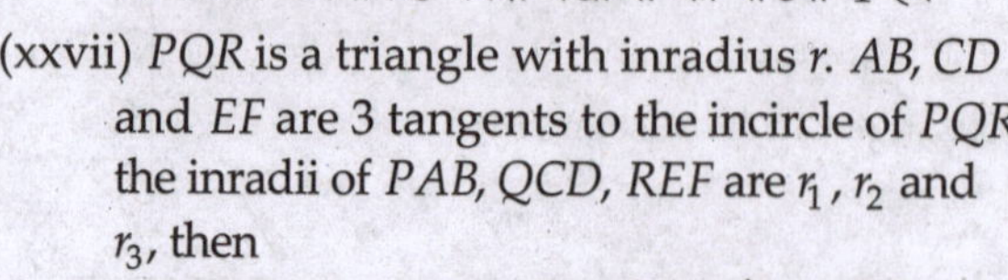

PQR अन्तःत्रिज्या r वाला एक त्रिभुज है। AB, CD और EF, PQR के अन्तःवृत्त की 3 स्पर्श रेखाएँ हैं, PAB, QCD, REF की अन्तःत्रिज्या r_1, r_2, r_3 हैं, तो

$$r = r_1 + r_2 + r_3$$

or/या $r = \sqrt{r_1 r_2} + \sqrt{r_2 r_3} + \sqrt{r_3 r_1}$

(xxviii) If the incircle of ΔABC touches the sides BC, AC and AB at D, E and F respectively, then

$AF + BD + CE = AE + BF + CD = \frac{1}{2}$ (Perimeter of ΔABC)

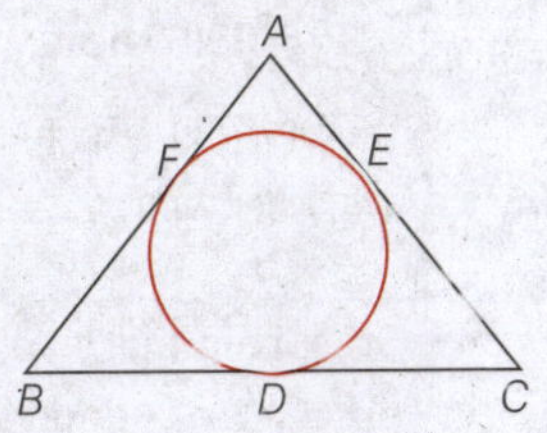

यदि ΔABC का अन्तःवृत्त क्रमशः D, E और F पर भुजाओं BC, AC और AB को स्पर्श करता है, तो

$AF + BD + CE = AE + BF + CD$
$= \frac{1}{2}$ (ΔABC का परिमाप)

Example **A circle is inscribed inside a triangle PQR. It touches the side PQ, QR and PR at the points D, E and F, respectively. If $DQ = 3$ cm, $RE = 5$ cm and $PF = 7$ cm, then the perimeter (in cm) of triangle PQR is**

एक वृत्त किसी त्रिभुज PQR के अन्तःनिर्मित है। यह भुजाओं PQ, QR और PR को क्रमशः D, E और F बिन्दुओं पर स्पर्श करती हैं। यदि $DQ = 3$ सेमी, $RE = 5$ सेमी तथा $PF = 7$ सेमी हैं, तब ΔPQR का परिमाप (सेमी में) ज्ञात कीजिए।

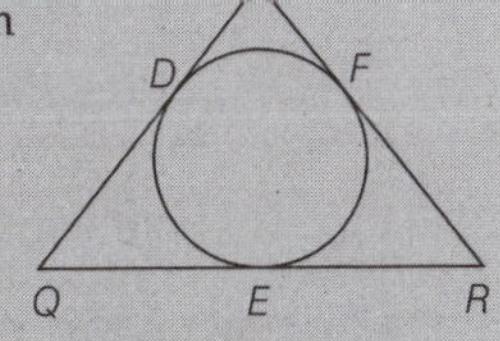

Sol. Given, $DQ = 3$ cm, $RE = 5$ cm and $PF = 7$ cm

Perimeter of $\Delta PQR = 2(DQ + ER + PF)$
$= 2(3 + 5 + 7)$
$= 2 \times 15 = 30$ cm

Lengths of Tangents स्पर्श रेखाओं की लम्बाइयाँ

(i) The lengths of two tangents drawn from the external point to the circle are equal

बाह्यः बिन्दु से एक वृत्त पर खींची गई दो स्पर्श रेखाएँ बराबर होती हैं **$PT = PQ$**

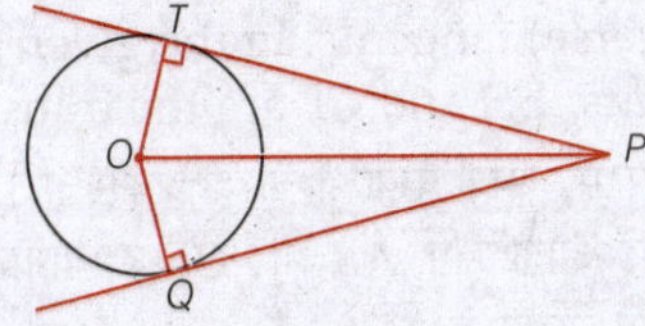

PT or $PQ = \sqrt{(OP)^2 - (OT)^2}$

Example **A is a point outside of a circle with centre O. AP and AQ are two tangents of the circle. If $AP = a^2 + 14$ and $AQ = 239$, then what is the value of a?**

केन्द्र O वाले वृत्त के बाहर एक बिन्दु A है। AP तथा AQ वृत्त की दो स्पर्श रेखाएँ हैं। यदि $AP = a^2 + 14$ तथा $AQ = 239$ हैं, तब a का मान ज्ञात कीजिए।

SSC (10+2) 2023

Sol. Given that,

$AP = a^2 + 14$ and $AQ = 239$

$\because AP = AQ$ [tangents drawn from an external point to a circle are equal.]

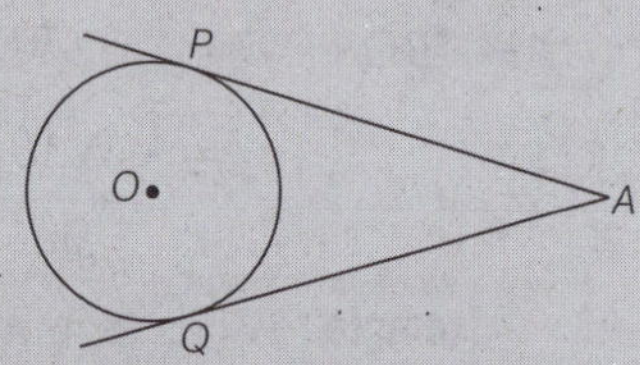

$\Rightarrow \quad a^2 + 14 = 239$

$\Rightarrow \quad a^2 = 239 - 14$

$\Rightarrow \quad a^2 = 225 \Rightarrow a = 15$

Example **If two tangents inclined at an angle 120° are drawn to a circle of radius 6 cm, then what is the length (in cm) of each tangent?**

यदि दो स्पर्श रेखाएँ, जो 120° पर झुकी हैं, 6 सेमी त्रिज्या वाले एक वृत्त पर खींची जाती हैं, तो प्रत्येक स्पर्श रेखा की लम्बाई (सेमी में) क्या है? *SSC (10+2) 2023*

Sol. In ΔPOR and ΔPOQ,

$OR = OQ$ (radii)

$PR = PQ$ (tangents)

$\angle R = \angle Q = 90°$

So, $\Delta POR \cong \Delta POQ$ [by SAS]

By CPCT, $\angle OPQ = \angle OPR = \dfrac{\angle QPR}{2} = \dfrac{120°}{2} = 60°$

In ΔPQR, $\tan 60° = \dfrac{OR}{PR} \Rightarrow \sqrt{3} = \dfrac{6}{PR}$

$\therefore \quad PR = \dfrac{6}{\sqrt{3}} = 2\sqrt{3}$ cm

(ii) In a given pair of circles, there are two types of tangents. The direct and cross (or transverse) tangents. In the given figure AB and CD are the direct tangents. EH and GF are the transverse tangents.

वृत्तों के दिए गए एक युग्म में, स्पर्श रेखाओं के दो प्रकार होते हैं। अनुस्पर्श रेखाएँ और तिर्यक स्पर्श रेखाएँ। दी गई आकृति में AB और CD उभयनिष्ठ अनुस्पर्श रेखाएँ हैं। EH और GF तिर्यक स्पर्श रेखाएँ हैं।

Length of the direct common tangent

उभयनिष्ठ अनुस्पर्श रेखा की लम्बाई $= \sqrt{d^2 - (r_1 - r_2)^2}$

Length of the transverse tangent

उभयनिष्ठ तिर्यक स्पर्श रेखा की लम्बाई $= \sqrt{d^2 - (r_1 + r_2)^2}$

Where, d is the distance between their centres.
जहाँ, d वृत्तों के केन्द्रों के बीच की दूरी है।

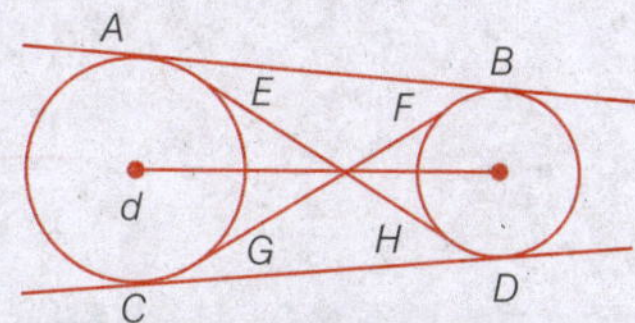

Example **Radii of two circles are 11 cm and 3 cm. Distance between their centres is 17 cm. What is the length of the direct common tangent?**

दो वृत्त की त्रिज्याएँ 11 सेमी तथा 3 सेमी हैं। उनके केन्द्रों के बीच की दूरी 17 सेमी है। उभयनिष्ठ स्पर्श रेखा की लम्बाई ज्ञात कीजिए। *SSC (10+2) 2023*

Sol. Length of direct common tangent

$$= \sqrt{D^2 - (r_1 - r_2)^2}$$

$$= \sqrt{17^2 - (11-3)^2} \qquad \{\because r_1 = 11, r_2 = 3\}$$

$$= \sqrt{289 - 64} = \sqrt{225} = 15 \text{ cm}$$

Common Chord उभयनिष्ठ जीवा

(i) When two circles of radii R and r intersect and AB is the common chord, then
जब R तथा r त्रिज्या के दो वृत्त एक-दूसरे को प्रतिच्छेद करते हों तथा AB उनकी उभयनिष्ठ जीवा हो, तो

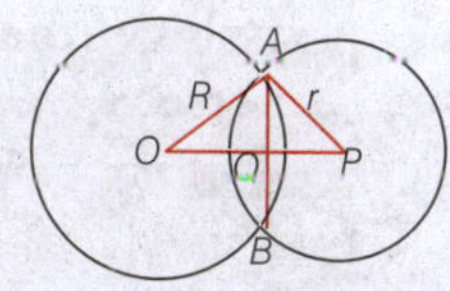

$$AQ = BQ = \frac{AB}{2}, \angle AQO = \angle AQP = 90°$$

$$OP = \sqrt{R^2 - \left(\frac{AB}{2}\right)^2} + \sqrt{r^2 - \left(\frac{AB}{2}\right)^2}$$

(ii) When radii of both the circles are equal, then/
जब दोनों वृत्तों की त्रिज्या समान हों, तब

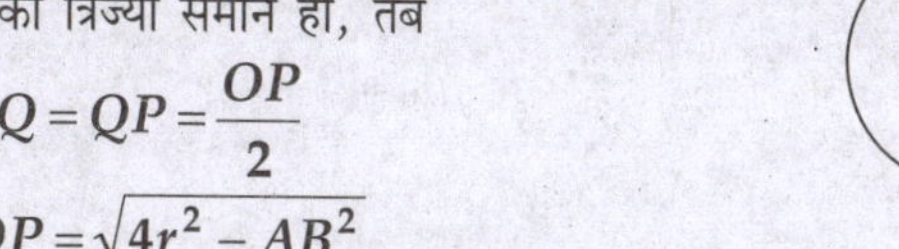

$$OQ = QP = \frac{OP}{2}$$

$$OP = \sqrt{4r^2 - AB^2}$$

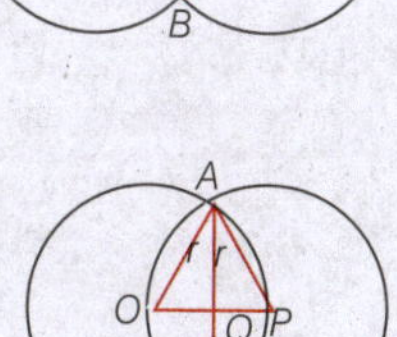

(iii) When radii of both the circles are equal and both circles pass through the centres of the other circle, then
जब दो वृत्तों की त्रिज्या बराबर हों तथा दोनों वृत्त एक-दूसरे के केन्द्र से होकर गुजरते हों, तब

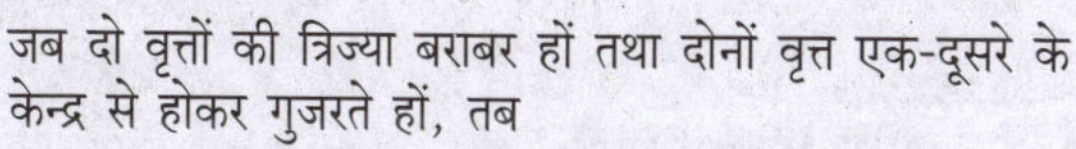

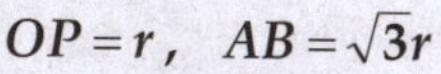

$$OP = r, \quad AB = \sqrt{3}r$$

(iv) $\Rightarrow$ Here, $PR = 2\sqrt{ab}$

$$PQ = 2\sqrt{ac}$$

$$QR = 2\sqrt{bc}$$

$$\sqrt{ab} = \sqrt{ac} + \sqrt{bc}$$

and $$\frac{1}{\sqrt{c}} = \frac{1}{\sqrt{b}} + \frac{1}{\sqrt{a}}$$

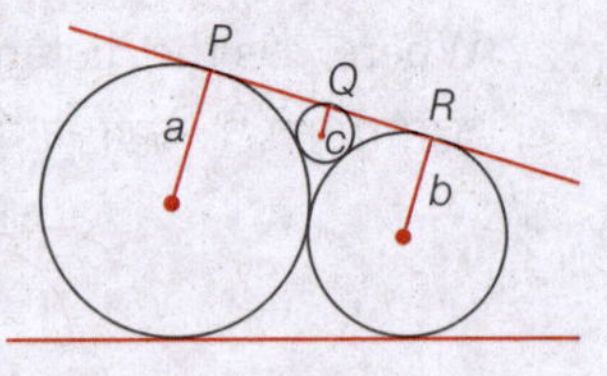

Example **Two circles of radius 13 cm and 15 cm intersect each other at points *A* and *B*. If the length of the common chord is 24 cm, then what is the distance between their centres?**

13 सेमी और 15 सेमी त्रिज्या वाले दो वृत्त एक-दूसरे को बिन्दु A और B पर प्रतिच्छेद करते हैं। यदि उभयनिष्ठ जीवा की लम्बाई 24 सेमी है, तो उनके केन्द्रों के बीच की दूरी क्या है? *SSC CGL (TIER II) 2022*

Sol. In ΔOAC,

$$OC = \sqrt{(13)^2 - (12)^2} = \sqrt{169 - 144} = \sqrt{25} = 5 \text{ cm}$$

In $\Delta O'AC$,

$$O'C = \sqrt{(15)^2 - (12)^2} = \sqrt{225 - 144} = \sqrt{81} = 9 \text{ cm}$$

$\therefore$ Distance between their centres $= 5 + 9 = 14$ cm

Example **ΔABC is an isosceles triangle with $AB = AC$. A circle passing through B touching AC at the middle point and intersects AB at M. The $AM : AB$ is**

ΔABC एक समद्विबाहु त्रिभुज है, जिसमें $AB = AC$ है। B से होकर जाने वाला एक वृत्त AC को मध्य-बिन्दु पर स्पर्श करता है और AB को M पर प्रतिच्छेद करता है। तब $AM : AB$ है

Sol. D is the mid point of AC

$\therefore$ $$AD = \frac{AC}{2} = \frac{AB}{2} \qquad (\because AB = AC)$$

Now $$AM \times AB = AD^2$$

$\Rightarrow$ $$AM \times AB = \frac{AB^2}{4} \Rightarrow AM : AB = 1 : 4$$

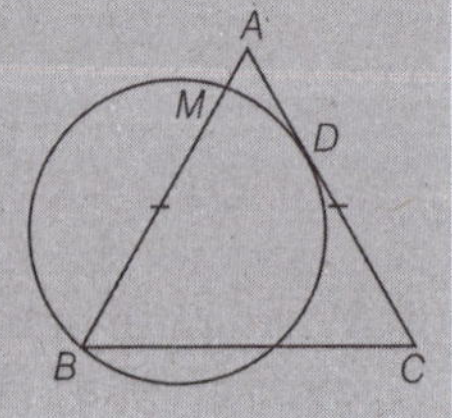

अध्याय 06

Mensuration-2D

द्विविमीय क्षेत्रमिति

- The literal meaning of mensuration is measure. It's generally used where geometrical figures are concerned and its parameters are area, length, volume, lateral surface area, etc.

  In mensuration 2D, we will study the measurements of area and perimeter of the figures.

  क्षेत्रमिति का शाब्दिक अर्थ मापना है। इसका उपयोग आमतौर पर वहाँ किया जाता है, जहाँ ज्यामितीय आकृतियों पर विचार होता है और इसके मापदण्ड क्षेत्रफल, लम्बाई, आयतन, वक्र पृष्ठ क्षेत्रफल इत्यादि हैं।

  द्वि-आयामी क्षेत्रमिति में, हम आकृतियों के क्षेत्रफल और परिमाप की माप का अध्ययन करेंगे।

Area क्षेत्रफल

- Area is defined as the total space taken up by a flat surface (2D) or shape of an object.

  क्षेत्रफल को किसी वस्तु की समतल (सपाट) सतह द्वारा घेरे जाने वाले कुल स्थान के रूप में परिभाषित किया गया है।

Perimeter परिमाप

- Perimeter is defined as the distance or length around the boundary of a two-dimensional shape.

  परिमाप को एक द्वि-आयामी आकृति की सीमा के चारों ओर की दूरी या लम्बाई के रूप में परिभाषित किया गया है।

1. **Scalene Triangle/विषमबाहु त्रिभुज**

   Perimeter/परिमाप $= a + b + c$

   Area/क्षेत्रफल $\Delta = \sqrt{s(s-a)(s-b)(s-c)}$,

   Where/जहाँ, $s = \dfrac{a+b+c}{2}$

   ($s \rightarrow$ Semi-perimeter/अर्द्ध-परिमाप)

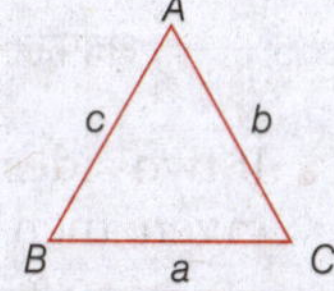

2. **Right Angle Triangle/समकोण त्रिभुज**

Perimeter/परिमाप $= P + B + H$

Area/क्षेत्रफल $= \frac{1}{2} PB$

3. **Isosceles Right Angle Triangle समकोण समद्विबाहु त्रिभुज**

Perimeter/परिमाप $= H + \sqrt{2}H$

Area/क्षेत्रफल $= \frac{H^2}{4}$

4. **Equilateral Triangle/समबाहु त्रिभुज**

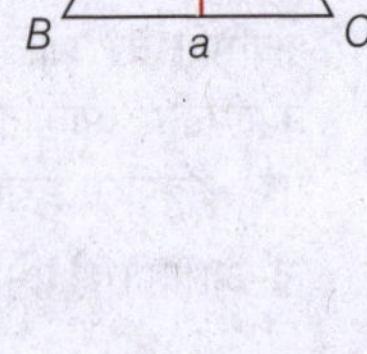

Perimeter/परिमाप $= 3a$

Area/क्षेत्रफल $= \frac{\sqrt{3}}{4} a^2$

Height/ऊँचाई $= \frac{\sqrt{3}}{2} a$

Inradius/अन्तः त्रिज्या $(r) = \frac{a}{2\sqrt{3}}$ or $\frac{h}{3}$

Circumradius/परित्रिज्या $(R) = \frac{a}{\sqrt{3}}$ or $\frac{2h}{3}$

5. **Isosceles Triangle/समद्विबाहु त्रिभुज**

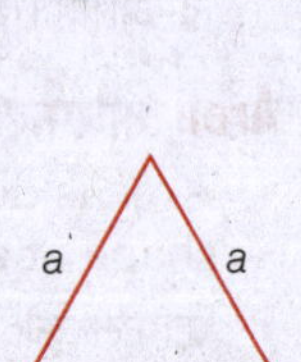

Perimeter/परिमाप $= 2a + b$

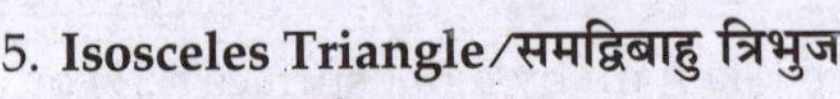

Area/क्षेत्रफल $= \frac{b}{4}\sqrt{4a^2 - b^2}$ M. Imp

Height/ऊँचाई $= \frac{1}{2}\sqrt{4a^2 - b^2}$

Some Important Properties of Triangle त्रिभुज के कुछ महत्त्वपूर्ण गुण

- If one side (base) and corresponding height (altitude) of triangle is given, then
  यदि त्रिभुज की एक भुजा (आधार) और संगत ऊँचाई (शीर्ष लम्ब) दिया गया हो, तब

  Area $= \frac{1}{2} \times$ side $\times$ corresponding height

  क्षेत्रफल $= \frac{1}{2} \times$ भुजा $\times$ संगत ऊँचाई

- If two sides of triangle and angle between them is given, then
  यदि त्रिभुज की दो भुजाएँ और उनके बीच का कोण दिया गया हो, तब

  Area/क्षेत्रफल $= \frac{1}{2} \times bc\sin\alpha = \frac{1}{2} \times ca \times \sin\beta = \frac{1}{2} \times ab \times \sin\gamma$

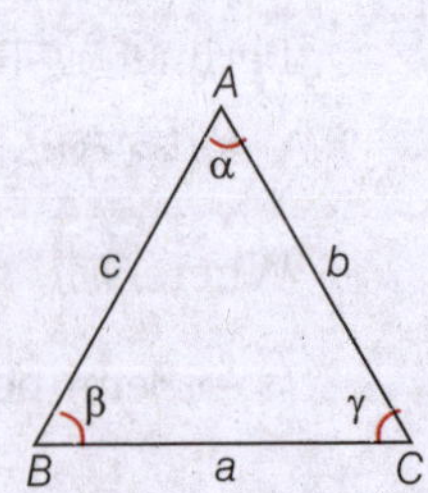

- If the length of the perpendicular drawn from a point inside an equilateral triangle to its sides are P_1, P_2 and P_3, then
यदि किसी समबाहु त्रिभुज के अन्दर किसी बिन्दु से इसकी भुजाओं पर डाले गए लम्बों की लम्बाई P_1, P_2 और P_3 हों, तब

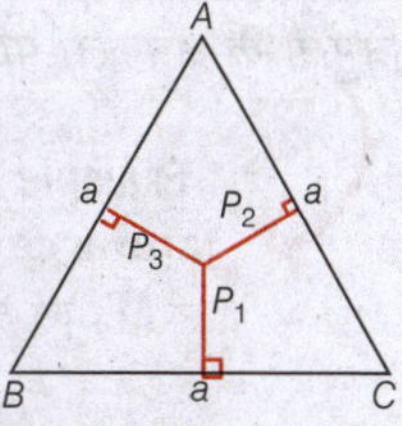

$$\therefore \textbf{ Side/भुजा}, a = \frac{2}{\sqrt{3}}(P_1 + P_2 + P_3)$$ *M. Imp*

यह सूत्र SSC परीक्षा की दृष्टि से महत्वपूर्ण है।

- If in ΔABC, the length of medians AD, BE and CF are a, b and c respectively, then
यदि ΔABC में, माध्यिकाएँ AD, BE और CE की लम्बाई क्रमशः a, b और c हैं, तब

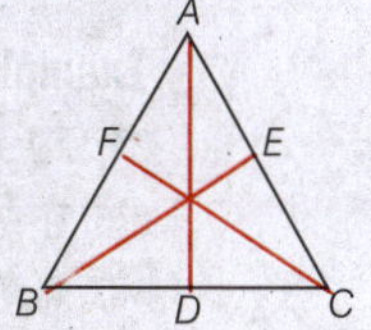

Area of ΔABC/ΔABC का क्षेत्रफल

$$= \frac{4}{3}\sqrt{s(s-a)(s-b)(s-c)}$$

Where/जहाँ, $s = \frac{a+b+c}{2}$

- If P, Q and R be the mid point of sides AB, BC and AC, respectively.
यदि P, Q और R क्रमशः भुजाओं AB, BC और AC के मध्य बिन्दु हो, तो

(i) Perimeter of $\Delta PQR = \frac{1}{2} \times$ Perimeter of ΔABC

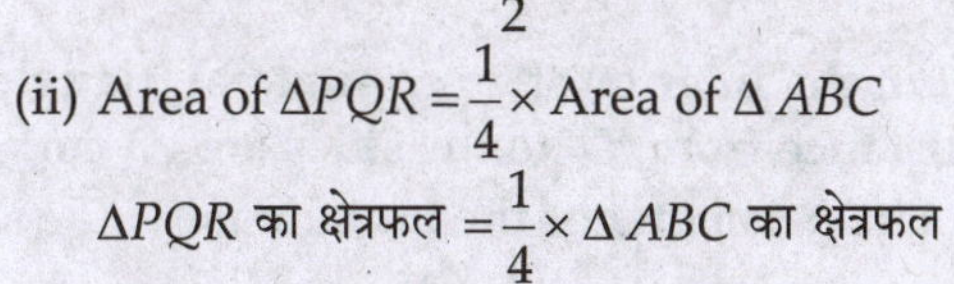

ΔPQR का परिमाप $= \frac{1}{2} \times \Delta ABC$ का परिमाप

(ii) Area of $\Delta PQR = \frac{1}{4} \times$ Area of ΔABC

ΔPQR का क्षेत्रफल $= \frac{1}{4} \times \Delta ABC$ का क्षेत्रफल

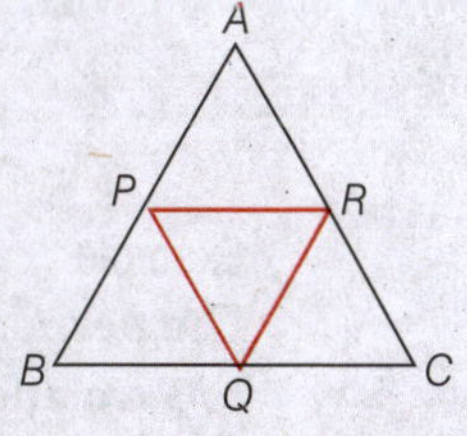

- Inradius of a triangle/ त्रिभुज के अन्तःवृत्त की त्रिज्या

$$= \frac{\text{Area of triangle}}{\text{Semi-perimeter of triangle}}$$

$$= \frac{\text{त्रिभुज का क्षेत्रफल}}{\text{त्रिभुज का अर्द्ध-परिमाप}} = \frac{\Delta}{S}$$

- Circumradius of triangle/त्रिभुज के परिवृत्त की त्रिज्या $= \frac{abc}{4\Delta}$

Where / जहाँ, a, b, c — side of triangle/ त्रिभुज की भुजाएँ

Δ — area of triangle/ त्रिभुज का क्षेत्रफल

त्रिभुजों के क्षेत्रफल, परिमाप तथा उनके अन्तःवृत्त, परिवृत्त पर आधारित उदाहरण निम्नलिखित हैं

Example **The area of the triangle, whose sides are 3 cm, 5 cm and 6 cm is**

उस त्रिभुज का क्षेत्रफल ज्ञात कीजिए, जिसकी भुजाएँ 3 सेमी, 5 सेमी और 6 सेमी हैं,

SSC SELECTION POST 2023

Sol. $S = \frac{3+5+4}{2} = 7$

Area of triangle $= \sqrt{7 \times (7-3) \times (7-5) \times (7-6)}$

$= \sqrt{7 \times 4 \times 2 \times 1} = 2\sqrt{14} \text{ cm}^2$

Example **Find the area of an isosceles triangle, whose base is 54 cm long and each equal side is 45 cm long.**

एक समद्विबाहु त्रिभुज का क्षेत्रफल ज्ञात कीजिए, जिसका आधार 54 सेमी लम्बा है और प्रत्येक बराबर भुजा 45 सेमी लम्बी है।

RRB NTPC (CBT II) 2022

Sol. Area of triangle $= \frac{b}{4}\sqrt{4a^2 - b^2}$

$= \frac{54}{4}\sqrt{4(45)^2 - (54)^2}$

$= \frac{27}{2}\sqrt{4 \times 2025 - 2916}$

$= \frac{27}{2}\sqrt{8100 - 2916}$

$= \frac{27}{2}\sqrt{5184} = \frac{27}{2} \times 72 = 972 \text{ cm}^2$

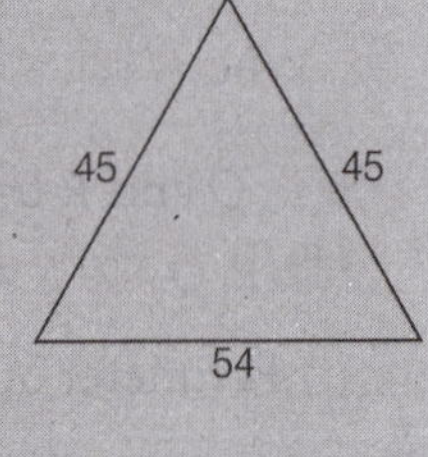

Example **O is a point in the interior of an equilateral triangle. The perpendicular distance from 'O' to the sides are $\sqrt{3}$ cm, $2\sqrt{3}$ cm and $5\sqrt{3}$ cm. The perimeter of the triangle is:**

'O' एक समबाहु त्रिभुज के अभ्यन्तर में स्थित एक बिन्दु है। 'O' से भुजाओं की लम्बवत दूरी $\sqrt{3}$ सेमी, $2\sqrt{3}$ सेमी और $5\sqrt{3}$ सेमी है। त्रिभुज का परिमाप है

SSC CGL (TIER I) 2022

Sol. Side of triangle $(a) = \frac{2}{\sqrt{3}}(P_1 + P_2 + P_3)$

$= \frac{2}{\sqrt{3}}(\sqrt{3} + 2\sqrt{3} + 5\sqrt{3})$

$= \frac{2}{\sqrt{3}} \times 8\sqrt{3} = 16 \text{ cm}$

$\therefore$ Perimeter of triangle $= 3 \times 16 = 48 \text{ cm}$

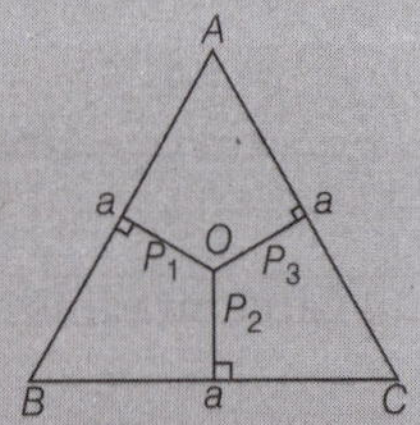

Example **ABC is a right angled triangle. If the length of two sides containing the right angle are 4 cm and 3 cm. The radius of its incircle is**

ABC एक समकोण त्रिभुज है। यदि समकोण वाली दो भुजाओं की लम्बाई 4 सेमी और 3 सेमी है। इसके भीतर वाले वृत्त की त्रिज्या है

RRB NTPC (CBT- I) 2021

Sol. $AC^2 = AB^2 + BC^2$

$\therefore \quad AC = \sqrt{3^2 + 4^2} = 5$

Area of triangle $(\Delta) = \frac{1}{2} \times 4 \times 3 = 6 \text{ cm}^2$

$$S = \frac{3+4+5}{2} = 6 \text{ cm}$$

$\therefore$ Inradius $(r) = \frac{\Delta}{S} = \frac{6}{6} = 1 \text{ cm}$

Alternate Method

Inradius $= \frac{a+b-c}{2} = \frac{3+4-5}{2} = 1 \text{ cm}$

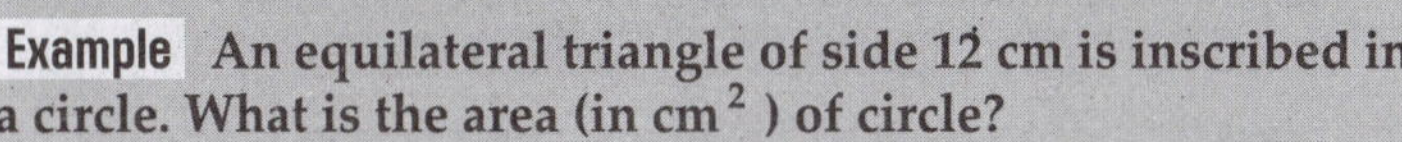

Example **An equilateral triangle of side 12 cm is inscribed in a circle. What is the area (in cm^2) of circle?**

भुजा 12 सेमी का एक समबाहु त्रिभुज एक वृत्त के अन्तर्गत बना हुआ है। वृत्त का क्षेत्रफल (सेमी^{2} में) क्या है? *RRB NTPC (CBT II) 2022*

Sol. Radius of outer circle

$$= \frac{a}{\sqrt{3}} = \frac{12}{\sqrt{3}} = 4\sqrt{3}$$

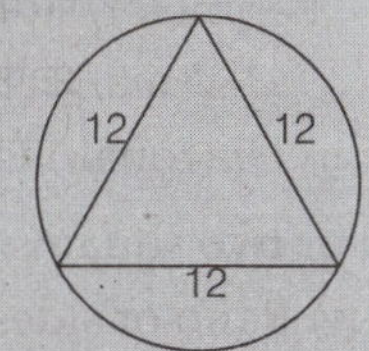

$\therefore$ Area of the circle

$$= \pi r^2 = \pi(4\sqrt{3})^2 = 48\pi \text{ cm}^2$$

Quadrilateral चतुर्भुज

1. Quadrilateral चतुर्भुज

Area/क्षेत्रफल $= \frac{1}{2} \times AC \times (h_1 + h_2)$

$= \frac{1}{2} \times$ Diagonal/विकर्ण

$\times$ Sum of perpendiculars drawn on this diagonal/इस विकर्ण पर खींचे गए लम्बों का योग

- The quadrilateral formed by joining the mid points of the adjacent sides of the quadrilateral will be the parallelogram of half area.

  चतुर्भुज की आसन्न भुजा के मध्य बिन्दु को मिलाने से बनने वाला चतुर्भुज आधे क्षेत्रफल का समान्तर चतुर्भुज होगा।

  If the area of the quadrilateral $ABCD$ is x, then area of the parallelogram $PQRS = \frac{x}{2}$

  यदि चतुर्भुज $ABCD$ का क्षेत्रफल x हो तो समान्तर चतुर्भुज $PQRS$ का क्षेत्रफल $= \frac{x}{2}$

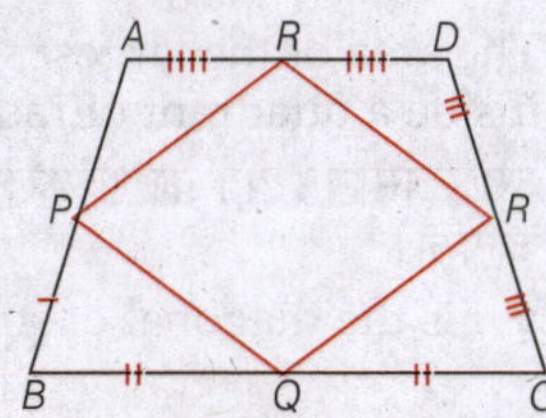

चतुर्भुज के क्षेत्रफल पर आधारित उदाहरण निम्नलिखित है

Example **In a quadrilateral $ABCD$, $AC = 24$ cm. If length of the perpendiculars drawn from B and D to line AC are 10 cm and 14 cm, then the area of the quadrilateral $ABCD$ is**

किसी चतुर्भुज $ABCD$ में, $AC = 24$ सेमी है। यदि B तथा D से विकर्ण AC पर डाले गए लम्बों की लम्बाइयाँ, 10 सेमी और 14 सेमी हो, तो चतुर्भुज $ABCD$ का क्षेत्रफल है

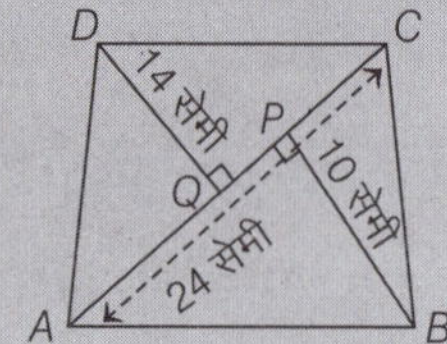

Sol. Area of quadrilateral $ABCD = \frac{1}{2} \times 24 \times (14 + 10) = 288 \text{ cm}^2$

2. Square वर्ग

If the side of a square is a, then/यदि एक वर्ग की भुजा a है, तब

Perimeter/परिमाप $= 4a$

Area/क्षेत्रफल $= a^2$

Diagonal/विकर्ण $= a\sqrt{2}$

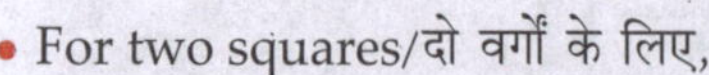

- For two squares/दो वर्गों के लिए,

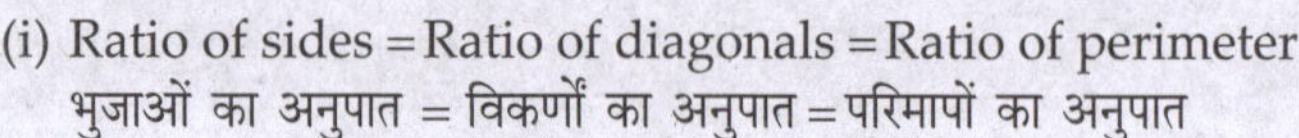

  (i) Ratio of sides = Ratio of diagonals = Ratio of perimeter
  भुजाओं का अनुपात = विकर्णों का अनुपात = परिमापों का अनुपात

  (ii) Ratio of areas $=$ (Ratio of sides)^2
  $=$ (Ratio of diagonals)^2 $=$ (Ratio of perimeters)^2
  क्षेत्रफलों का अनुपात $=$ (भुजाओं का अनुपात)^2 $=$ (विकर्णों का अनुपात)^2
  $=$ (परिमापों का अनुपात)^2

- If one of the diagonals or the perimeter becomes x times then the area becomes x^2 times or increases by $(x^2 - 1)$ times.
  यदि एक विकर्ण या परिमाप x गुना हो जाए, तब क्षेत्रफल x^2 गुना हो जाता है या $(x^2 - 1)$ गुना बढ़ जाता है।

- The area of the largest square that can be drawn inside a semi-circle of radius r.
  उस सबसे बड़े वर्ग का क्षेत्रफल, जिसे r त्रिज्या के अर्द्धवृत्त के अन्दर बनाया जा सकता है।

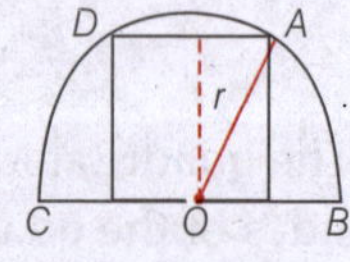

$\therefore$ Area of square/वर्ग का क्षेत्रफल $= \frac{4}{5} r^2$

- The area of the largest square that can be drawn inside a quadrant of radius r.
  उस सबसे बड़े वर्ग का क्षेत्रफल, जिसे r त्रिज्या के चतुर्थांश में बनाया जाता है।
  $\because$ r is the diagonal of square/r वर्ग का विकर्ण है।

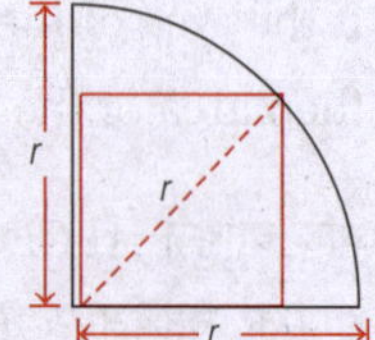

$\therefore$ Area of square/वर्ग का क्षेत्रफल $= \frac{r^2}{2}$

- Area of the square inscribed in the quadrant of a circle of radius 'r' (As shown in figure)
  r त्रिज्या वाले वृत्त के चतुर्थांश के अन्दर बनाए गए वर्ग का क्षेत्रफल (जैसा कि चित्र में दिखाया गया है।)

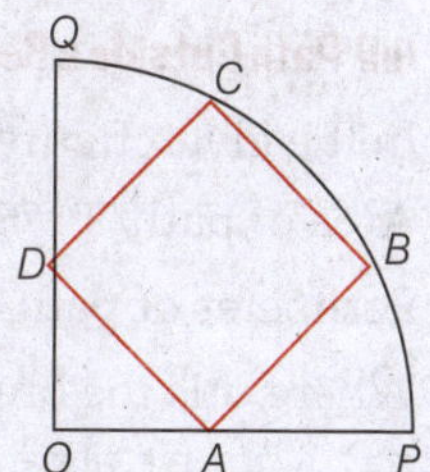

∴ Area of square/वर्ग का क्षेत्रफल $= \frac{2}{5}r^2$

3. Rectangle आयत

If the length and breadth of a rectangle is a and b, respectively.

यदि आयत की लम्बाई व चौड़ाई क्रमशः a और b हैं।

$$\text{Perimeter} = 2(a+b)$$
$$= 2(\text{Length} + \text{Breadth})$$
$$\text{Area} = ab = \text{Length} \times \text{Breadth}$$
$$\text{Diagonal} = \sqrt{a^2 + b^2} = \sqrt{(\text{Length})^2 + (\text{Breadth})^2}$$

परिमाप $= 2(a+b) = 2$ (लम्बाई + चौड़ाई)

क्षेत्रफल $= ab =$ लम्बाई × चौड़ाई

विकर्ण $= \sqrt{a^2 + b^2} = \sqrt{(\text{लम्बाई})^2 + (\text{लम्बाई})^2}$

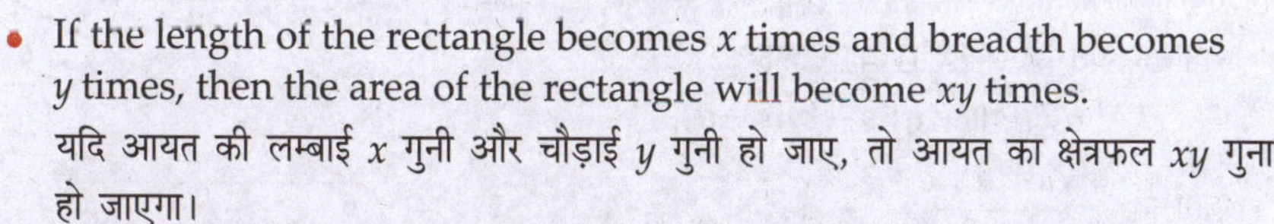

- If the length of the rectangle becomes x times and breadth becomes y times, then the area of the rectangle will become xy times.
  यदि आयत की लम्बाई x गुनी और चौड़ाई y गुनी हो जाए, तो आयत का क्षेत्रफल xy गुना हो जाएगा।

Path Around or Inside a Rectangle एक आयत के चारों ओर या अन्दर पथ

(i) Crossing Road Inside Rectangle आयत के अन्दर क्रॉसिंग सड़क

In the given figure/दी गई आकृति में,

Area of crossing road/क्रॉसिंग सड़क का क्षेत्र $= x(l + b - x)$ *M. Imp.*

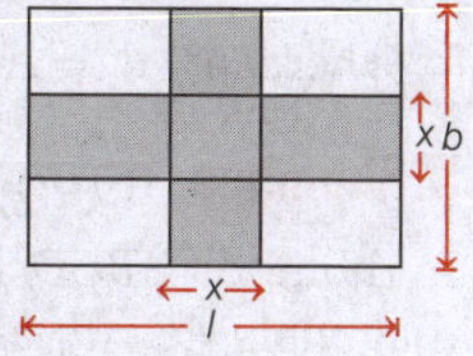

Perimeter of path/पथ का परिमाप $= \mathbf{2(l + b - 2x)}$

Where, x is the width of path/जहाँ, x पथ की चौड़ाई है।

(ii) Path Inside a Rectangle एक आयत के अन्दर पथ

In the given figure/दी गई आकृति में,

Area of path/पथ का क्षेत्र $= \mathbf{2x(l + b - 2x)}$

Perimeter of path/पथ का परिमाप $= \mathbf{4(l + b - 2x)}$

Where, x is the width of path.

जहाँ, x पथ की चौड़ाई है।

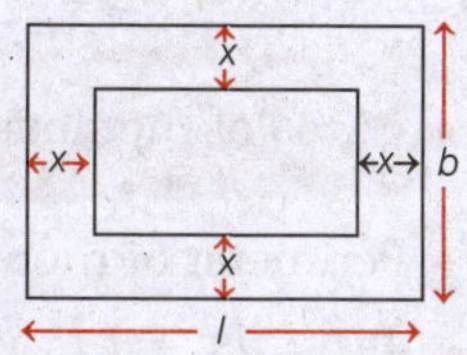

(iii) Path Outside a Rectangular Field एक आयताकार मैदान के बाहर पथ

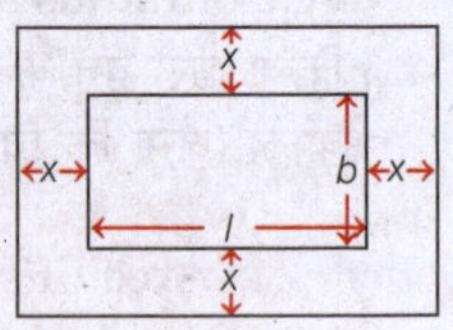

In the given figure/दी गई आकृति में,

Area of path/पथ का क्षेत्रफल $= \mathbf{2x(l + b + 2x)}$

Perimeter of path/पथ का परिमाप $= \mathbf{4(l + b + 2x)}$

Where, x is the breadth of path

जहाँ, x पथ की चौड़ाई है।

4. Parallelogram समान्तर चतुर्भुज

In a parallelogram $ABCD$, let sides $AB = a$ cm, and $BC = b$ cm, then

समान्तर चतुर्भुज $ABCD$ में,

भुजाएँ $AB = a$ सेमी और $BC = b$ सेमी, तब

- Area = Base × Height $= ah_1 = bh_2$

  क्षेत्रफल = आधार × ऊँचाई $= ah_1 = bh_2$
- Perimeter/परिमाप $= 2(a + b)$
- The length of one diagonal is d, then

  Area of parallelogram $ABCD = \mathbf{2\sqrt{s(s-a)(s-b)(s-d)}}$

  Where, $s = \dfrac{a+b+d}{2}$

  एक विकर्ण की लम्बाई d है, तब

  समान्तर चतुर्भुज $ABCD$ का क्षेत्रफल

  $= \mathbf{2\sqrt{s(s-a)(s-b)(s-d)}}$

  जहाँ, $s = \dfrac{a+b+d}{2}$,

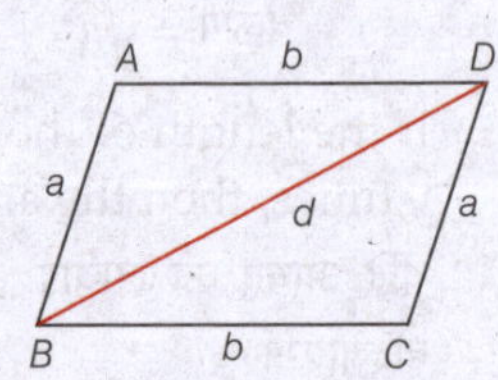

$$AC^2 + BD^2 = 2(a^2 + b^2)$$

$$d_1^2 + d_2^2 = 2(a^2 + b^2)$$

where/जहाँ $d_1 = AC, d_2 = BD$

5. Rhombus समचतुर्भुज

In rhombus $ABCD$, let each side is a and diagonals $AC = d_1$ and $BD = d_2$, then

माना, समचतुर्भुज $ABCD$ में, प्रत्येक भुजा a तथा विकर्ण $AC = d_1$ तथा $BD = d_2$ हैं, तब

ये सूत्र SSC, Railway आदि परीक्षा के लिए M. Imp है। इन पर आधारित 1 प्रश्न अवश्य पूछा जाता है।

- Area of rhombus/समचतुर्भुज का क्षेत्रफल $= \dfrac{1}{2} \times d_1 \times d_2$ *M. Imp.*
- Perimeter of rhombus/समचतुर्भुज का परिमाप $= 4a$
- $\mathbf{4a^2 = d_1^2 + d_2^2}$

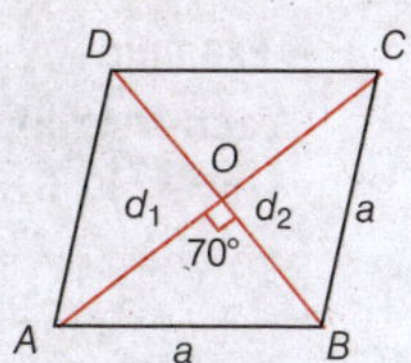

- Area/क्षेत्रफल = Base (आधार) × Height (ऊँचाई)
- Area of ΔAOB = Area of ΔBOC
  = Area of ΔCOD
  = Area of ΔAOD

  क्षेत्रफल ΔAOB = क्षेत्रफल ΔBOC = क्षेत्रफल ΔCOD
  = क्षेत्रफल ΔAOD

6. Trapezium समलम्ब चतुर्भुज

In the given figure, $ABCD$ is a trapezium, then
दी गई आकृति में, $ABCD$ एक समलम्ब चतुर्भुज है, तब

- Perimeter/परिमाप $= a + b + c + d$
- Area/क्षेत्रफल $= \frac{1}{2}$ (Sum of parallel sides) × Distance between them
  $= \frac{1}{2}$ (समान्तर भुजाओं का योग) × उनके बीच की दूरी
  $= \frac{1}{2}(a+b) \times h$
- $d_1^2 + d_2^2 = c^2 + d^2 + 2ab$

वर्ग, आयत, समान्तर चतुर्भुज, समचतुर्भुज एवं समलम्ब पर आधारित उदाहरण निम्न हैं

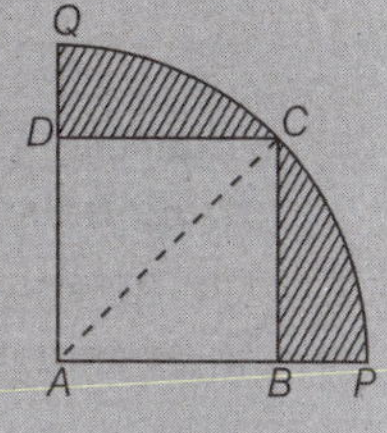

Example **In the given figure, a square $ABCD$ is inscribed in a quadrant $APCQ$. If $AB = 16$ cm, then find the area of the shaded region (take $\pi = 3.14$) (correct to two places of decimal).**

दी गई आकृति में, एक चतुर्थांश $APCQ$ में एक वर्ग $ABCD$ उत्कीर्णित है। यदि $AB = 16$ सेमी. है, तो छायांकित क्षेत्र का क्षेत्रफल ज्ञात कीजिए।
($\pi = 3.14$ लें) (दशमलव के दो स्थानों तक सही उत्तर दीजिए) **SSC CGL 2022**

Sol. Area of square $= (16)^2 = 256 \text{ cm}^2$

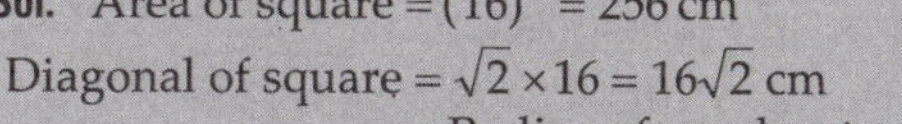
Diagonal of square $= \sqrt{2} \times 16 = 16\sqrt{2}$ cm
= Radius of quadrant

So, area of the quadrant $= \frac{1}{4}\pi r^2$

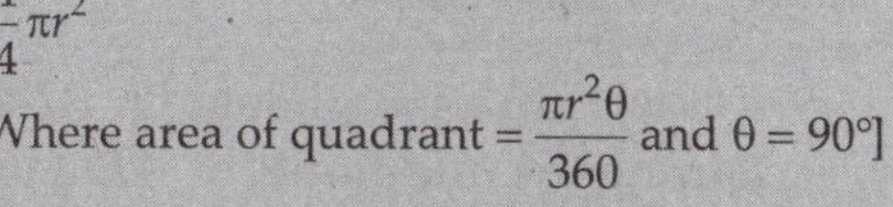
[Where area of quadrant $= \frac{\pi r^2 \theta}{360}$ and $\theta = 90°$]

$= \frac{1}{4} \times 3.14 \times (16\sqrt{2})^2 = 401.92 \text{ cm}^2$

Now, area of shaded portion = Area of quadrant − Area of square
$= 401.92 - 256 = 145.92 \text{ cm}^2$

Example **A path 6m wide runs around and outside of a rectangular plot of length 10 m and breadth 8 m. The area (in m^2) of the path is**

10 मी लम्बाई और 8 मी चौड़ाई वाले आयताकार भूखण्ड के चारों ओर 6 मी चौड़ा एक पथ बनाया गया है। पथ का क्षेत्रफल (मी^{2} में) है *SSC MTS 2021*

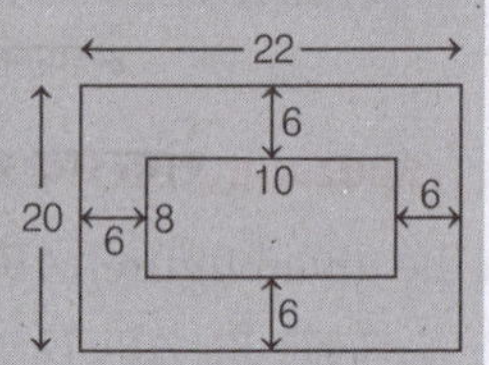

Sol. Area of the path $= 2 \times 6(10 + 8 + 2 \times 6)$

$= 12 \times 30 = 360\ m^2$

Alternate Method

Area of the path $= 22 \times 20 - 10 \times 8$

$= 440 - 80$

$= 360\ m^2$

Example **A rectangular park is 110 m long and 70 m wide. Two cross paths; each 6 m wide, are constructed parallel to the sides. If the cost of construction of these paths is ₹ 175/m^2, then what is the total cost (in ₹) of construction of these paths?**

एक आयताकार पार्क 110 मी लम्बा और 70 मी चौड़ा है। दो क्रॉस पथ, प्रत्येक 6 मी चौड़े, भुजाओं के समान्तर बनाए गए हैं। यदि इन पथों के निर्माण की लागत ₹ 175/मी^{2} है, तो इन पथों के निर्माण की कुल लागत (₹ में) क्या है?

SSC CONSTABLE (GD) 2021

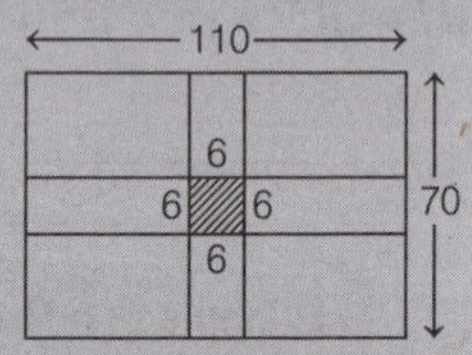

Sol. Total area of the two crosspaths

$= 6(110 + 70 - 6)$

$= 6 \times 174 = 1044\ m^2$

Now the cost of construction the paths

$= 1044 \times 175 =$ ₹ 182700

Example **One side of a rhombus is 13 cm and one of its diagonals is 24 cm. What is the area of the rhombus (in cm^2)?**

एक समचतुर्भुज की एक भुजा 13 सेमी है, और इसके विकर्णों में से एक 24 सेमी है, तो समचतुर्भुज का क्षेत्रफल (सेमी^{2} में) क्या है? *SSC CPO 2020*

Sol. Given, $a = 13$ cm and $d_1 = 24$ cm

$\therefore \quad 4a^2 = d_1^2 + d_2^2$

$4 \times 169 = 576 + d_2^2 \Rightarrow d_2^2 = 676 - 576$

$\Rightarrow \quad d_2^2 = 100 \Rightarrow d_2 = 10$ cm

$\therefore$ Area of rhombus $= \frac{1}{2} \times d_1 \times d_2 = \frac{1}{2} \times 24 \times 10 = 120\ cm^2$

Example **In a parallelogram *ABCD*, the length of the line joining the midpoints of *AD* and *AC* is 2 units. If the perimeter of the parallelogram is 20 units, then find the length of *AD*.**

एक समान्तर चतुर्भुज *ABCD* में, *AD* और *AC* में मध्य बिन्दुओं को मिलाने वाली रेखा की लम्बाई 2 इकाई है। यदि समान्तर चतुर्भुज का परिमाप 20 इकाई है, तो *AD* की लम्बाई ज्ञात कीजिए। *RRC GROUP D 2022*

Sol. Given, $PQ = 2$ units

Using mid point theorem

$$DC = 2 \times PQ = 2 \times 2 = 4 \text{ units}$$

Perimeter of parallelogram

$$= 2 \times (\text{sum of adjacent sides})$$
$$= 2 \times (AD + DC) = 20 \text{ units}$$
$$\Rightarrow \quad 2 \times (AD + 4) = 20 \Rightarrow AD = 6 \text{ units}$$

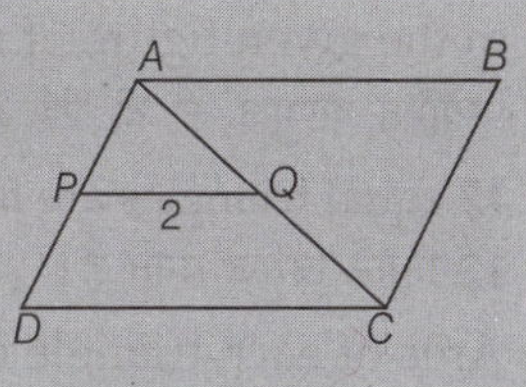

Polygon बहुभुज

Regular Polygon समबहुभुज

Area/ क्षेत्रफल $= \frac{1}{2} \times a \times r \times n$ or/ या क्षेत्रफल/Area $= \frac{na^2}{4} \cot \frac{(180°)}{n}$

- Circumradius /परित्रिज्या, $R = \frac{a}{2} \operatorname{cosec}\left(\frac{180°}{n}\right)$
- Inradius/अन्तः त्रिज्या, $r = \frac{a}{2} \cot\left(\frac{180°}{n}\right)$

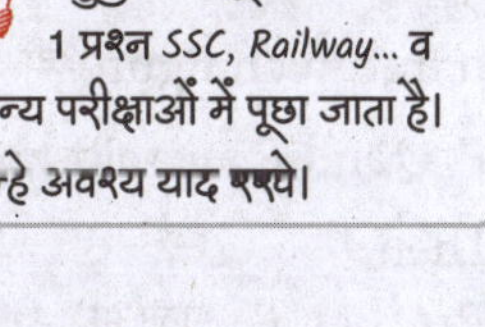

Where, a – length of sides
जहाँ, a = भुजा की लम्बाई
r = inradius and n = number of sides
r = अन्तः त्रिज्या और n = भुजाओं की संख्या

(i) Area of regular pentagon

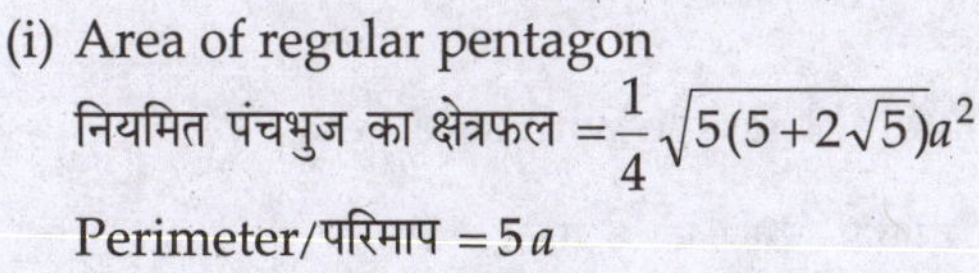

नियमित पंचभुज का क्षेत्रफल $= \frac{1}{4}\sqrt{5(5+2\sqrt{5})}a^2$

Perimeter/परिमाप $= 5a$

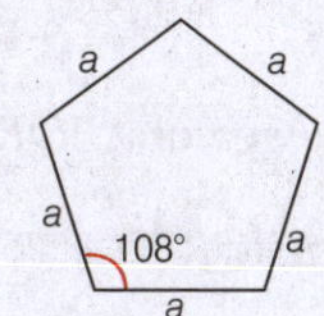

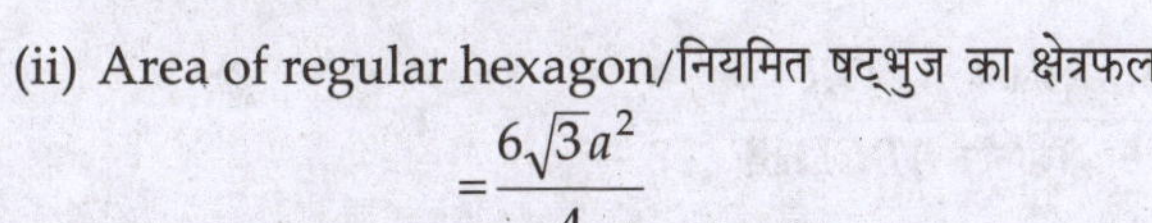

(ii) Area of regular hexagon/नियमित षट्भुज का क्षेत्रफल

$$= \frac{6\sqrt{3}a^2}{4}$$

- Perimeter/परिमाप $= 6a$

  Large diagonal/बड़ा विकर्ण

  $FC = AD = BE = 2a$

- Short diagonal/छोटा विकर्ण

  $FD = DB = EF = \sqrt{3}a$

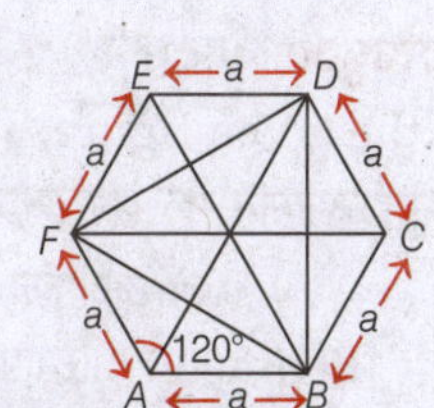

- In the given figure of regular hexagon,
  नियमित षट्भुज की दी गई आकृति में,
  12 equal triangles are formed.
  12 समान त्रिभुज बनते हैं।
  Area of each triangle is same.
  प्रत्येक त्रिभुज का क्षेत्रफल समान है।

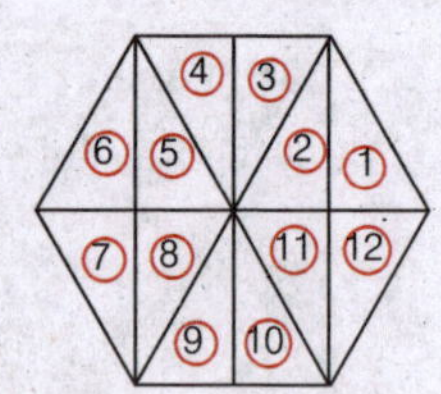

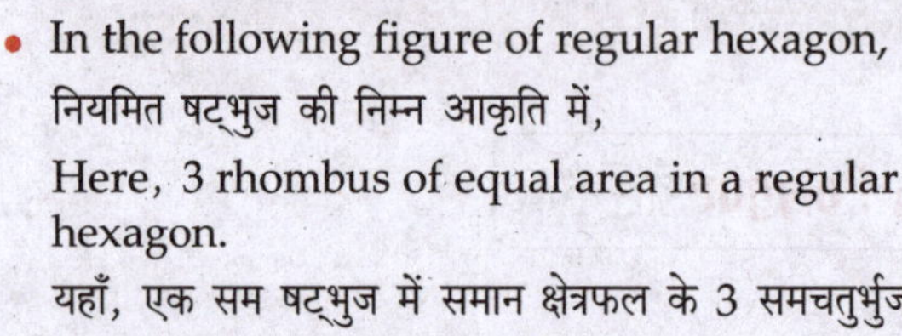

- In the following figure of regular hexagon,
  नियमित षट्भुज की निम्न आकृति में,
  Here, 3 rhombus of equal area in a regular hexagon.
  यहाँ, एक सम षट्भुज में समान क्षेत्रफल के 3 समचतुर्भुज।

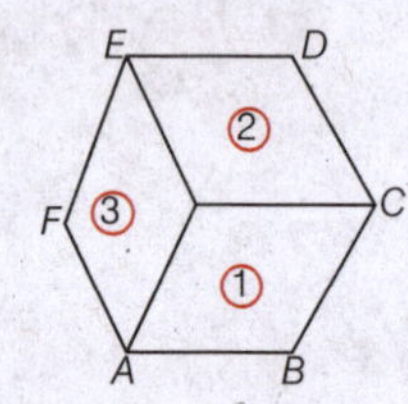

- In the given figure/दी गई आकृति में,
  ΔEAC is an equilateral triangle of side $\sqrt{3}a$.
  ΔEAC, $\sqrt{3}a$ भुजा का एक समबाहु त्रिभुज है।

$$\frac{\textbf{Area / क्षेत्रफल } \Delta EAC}{\textbf{Area / क्षेत्रफल } ABCDEF} = \frac{1}{2}$$

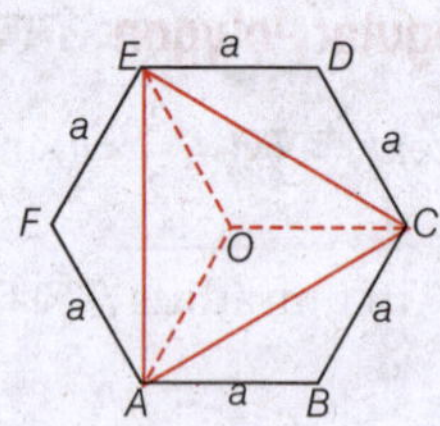

- In the given figure/दी गई आकृति में,
  P, Q and R are mid-points of sides AF, BC and ED.
  Then,
  P, Q और R, भुजाओं AF, BC और ED के मध्य-बिन्दु हैं।
  तब, $PQ = \dfrac{2a + a}{2} = \dfrac{3a}{2}$

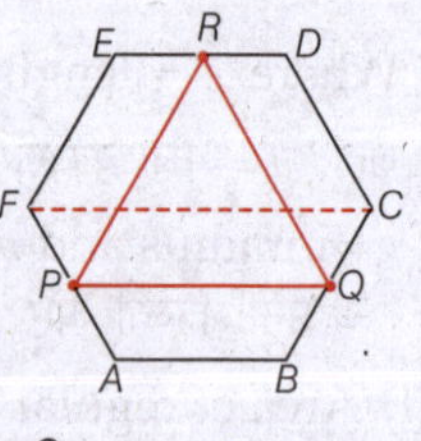

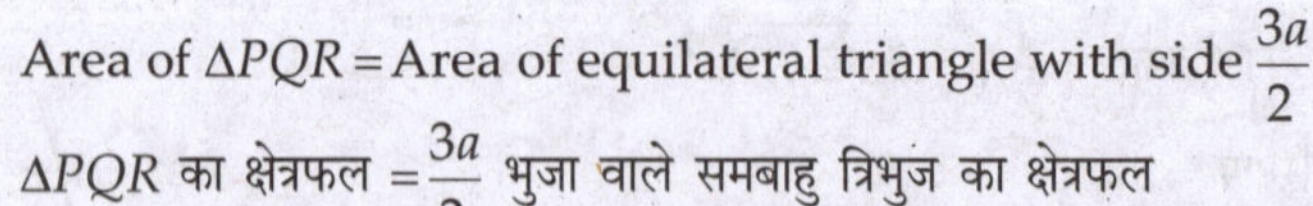

Area of ΔPQR = Area of equilateral triangle with side $\dfrac{3a}{2}$

ΔPQR का क्षेत्रफल $= \dfrac{3a}{2}$ भुजा वाले समबाहु त्रिभुज का क्षेत्रफल

$$\frac{\textbf{Area of } \Delta PQR\textbf{/ क्षेत्रफल } \Delta PQR}{\textbf{Area of } ABCDEF\textbf{/ क्षेत्रफल } ABCDEF} = \frac{3}{8}$$

Octagon अष्टभुज

In the given figure, $ABCDEFGH$ is a regular octagon with side a
दी गई आकृति में, $ABCDEFGH$ एक भुजा a वाला नियमित अष्टभुज है।

Area/क्षेत्रफल $= 2(\sqrt{2} + 1)a^2$

Perimeter/परिमाप $= 8a$

Inradius/अन्तःत्रिज्या $= \dfrac{a}{2\sqrt{2} - 2}$

Circumradius/परित्रिज्या $= \dfrac{\sqrt{2 + \sqrt{2}}}{2}a$

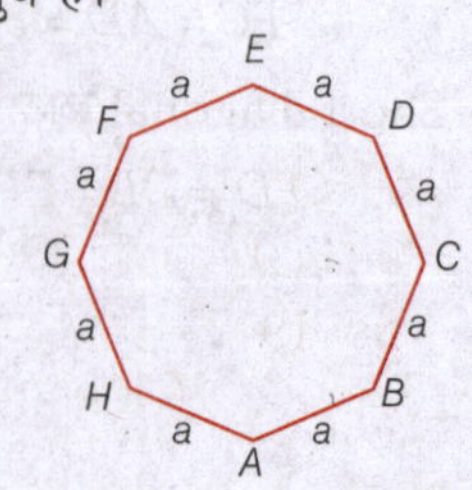

बहुभुज पर आधारित उदाहरण निम्न हैं

Example **A circle is inside a hexagon, perimeter of hexagon is 84 cm. Find the difference between the area of the circle and hexagon. Take ($\sqrt{3} = 1.73$)**

एक षट्भुज के अन्दर एक वृत्त है, षट्भुज का परिमाप 84 सेमी है। वृत्त और षट्भुज के क्षेत्रफल के बीच का अन्तर ज्ञात कीजिए। ($\sqrt{3} = 1.73$) लीजिए।

IBPS CLERK (MAINS) 2021

Sol. Perimeter of hexagon = 84 cm

$\therefore \quad 6 \times \text{Side} = 84 \Rightarrow \text{Side} = 14 \text{ cm}$

Area of hexagon $= 6 \times \frac{\sqrt{3}}{4} (\text{side})^2$

$= 6 \times \frac{\sqrt{3}}{4} \times (14)^2 = 294\sqrt{3}$

$= 508.62 \text{ cm}^2$

Inradius (r) of regular hexagon $= \frac{\sqrt{3}}{2} a = \frac{\sqrt{3}}{2} \times 14 = 7\sqrt{3}$ cm

Area of circle $= \pi r^2 = \frac{22}{7} \times 7\sqrt{3} \times 7\sqrt{3} = 462 \text{ cm}^2$

$\therefore$ Required difference $= 508.62 - 462 = 46.62 \text{ cm}^2$

Example **In the given figure, *ABCDEF* is a regular hexagon of side 12 cm. *P, Q* and *R* are the mid point of the sides *AB, CD* and *EF*, respectively. What is the area (in cm^2) of triangle *PQR*?**

दी गई आकृति में *ABCDEF*, 12 सेमी भुजा वाला एक सम षट्भुज है। *P, Q* और *R* क्रमशः *AB, CD* और *EF* भुजाओं के मध्य-बिन्दु हैं। ΔPQR का क्षेत्रफल (सेमी^{2} में) क्या है?

SSC CGL (TIER II) 2018

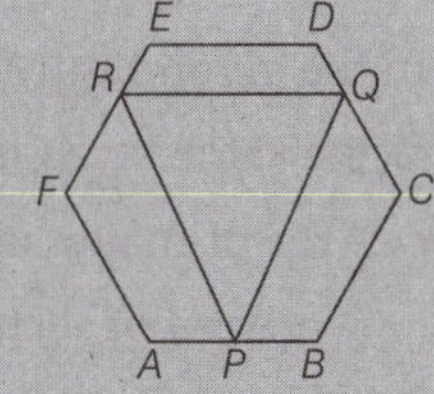

Sol. $a = 12$ cm

$\therefore \quad PQ = \frac{3a}{2} = \frac{3 \times 12}{2} = 18 \text{ cm}$

Area of ΔPQR = Area of equilateral triangle having side PQ

$= \frac{\sqrt{3}}{4} \times (18)^2 = 81\sqrt{3} \text{ cm}^2$

In the given figure, radius of circle = r, then

दी गई आकृति में, वृत्त की त्रिज्या = r, तब

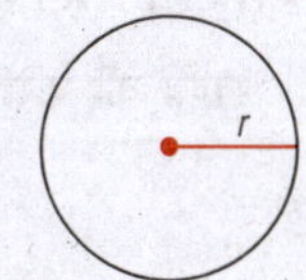

Circumference/परिधि $= 2\pi r$

Area/क्षेत्रफल $= \pi r^2$

$$\frac{\text{Circumference / परिधि}}{\text{Diameter / व्यास}} = \pi$$

Sector of a Circle
त्रिज्यखण्ड

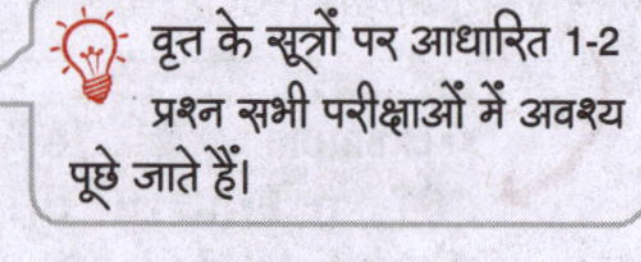

In the given figure/दी गई आकृति में,

$$\theta = \text{Central angle}/\theta \text{ केन्द्रीय कोण} = \frac{l}{r}$$

(Sector, is a part of circle)(त्रिज्यखण्ड, वृत्त का एक भाग है)

Area of Sector त्रिज्यखण्ड का क्षेत्रफल

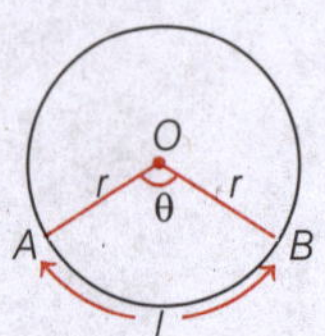

$$= \frac{\theta}{360°} \times \pi r^2 = \frac{lr}{2}$$ M. Imp

Length of $\widehat{AB}/\widehat{AB}$ की लम्बाई $= l = \frac{\theta}{360°} \times 2\pi r$f

Segment वृत्तखण्ड

1. Minor Segment लघु वृत्तखण्ड

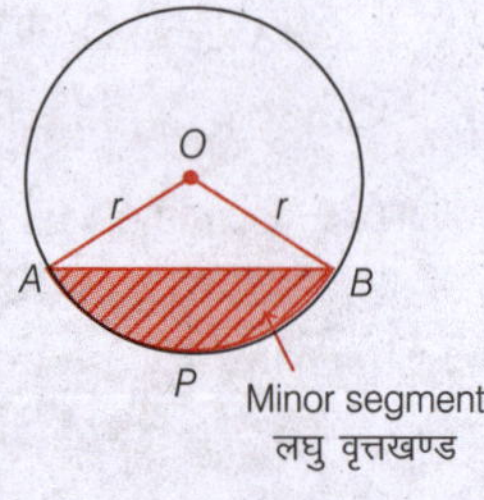

In the given figure, $APBA$ is minor segment.

दी गई आकृति में, $APBA$ लघु वृत्तखण्ड है।

- Area of minor segment
  = Area of sector – Area of ΔOAB
  लघु वृत्तखण्ड का क्षेत्रफल
  = त्रिज्यखण्ड $OAPB$ का क्षेत्रफल $-\Delta OAB$ का क्षेत्रफल

2. Major Segment दीर्घ वृत्तखण्ड

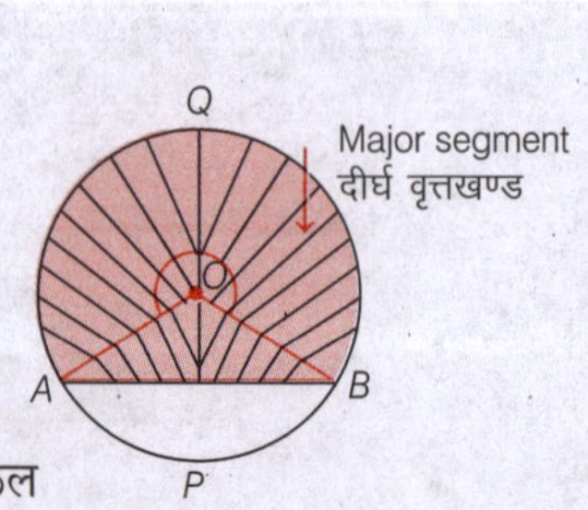

In the given figure, $AQBA$ is major segment.

दी गई आकृति में, $AQBA$ दीर्घ वृत्तखण्ड है।

- Area of major segment
  = Area of sector $OAQB$ + Area of ΔAOB
  दीर्घ वृत्तखण्ड का क्षेत्रफल
  = त्रिज्यखण्ड $O\,AQB$ का क्षेत्रफल + ΔAOB का क्षेत्रफल

  Area of major segment = Area of circle – Area of minor segment
  दीर्घ वृत्तखण्ड का क्षेत्रफल = वृत्त का क्षेत्रफल – लघु वृत्तखण्ड का क्षेत्रफल
- Perimeter (circumference) of segment
  खण्ड की परिधि (परिमाप) $= 2r\left(\frac{\pi\theta}{360°} + \sin\frac{\theta}{2}\right)$

Semi-circle अर्द्धवृत्त

In the given figure, there is a semi-circle with radius r.
दी गई आकृति में, r त्रिज्या वाला एक अर्द्धवृत्त है।

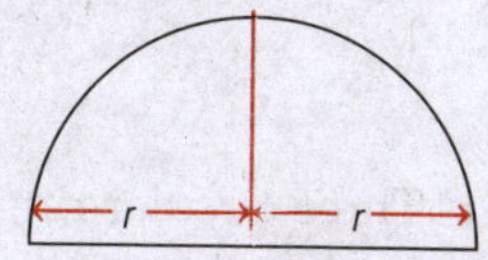

- Perimeter/परिधि $= \pi r +$ Diameter/व्यास $= \pi r + 2r$
- Area/क्षेत्रफल $= \dfrac{\pi r^2}{2}$

For a quadrant/एक चतुर्थांश के लिए,

In the given figure, there is a quadrant with radius r.
दी गई आकृति में, r त्रिज्या का एक चतुर्थांश है।

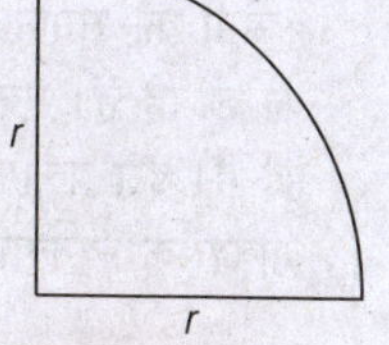

Circumference or Perimeter/परिधि या परिमाप

$$= r + r + \frac{2\pi r}{4} = 2r + \frac{\pi r}{2}$$

Area/क्षेत्रफल $= \dfrac{\pi r^2}{4}$

- According to the figure two semicircles are inscribed in a quadrant.
चित्रानुसार, दो अर्द्धवृत्तों को एक चतुर्थांश के अन्दर निर्मित किया गया है।

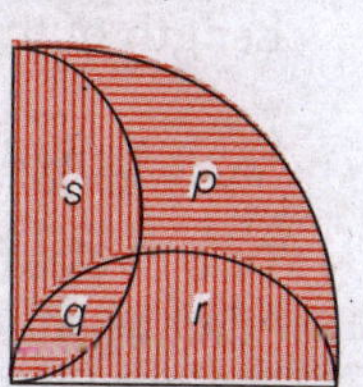

(i) Area of shaded region (p)/छायांकित क्षेत्र (p) का क्षेत्रफल
= Area of shaded region (q)/छायांकित क्षेत्र (q) का क्षेत्रफल

(ii) Area of shaded region (r)/छायांकित क्षेत्र (r) का क्षेत्रफल
= Area of shaded region (s)/छायांकित क्षेत्र (s) का क्षेत्रफल

(iii) Ratio of the area of the region (p), (q), (r), (s) respectively/(p), (q), (r), (s) के क्षेत्रफलों का अनुपात,

$$p : q : r : s = 4 : 4 : 7 : 7$$

- In the given figure ΔABC is an equilateral triangle.

$$OB = \frac{\text{side of equilateral triangle}}{\sqrt{3}} = \frac{2r}{\sqrt{3}}$$

दी गई आकृति में, ΔABC एक समबाहु त्रिभुज है।

$$OB = \frac{\text{समबाहु त्रिभुज की भुजा}}{\sqrt{3}} = \frac{2r}{\sqrt{3}}$$

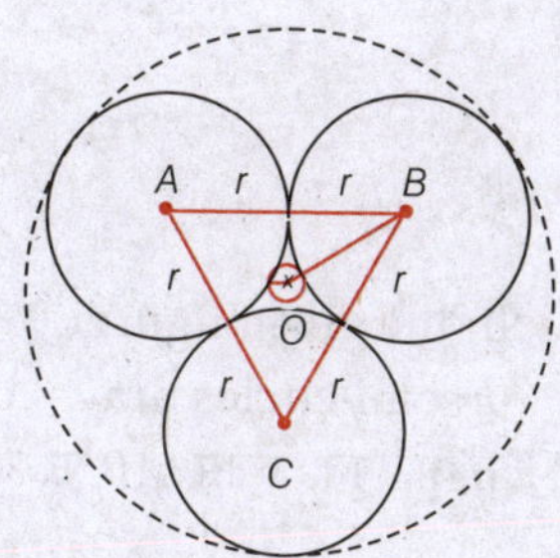

Radius of small circle/ छोटे वृत्त की त्रिज्या,

$$x = \frac{2r}{\sqrt{3}} - r = \frac{(2-\sqrt{3})r}{\sqrt{3}}$$

If we draw a big circle around these 3 circles, then radius of that big circle $= \frac{2r}{\sqrt{3}} + r$

यदि हम इन 3 वृत्तों के चारों ओर एक बड़ा वृत्त बनाते हैं, तो उस बड़े वृत्त की त्रिज्या $= \frac{2r}{\sqrt{3}} + r$

- Three circular rings of equal radii of r cm each are touching each other. A string runs all around the set of rings very trightly. What is the minimum length of string required to bind all the three rings in the given manner?

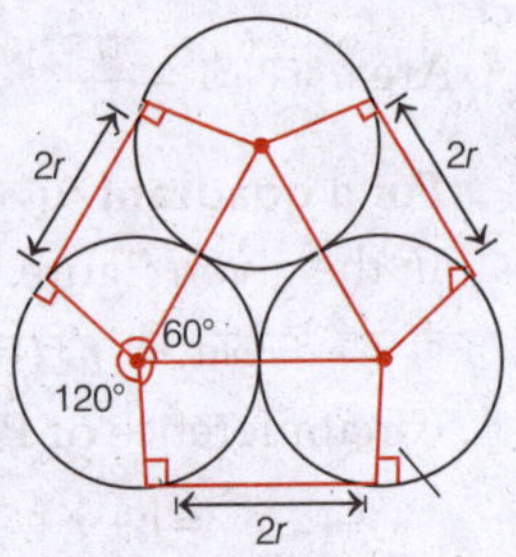

r सेमी की समान त्रिज्या के तीन गोलाकार छल्ले एक-दूसरे को छू रहे हैं। एक डोरी छल्लों के समूह के चारों ओर बन्धी हुई है। इस प्रकार से तीनों छल्लों को बाँधने के लिए आवश्यक न्यूनतम डोरी की लम्बाई है

Length of three arc/तीनों चापों की लम्बाई $= 3 \times \left(\frac{2\pi r}{360} \times 120\right) = 2\pi r$

$\therefore$ Length of string/रस्सी की लम्बाई $= 2\pi r + 3 \times 2r = 2r(\pi + 3)$

- Two circles of equal radius are touching each other as shown in figure, a square of side 'a' is placed between two circles. One side of square is on direct common tangent of both circles, then

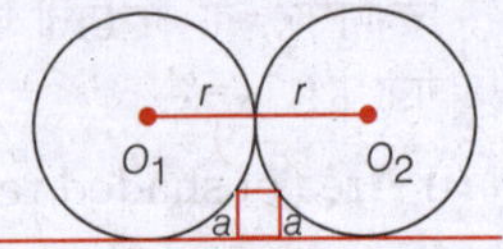

समान त्रिज्या r वाले दो वृत्त एक-दूसरे को स्पर्श करते हैं। इन वृत्तों के मध्य a भुजा वाला एक वर्ग इस प्रकार स्थित है, कि वर्ग की एक भुजा दोनों वृत्तों की उभयनिष्ठ अनुस्पर्श रेखा पर स्थित है, तब $\frac{r}{a} = \frac{5}{2}$

- Study the given figure/दी गई आकृति का अध्ययन करें।

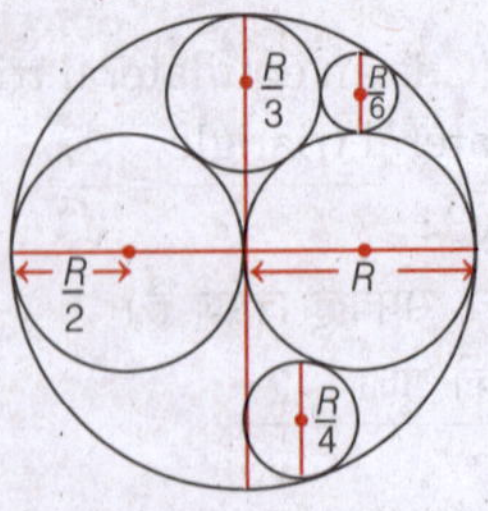

- In the given figure/दी गई आकृति में, n semi-circles drawn on diameter AB

n अर्द्धवृत्त, व्यास AB पर खींचे गए हैं।

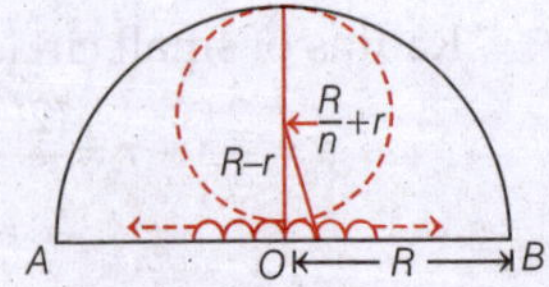

Radius of each small semi-circle $= \frac{R}{n}$, where n is even.

प्रत्येक छोटे अर्द्धवृत्त की त्रिज्या $= \frac{R}{n}$, जहाँ n सम है।

$\because \quad \left(\frac{R}{n} + r\right)^2 = \left(\frac{R}{n}\right)^2 + (R - r)^2$, then/तब

Radius of small circle/छोटे वृत्त की त्रिज्या, $r = \frac{nR}{2(n+1)}$

- In the given figure/दी गई आकृति में,

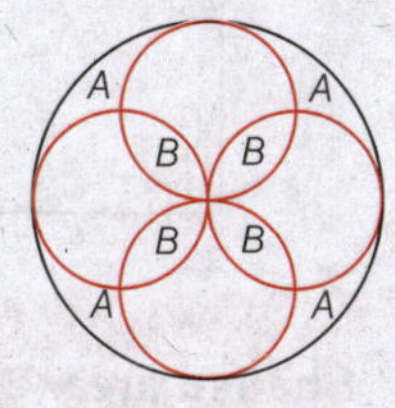

Area of A = Area of B

A का क्षेत्रफल = B का क्षेत्रफल

$$= \frac{\text{Area of } A/A \text{ का क्षेत्रफल}}{\text{Area of } B/B \text{ का क्षेत्रफल}} = \frac{1}{1} = 1 : 1$$

- In given figure

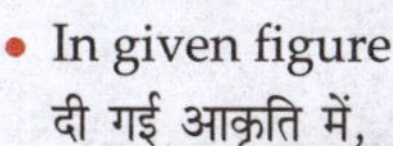

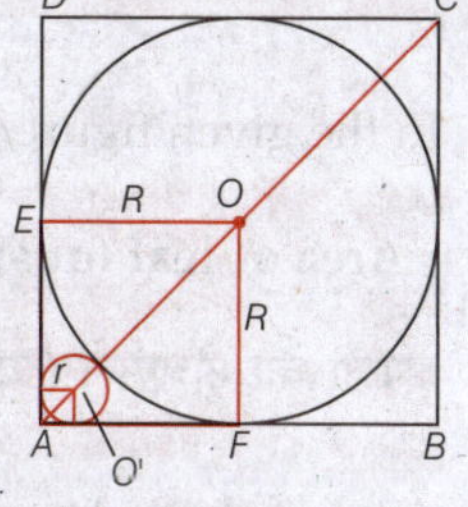

दी गई आकृति में,

$$AO = \sqrt{2}\, R$$

$$AO' = \sqrt{2} r$$

$$\therefore \quad \frac{R}{r} = \frac{\sqrt{2}+1}{\sqrt{2}-1}$$

- Area enclosed by two concentric circles

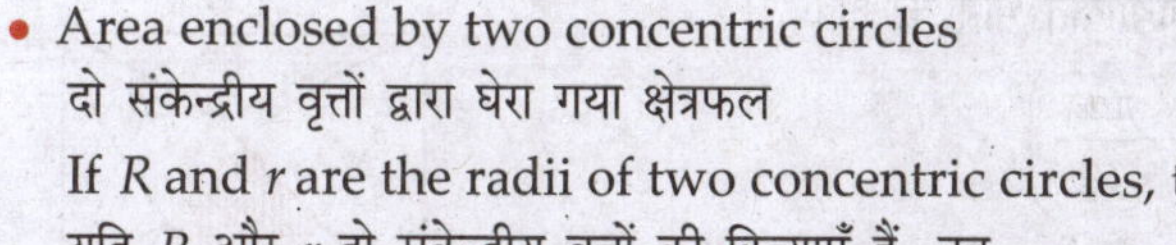

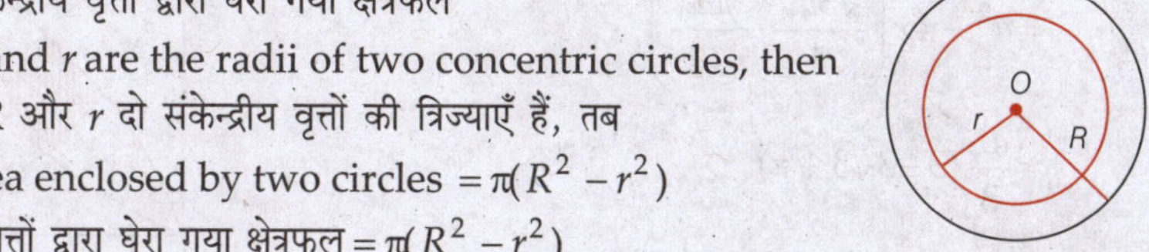

दो संकेन्द्रीय वृत्तों द्वारा घेरा गया क्षेत्रफल

If R and r are the radii of two concentric circles, then

यदि R और r दो संकेन्द्रीय वृत्तों की त्रिज्याएँ हैं, तब

(i) Area enclosed by two circles $= \pi(R^2 - r^2)$

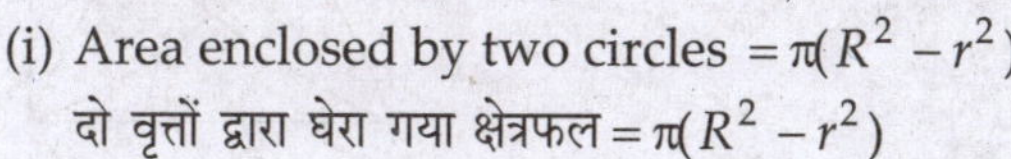

दो वृत्तों द्वारा घेरा गया क्षेत्रफल $= \pi(R^2 - r^2)$

(ii) Width of path/रास्ते की चौड़ाई $= R - r$

At each corner of an equilateral triangular field of side 'a' cm, a cow is tethered by a rope of length r_1, r_2 and r_3 cm, then the area grazed by the cow

$$= \frac{\pi}{6}(r_1^2 + r_2^2 + r_3^2) \quad \textit{M. Imp}$$

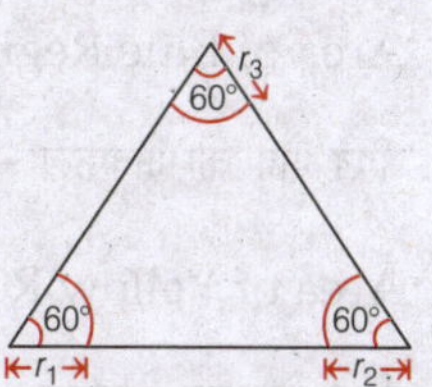

a सेमी भुजा वाले समबाहु त्रिभुजाकार खेत के प्रत्येक कोने पर r_1, r_2 तथा r_3 सेमी लम्बाई की रस्सियों से एक गाय को बाँधा जाता है, तब गाय द्वारा चरे गए खेत का क्षेत्रफल

$$= \frac{\pi}{6}(r_1^2 + r_2^2 + r_3^2)$$

- In the given figure/दी गई आकृति में,
  Area of A, B, C, D, E, F, G and H is equal.
  A, B, C, D, E, F, G और H का क्षेत्रफल बराबर है।

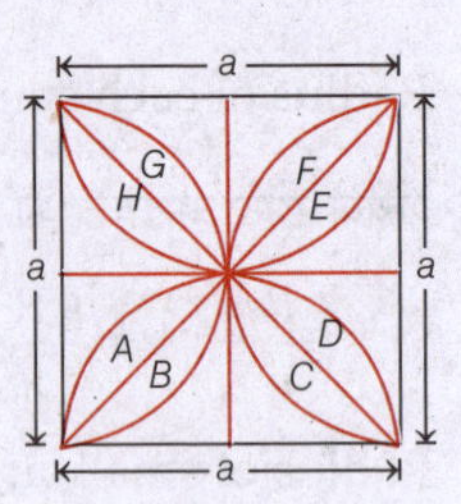

- In the given figure/दी गई आकृति में,

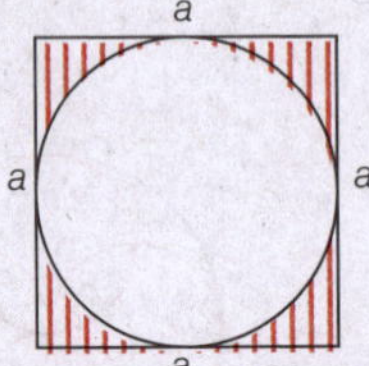

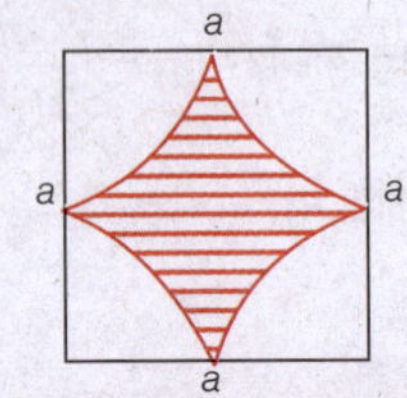

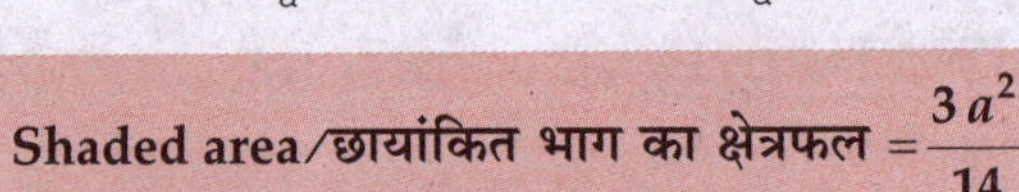

Shaded area/छायांकित भाग का क्षेत्रफल $= \dfrac{3a^2}{14}$

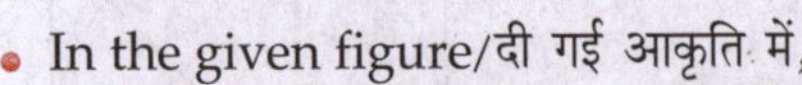

- In the given figure/दी गई आकृति में,

Area of leaf (or shaded area) $= \dfrac{4}{7}a^2$

पत्ती का क्षेत्रफल (या छायांकित क्षेत्रफल) $= \dfrac{4}{7}a^2$

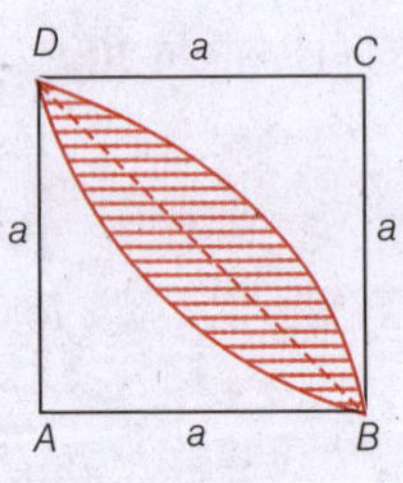

- Area of shaded part/छायांकित भाग का क्षेत्रफल

$$= a^2 - 4\times\left(\frac{\sqrt{3}a^2}{4} - \frac{\pi a^2}{12}\right)$$

$$= \frac{a^2}{3}(3 - 3\sqrt{3} + \pi)$$

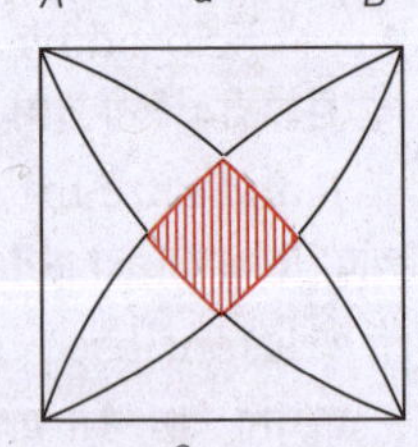

- In the given figure/दी गई आकृति में,

$$r_1 = \frac{a}{16},\ r_2 = \frac{3a}{8},\ r_3 = \frac{a}{6}$$

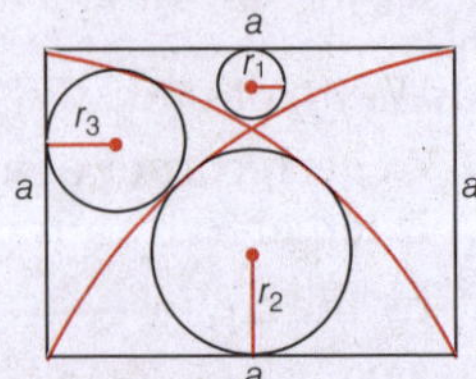

- In the given figure/दी गई आकृति में,
  Area of Blue Region
  नीले क्षेत्र का क्षेत्रफल $= \dfrac{\pi a^2}{3} - \dfrac{\sqrt{3}}{4}a^2$
  Area of Yellow Region
  पीले क्षेत्र का क्षेत्रफल $= \dfrac{\sqrt{3}}{4}a^2 - \dfrac{\pi a^2}{12}$

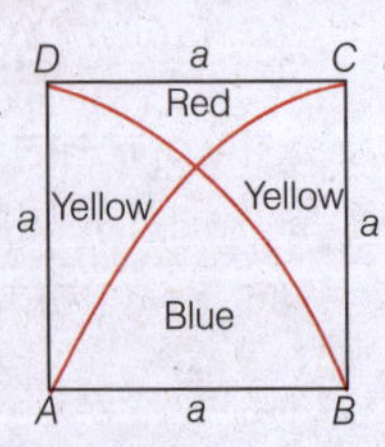

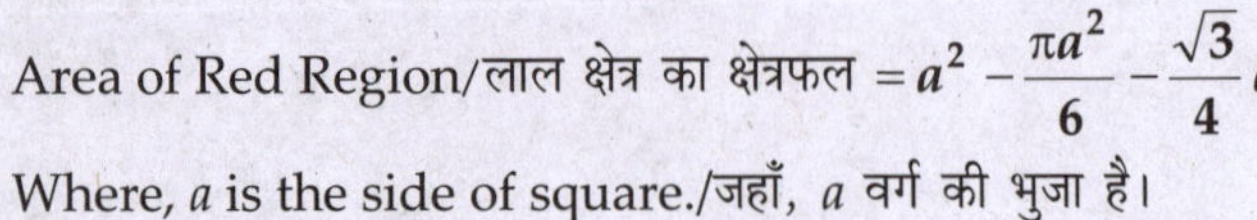

Area of Red Region/लाल क्षेत्र का क्षेत्रफल $= a^2 - \dfrac{\pi a^2}{6} - \dfrac{\sqrt{3}}{4}a^2$

Where, a is the side of square./जहाँ, a वर्ग की भुजा है।

वृत्त, त्रिज्यखण्ड, अर्द्धवृत्त पर आधारित उदाहरण निम्न हैं

Example **A chord of a circle of radius 21 cm makes a right angle at the center. Find the area of major segment $\left(\text{take } \pi = \frac{22}{7}\right)$ closest to nearest integer.**

21 सेमी त्रिज्या वाले वृत्त की जीवा केन्द्र पर समकोण बनाती है। दीर्घ वृत्तखण्ड का क्षेत्रफल ज्ञात कीजिए। $\left(\pi = \frac{22}{7} \text{ लीजिए}\right)$ और अपना उत्तर निकटतम तक पूर्णांकित कीजिए। **SSC SELECTION POST 2022**

Sol. Area of minor sector (OBC) $= \frac{90}{360} \times \frac{22}{7} \times 21 \times 21 = 346.5 \text{ cm}^2$

Area of $\Delta BOC = \frac{1}{2} \times 21 \times 21 = 220.5 \text{ cm}^2$

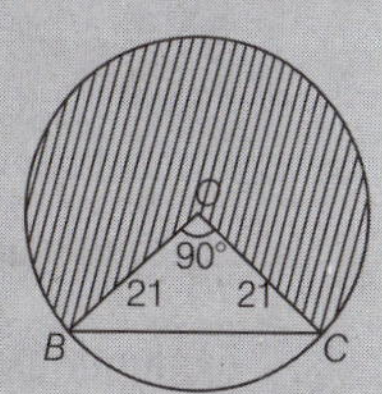

$\therefore$ Area of minor segment $= 346.5 - 220.5 = 126 \text{ cm}^2$

Now, Area of major segment

= Area of circle – Area of minor segment

$= \frac{22}{7} \times 21^2 - 126 = 1386 - 126 = 1260 \text{ cm}^2$

Example **Three circles of radius 7 cm are placed in such a way that each circle touches the other two. What will be the area of the portion enclosed by these three circles?**

7 सेमी त्रिज्या वाले तीन वृत्त इस प्रकार रखे गए हैं कि प्रत्येक वृत्त अन्य दो को स्पर्श करता है। इन तीन वृत्त द्वारा घिरे क्षेत्र का क्षेत्रफल क्या होगा? **SSC (10+2) 2023**

Sol. Side of equilateral triangle $= 2r = 14$ cm

$\therefore$ Area of equilateral triangle

$= \frac{\sqrt{3}}{4} \times 14 \times 14 = 49\sqrt{3} \text{ cm}^2$

A 7 7 B
60° 60°
7 7
7 60° 7
C

Now, the area of the each sector

$= 3 \times \frac{60}{360} \times \pi r^2 = 3 \times \frac{1}{6} \times \frac{22}{7} \times (7)^2 = 77 \text{ cm}^2$

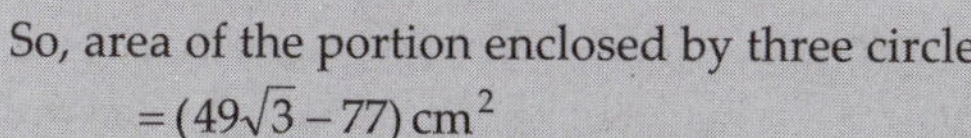

So, area of the portion enclosed by three circle

$= (49\sqrt{3} - 77) \text{ cm}^2$

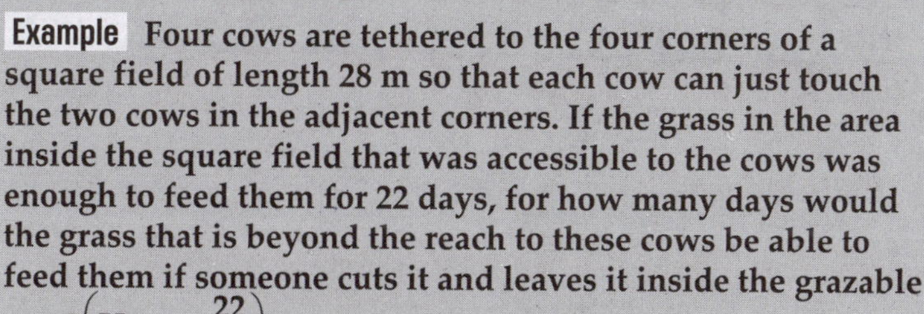

Example **Four cows are tethered to the four corners of a square field of length 28 m so that each cow can just touch the two cows in the adjacent corners. If the grass in the area inside the square field that was accessible to the cows was enough to feed them for 22 days, for how many days would the grass that is beyond the reach to these cows be able to feed them if someone cuts it and leaves it inside the grazable part?** $\left(\text{Use } \pi = \frac{22}{7}\right)$

चार गायों को 28 मी लम्बे एक वर्गाकार खेत के चारों कोनों से बांधा गया है ताकि प्रत्येक गाय बगल के कोनों में दो गायों को छू सके। यदि वर्गाकार खेत के अन्दर के क्षेत्र में गायों के लिए उपलब्ध घास, 22 दिनों के लिए, उन्हें खिलाने के लिए पर्याप्त थी, तो इन गायों की पहुँच से बाहर की घास कितने दिनों तक उन्हें खिलाने में सक्षम होगी यदि कोई इसे काटता है और इसे चरने योग्य भाग के अन्दर छोड़ देता है? $\left(\pi = \frac{22}{7}\right.$ का प्रयोग करें$\left.\right)$ ***RRB NTPC (CBT I) 2021***

Sol. Area of square $= (28)^2 = 784\ \text{m}^2$

Area of the field that can be grazed by the cows

$$= \text{Area of one circle}$$

$$= \frac{22}{7} \times 14 \times 14 = 616\ \text{cm}^2$$

Area of the remaining field $= 784 - 616 = 168\ \text{m}^2$

According to the question,

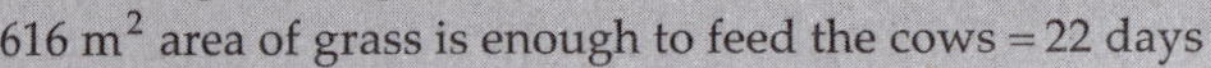

$616\ \text{m}^2$ area of grass is enough to feed the cows = 22 days

$\therefore$ $168\ \text{m}^2$ area of grass is enough to feed the cows $= \frac{22}{616} \times 168 = 6$ days

Example **In the given figure, four identical semicircles are drawn in quadrant, $XA = 7$ cm. What is the area (in cm^2) of shaded region?**

दी गई आकृति में, चतुर्थांश में चार समरूप अर्द्धवृत्त बनाए गए हैं। $XA = 7$ सेमी है। छायांकित भाग का क्षेत्रफल (सेमी^{2} में) क्या है? ***SSC CGL (TIER II) 2022***

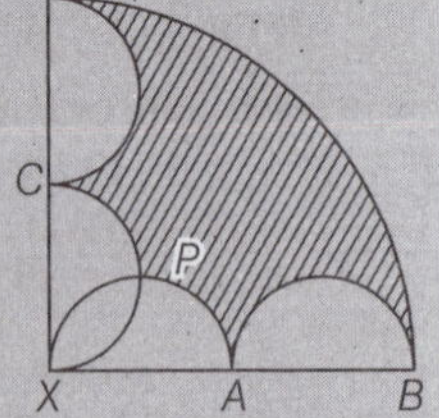

Sol. Area of shaded region

= Area of larger quadrant – Area of 4 smaller semicircle + Common area between semicircle XPA and XPC

Area of 4 smaller semicircle

$$= 4 \times \frac{1}{2} \times \frac{22}{7} \times \left(\frac{7}{2}\right)^2 = 77\ \text{cm}^2$$

Area of larger quadrant $= \frac{1}{4} \times \frac{22}{7} \times (14)^2 = 154\ \text{cm}^2$

Common area between semicircle XPA and XPC

$$= \frac{4}{(4+7)} \times \text{Area of semicircle } XPA = \frac{4}{11} \times \frac{77}{4} = 7\ \text{cm}^2$$

$\therefore$ Area of shaded region $= 154 - 77 + 7 = 84\ \text{cm}^2$

अध्याय 07

Mensuration-3D

त्रिविमीय क्षेत्रमिति

Three-dimensional shapes, also known as 3D shapes, are objects having length, width and height.

त्रि-आयामी आकृतियाँ, जिन्हें 3D आकृतियाँ भी कहा जाता है, वे वस्तुएँ हैं जिनकी लम्बाई, चौड़ाई और ऊँचाई होती हैं।

Primary Components of 3D Shapes

3D आकृतियों के प्राथमिक घटक

The primary components of 3D shapes are faces, edges and vertices.

(i) **Faces** are the flat surface that make up the shape.

(ii) **Edges** are the line where two faces meet.

(iii) **Vertices** are the point where three or more edges meet.

3D आकृतियों के प्राथमिक घटक सतह (फलक), किनारे और शीर्ष (कोने) हैं।

(i) **सतह** (फलक) सपाट/समतल बाहरी सतह है जो आकृति या आकार बनाते हैं।

(ii) **किनारे** वे रेखाएँ हैं जहाँ दो सतहें मिलती हैं।

(iii) **शीर्ष** वे बिन्दु हैं, जहाँ तीन या अधिक किनारे मिलते हैं।

Surface Area पृष्ठीय क्षेत्रफल

The total area covered by the outer surfaces of 3D object is called its surface area. It is measured in square units.
3D वस्तुओं की बाहरी सतहों के द्वारा घेरे गए कुल क्षेत्रफल को उसका पृष्ठीय क्षेत्रफल कहा जाता है। इसे वर्ग इकाई में मापा जाता है।

The following are the types of surface area

पृष्ठीय क्षेत्रफल के प्रकार निम्न हैं

(i) **Total Surface Area/कुल पृष्ठीय क्षेत्रफल** Total surface area occupied by the base(s) and the remaining surfaces of the object.
It is the total area covered by all the surfaces of the object.
कुल पृष्ठीय क्षेत्रफल वस्तु के आधार(ओं) और शेष सतहों द्वारा घेरा गया क्षेत्र है। यह वस्तु की सभी सतहों द्वारा घेरा गया कुल क्षेत्र है।

(ii) **Curved/Lateral Surface Area/वक्रीय/पार्श्व पृष्ठीय क्षेत्रफल**
Curved/Lateral surface area is the area occupied by all the remaining surfaces of the object other than its base(s).
वक्र/पार्श्व पृष्ठीय क्षेत्रफल, वस्तु के आधार(ओं) के अतिरिक्त शेष सभी सतहों द्वारा घेरा गया क्षेत्रफल है।

Volume आयतन

The amount of space occupied by a three dimensional object is called the 'Volume' of that object and it is measured in cubic units.

किसी त्रिविमीय वस्तु द्वारा घेरे गए स्थान की मात्रा को उस वस्तु का 'आयतन' कहा जाता है और इसे घन इकाई में मापा जाता है।

Euler's Theorem यूलर प्रमेय

Euler's formula deals with shapes called polyhedron. A polyhedron is a closed solid shape which has flat faces and straight edges.

यूलर का सूत्र बहुफलक कही जाने वाली आकृतियों से सम्बन्धित है। बहुफलक एक बन्द ठोस आकृति है जिसके समतल सतह और सीधे किनारे होते हैं।

By Euler's theorem/यूलर प्रमेय से, $F + V - E = 2$

Where/जहाँ, F = Number of faces/फलकों की संख्या

V = Number of vertices/शीर्षों की संख्या

E = Number of edges/किनारों की संख्या

यह सूत्र मुख्यतः सभी परीक्षाओं की दृष्टि से महत्त्वपूर्ण है। इस सूत्र का प्रयोग प्रश्न पूछे जाने की सम्भावना रहती है।

A solid figure in which length, breadth and height are equal. If a is the side of a cube, then

एक ठोस आकृति जिसकी लम्बाई, चौड़ाई और ऊँचाई बराबर होती है। यदि घन की भुजा a है, तब

(i) Lateral surface area
पार्श्व पृष्ठीय क्षेत्रफल $= 4a^2$

(ii) Total surface area
कुल पृष्ठीय क्षेत्रफल $= 6a^2$

(iii) Volume/आयतन $= a^3$

(iv) Diagonal/विकर्ण $= \sqrt{3}a$

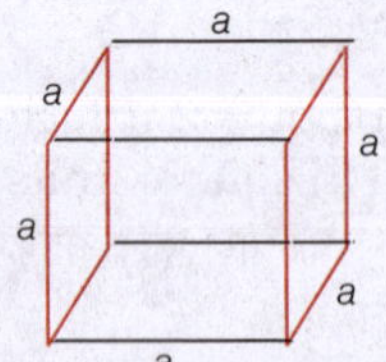

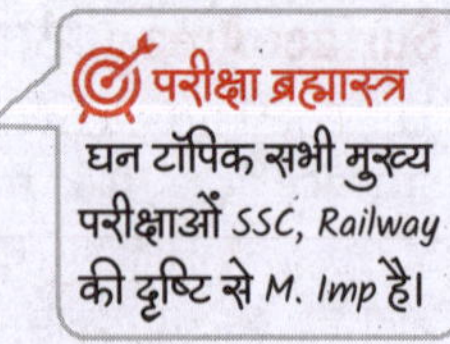

घन पर आधारित उदाहरण निम्नलिखित है

Example **If the total surface area of a cube is 24 sq. units, what is the volume of the cube?**

यदि घन का कुल पृष्ठीय क्षेत्रफल 24 वर्ग इकाई है, तो घन का आयतन क्या है?

SSC CGL (TIER-I) 2023

Sol. Side of cube $= \sqrt{\frac{24}{6}} = \sqrt{4} = 2$ units

Volume of the cube $= 2^3 = 8$ cu. units

Cuboid घनाभ

If l, b and h are the length, breadth and height of a cuboid respectively, then
यदि एक घनाभ की लम्बाई, चौड़ाई और ऊँचाई क्रमशः l, b और h हैं, तब

(i) Area of adjacent faces $= lb, bh, hl$
संलग्न फलकों का क्षेत्रफल $= lb, bh, hl$

(ii) Lateral surface area/पार्श्व पृष्ठीय क्षेत्रफल
$= 2(l+b)h$

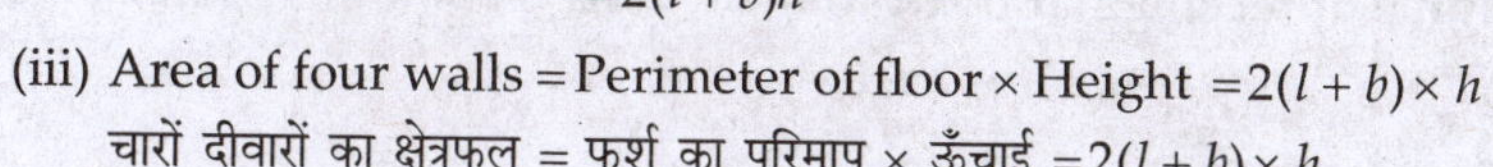

(iii) Area of four walls = Perimeter of floor × Height $= 2(l+b)\times h$
चारों दीवारों का क्षेत्रफल = फर्श का परिमाप × ऊँचाई $= 2(l+b)\times h$

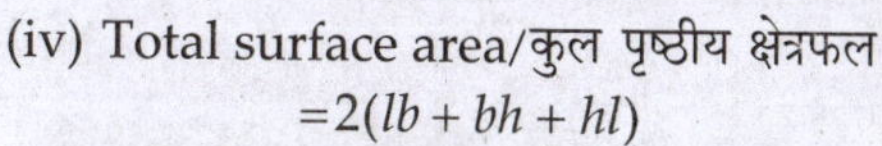

(iv) Total surface area/कुल पृष्ठीय क्षेत्रफल
$= 2(lb+bh+hl)$

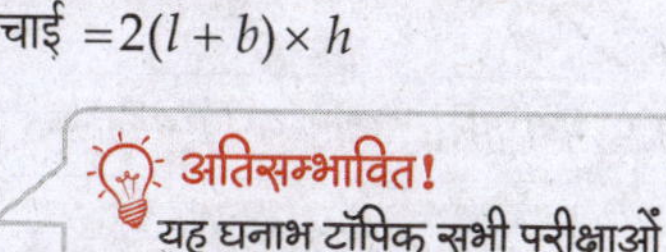

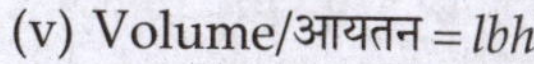

(v) Volume/आयतन $= lbh$

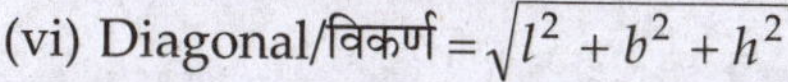

(vi) Diagonal/विकर्ण $= \sqrt{l^2+b^2+h^2}$

Note *Length of longest rod that can be placed in the room is equal to the length of diagonals.*
किसी कमरे में रखी जा सकने वाली सबसे बड़ी छड़ की लम्बाई, विकर्ण की लम्बाई के बराबर होती है।

(vii) If the area of 3 adjacent faces of a cuboid are x, y and z respectively, then
यदि एक घनाभ के 3 आसन्न (संलग्न) फलकों के क्षेत्रफल क्रमशः x, y और z हैं, तब

$$x = lb,\ y = bh,\ z = hl$$

and/तथा Volume of cuboid/घनाभ का आयतन $= \sqrt{xyz}$

(viii) If x, y and z are diagonals of three adjacent faces of a cuboid, then
यदि एक घनाभ के 3 आसन्न (संलग्न) फलकों के विकर्ण x, y और z हैं, तब,

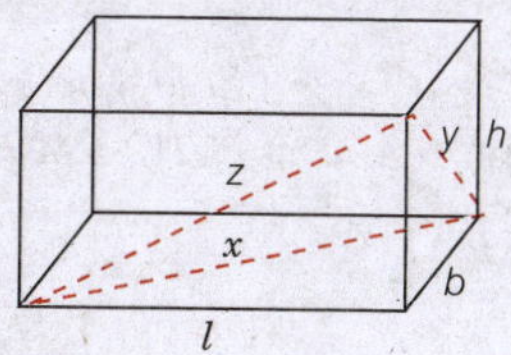

$$x^2 = l^2+b^2;\ y^2 = b^2+h^2;\ z^2 = l^2+h^2$$

Volume of cuboid/घनाभ का आयतन $= lbh$

$$= \frac{\sqrt{(x^2+y^2-z^2)(x^2-y^2+z^2)(y^2+z^2-x^2)}}{2\sqrt{2}}$$

(ix) Volume of hollow cuboid/खोखले घनाभ का आयतन

$$= lbh - (l-2x)(b-2x)(h-2x)$$

Where, x is the thickness of walls of the cuboid
जहाँ, x घनाभ की दीवार की मोटाई है।

(x) Making open rectangular box by cutting 4 corners of a rectangular sheet.
एक आयताकार शीट के 4 कोनों को काटकर, खुला आयताकार बॉक्स बनाना।

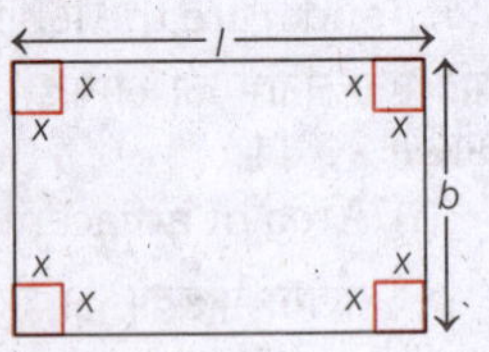

Volume of rectangular box आयताकार बॉक्स का आयतन M. Imp

$$= (l - 2x)(b - 2x)x$$

घनाभ पर आधारित उदाहरण निम्नलिखित हैं

Example **A cuboid is formed by joining 3 cubes of side 15 cm. What is the volume of the cuboid?**

15 सेमी भुजा के 3 घनों को मिलाकर एक घनाभ बनाया गया है। घनाभ का आयतन कितना है? *DSSSB 2022*

Sol. Length of cuboid $= 3 \times 15 = 45$ cm

Breadth of cuboid $= 15$ cm

Height of cuboid $= 15$ cm

$\therefore$ Volume of cuboid $= 45 \times 15 \times 15 = 10125 \text{ cm}^3$

Example **The length of a diagonal of cuboid is 11 cm. The surface area is 240 cm². What is the sum of its length, breadth and height?**

एक घनाभ के विकर्ण की लम्बाई 11 सेमी है। इसका पृष्ठीय क्षेत्रफल 240 वर्ग सेमी है। इसकी लम्बाई, चौड़ाई और ऊँचाई का योग कितना है?

Sol. $\because \sqrt{L^2 + B^2 + H^2} = 11 \text{ cm}$

and $2(LB + BH + HL) = 240 \text{ cm}^2$

We know that,

$$(a + b + c)^2 = a^2 + b^2 + c^2 + 2(ab + bc + ac)$$

$$\therefore \quad (L + B + H)^2 = (11)^2 + 2 \times 120 = 121 + 240 = 361$$

$$L + B + H = \sqrt{361} = 19 \text{ cm}$$

Example **An aquarium is in the form of cuboid whose external measures are 70 cm × 15 cm × 20 cm. The base, side faces and back face are to be covered with coloured paper. What is the area of paper needed?**

एक मछली घर एक घनाभ के आकार का है जिसकी बाहरी माप 70 सेमी × 15 सेमी × 20 सेमी है। आधार फलक, पार्श्व फलकों और पीछे के फलकों को रंगीन कागज से ढ़कना है। आवश्यक कागज का क्षेत्रफल कितना है?

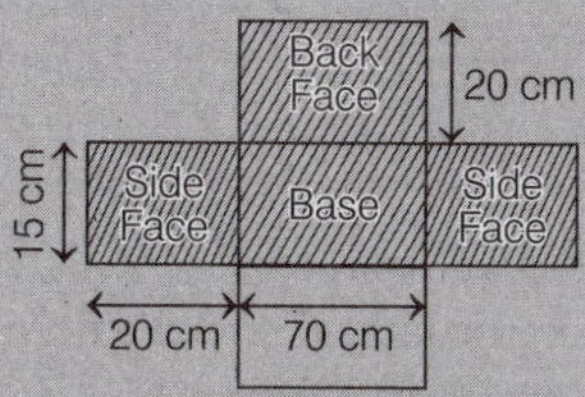

Sol. Base area $= 70 \times 15 = 1050\ \text{cm}^2$

Side face area $= 2 \times (15 \times 20) = 600\ \text{cm}^2$

Back face area $= 70 \times 20 = 1400\ \text{cm}^2$

$\therefore$ Total paper area $= 1050 + 600 + 1400 = 3050\ \text{cm}^2$

Example **How many 25 cm × 11.25 cm × 6 cm bricks will be required to construct an 8 m × 6 m × 22.5 cm wall? (Ignoring other material)**

एक 8 मी × 6 मी × 22.5 सेमी की दीवार बनाने के लिए कितनी 25 सेमी × 11.25 सेमी × 6 सेमी की ईंटों की आवश्यकता होगी? (प्रयुक्त अन्य सामग्री को नगण्य मानिए)

Sol. $\because$ 1 m = 100 cm

$$\therefore \text{Required bricks} = \frac{800 \times 600 \times 22.5}{25 \times 11.25 \times 6} = 6400$$

Example **From the four corners of a rectangular sheet of dimensions 25 cm × 20 cm, square of side 2 cm is cut off from corners and a box is made. The volume of the box is**

25 सेमी × 20 सेमी विमाओं वाली आयताकार शीट के चार कोनों में से 2 सेमी भुजा वाले वर्ग को चारों कोनों से काटकर एक डिब्बा बनाया जाता है। डिब्बे का आयतन है

SSC CGL 2021

Sol. Volume of box $= (25 - 2 \times 2)(20 - 2 \times 2) \times 2$

$= 21 \times 16 \times 2 = 672\ \text{cm}^3$

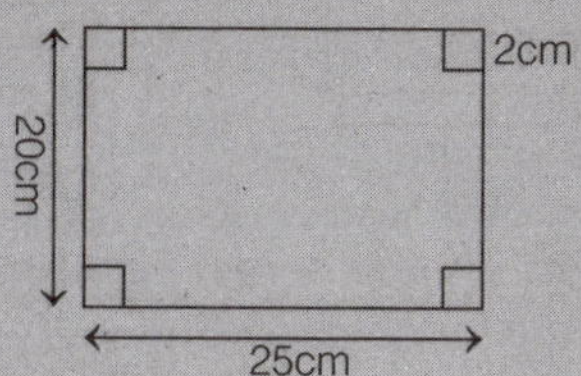

If radius of cylinder is r and height of cylinder is h.
यदि बेलन की त्रिज्या r और बेलन की ऊँचाई h है।

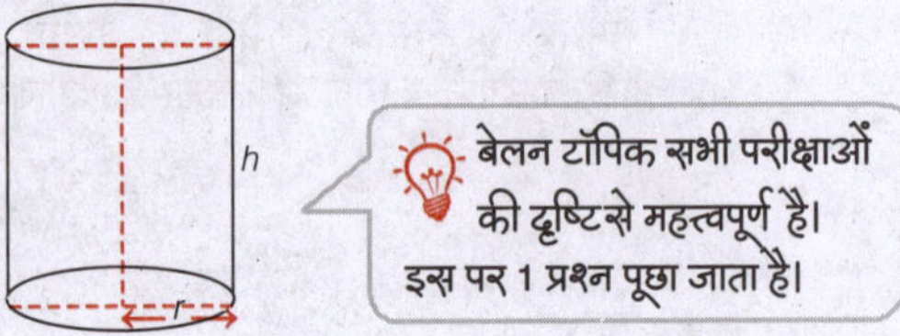

(i) Volume/आयतन $= \pi r^2 h$

(ii) Curved surface area or lateral surface area
वक्र पृष्ठीय क्षेत्रफल या पार्श्व पृष्ठीय क्षेत्रफल $= 2\pi rh$

(iii) Total surface area/कुल पृष्ठीय क्षेत्रफल $= 2\pi rh + 2\pi r^2 = 2\pi r(h + r)$

(iv) Folding of a rectangular sheet to form a cylinder
एक बेलन बनाने के लिए आयताकार शीट को मोड़ना

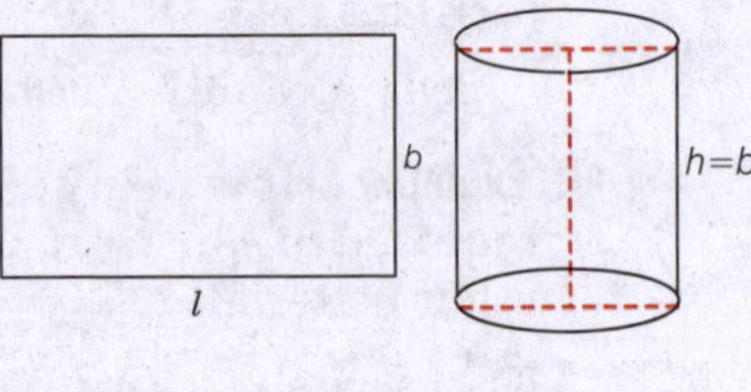

When folding along length, then
जब लम्बाई के अनुदिश मोड़ना है, तब

$$h = b \text{ and/और } r = \frac{l}{2\pi}$$

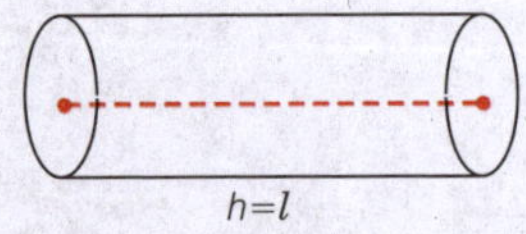

When folding along breadth, then
जब चौड़ाई के अनुदिश मोड़ना है, तब

$$h = l \text{ and/और } r = \frac{b}{2\pi}$$

बेलन पर आधारित उदाहरण निम्नलिखित हैं

Example **A rectangular paper of length 56 cm and breadth 22 cm is folded along the length to make a cylinder. The volume of the cylinder so formed.**

56 सेमी लम्बाई और 22 सेमी चौड़ाई के आयताकार कागज को एक बेलन बनाने के लिए लम्बाई के समानान्तर मोड़ा जाता है। इस प्रकार बने बेलन का आयतन ज्ञात कीजिए।

Sol.

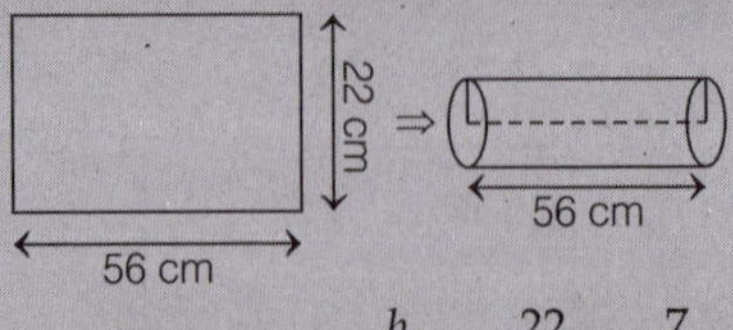

$h = l = 56$ cm, $\quad r = \dfrac{b}{2\pi} = \dfrac{22}{2 \times \dfrac{22}{7}} = \dfrac{7}{2}$ cm

$\therefore$ Volume of cylinder $= \pi r^2 h = \dfrac{22}{7} \times \left(\dfrac{7}{2}\right)^2 \times 56 = 2156 \text{ cm}^3$

Example **A cylindrical drum is open from the top having a radius of 15 cm and height of 12 cm is to be painted from inside as well as outside. The rate of painting is ₹ 2/cm^2. Find the cost of painting the drum.**

एक बेलनाकार ड्रम ऊपर से खुला है जिसकी त्रिज्या 15 सेमी है और ऊँचाई 12 सेमी है। उसे अन्दर और बाहर से पेण्ट किया जाना है। पेण्टिग की दर ₹ 2 /सेमी^{2} है। ड्रम को पेण्ट करने की लागत ज्ञात कीजिए। *RRC GROUP D 2022*

Sol. Curved surface area $= 2 \times \frac{22}{7} \times 15 \times 12 = \frac{7920}{7}$ cm^2

Area of one base $= \frac{22}{7} \times (15)^2 = \frac{4950}{7}$ cm^2

∴ Total area to be painted (inside + outside)

$$= 2 \times \left(\frac{7920}{7} + \frac{4950}{7}\right) = \frac{25740}{7} \text{ cm}^2$$

∴ Cost of painting $= \frac{25740}{7} \times 2 = ₹\, 7354.3$

Hollow Cylinder खोखला बेलन

If/यदि r = Inner radius/आन्तरिक त्रिज्या

R = Outer radius/बाह्य त्रिज्या

h = Height/ऊँचाई

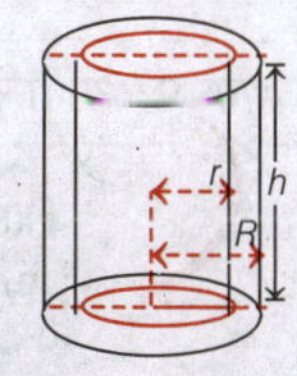

(i) Volume/आयतन $= \pi R^2 h - \pi r^2 h = \pi(R^2 - r^2)h$

(ii) Curved surface area/वक्रपृष्ठीय क्षेत्रफल $= 2\pi(R + r)h$

(iii) Total surface area/कुल पृष्ठीय क्षेत्रफल $= 2\pi(R + r)(h + R - r)$

खोखले बेलन पर आधारित उदाहरण निम्नलिखित है

Example **A well of radius 2 m is dug 16.8 m deep. The soil taken out of it is spread evenly all around the well of form a 48 cm high embankment. What is the width (in cm) of the embankment?**

2 मी त्रिज्या का एक कुआँ 16.8 मी गहराई तक खोदा जाता है। इसमें से खोदी गई मिट्टी को कुएँ के चारों ओर समान रूप से फैलाकर 48 सेमी ऊँचा तटबन्ध बनाया जाता है। तटबन्ध की चौड़ाई (मीटर में) क्या है? *RRB NTPC 2022*

Sol. Let width of the embankment be x m.

According to the question,

$$\pi (2)^2 \times 16.8 = \pi [(2 + x)^2 - 2^2] \times \frac{48}{100}$$

$$\Rightarrow \quad 4 \times 16.8 \times \frac{100}{48} = 4 + 4x + x^2 - 4 \Rightarrow 140 = x^2 + 4x$$

$$\Rightarrow \quad x^2 + 14x - 10x - 140 = 0 \Rightarrow x(x + 14) - 10(x + 14) = 0$$

$$\Rightarrow \quad (x + 14)(x - 10) = 0$$

$$x = 10 \text{ m} \qquad [\because x \neq -14]$$

If the radius, height and slant height of the cone are r, h and l respectively, then

यदि एक शंकु की त्रिज्या, ऊँचाई और तिर्यक ऊँचाई क्रमशः r, h और l हैं, तब

(i) Volume/आयतन $= \frac{1}{3}\pi r^2 h$

(ii) Curved surface area/वक्रपृष्ठीय क्षेत्रफल

$$= \frac{1}{2} \times 2\pi r \times l = \pi r l$$

अतिसम्भावित! यह शंकु टॉपिक सभी परीक्षाओं की दृष्टि से *M. Imp* है। उस पर आधारित 1 प्रश्न सभी परीक्षाओं में पूछा जाता है।

(iii) Total surface area/कुल पृष्ठीय क्षेत्रफल

$$= \pi r l + \pi r^2 = \pi r(l + r)$$

(iv) Slant height/तिरछी ऊँचाई $l = \sqrt{h^2 + r^2}$

(v) If S denotes the curved surface area of a right circular cone of height h and semivertical angle α, then

यदि S किसी शंकु के वक्रपृष्ठ क्षेत्रफल को व्यक्त करता है, h ऊँचाई को और α अर्द्धशीर्ष कोण को व्यक्त करता है, तब

$$S = \pi h^2 \sec\alpha \tan\alpha$$

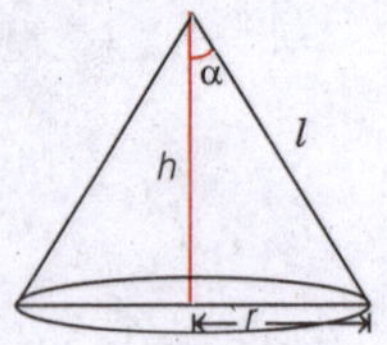

शंकु पर आधारित उदाहरण निम्नलिखित हैं

Example **A conical tent can accommodate 25 person. Each person must have 4 m^2 of space on the ground and 80 m^3 of air to breathe. Find the height of the tent.**

एक शंक्वाकार तम्बू में 25 व्यक्ति रह सकते हैं। प्रत्येक व्यक्ति को जमीन पर 4 मी^{2} जगह और साँस लेने के लिए 80 मी^{3} हवा होनी चाहिए। तम्बू की ऊँचाई ज्ञात कीजिए। *SSC CGL 2020*

Sol. Total base are required = $4 \times 25 = 100 \text{ m}^2$

$\therefore \quad \pi r^2 = 100$

Total volume of air required = $25 \times 80 = 2000 \text{ m}^3$

Let the height of the cone be h m

$\therefore \quad \frac{1}{3}\pi r^2 h = 2000$

$\Rightarrow \quad \frac{1}{3} \times 100 \times h = 2000$

$\Rightarrow \quad h = \frac{2000 \times 3}{100} \Rightarrow h = 60 \text{ m}$

Example **How many meters of 2 m wide cloth will be required to make a conical tent with a diameter of the base as 14 m and slant height as 9 m? (Ignore wastage)**

एक शंक्वाकार तम्बू जिसके आधार का व्यास 14 मी और तिर्यक ऊँचाई 9 मी हो, को बनाने के लिए 2 मी चौड़े कपड़े के कितने मीटर की आवश्यकता होगी? *MP PATWARI 2023*

Sol. Radius of the base of tent $= \frac{14}{2} = 7$ m and slant height = 9 m

$\therefore$ Curved surface area of tent $= \frac{22}{7} \times 7 \times 9 = 198\ m^2$

$\therefore$ Area of the cloth = $198\ m^2$ $\Rightarrow$ Length of cloth $= \frac{198}{2} = 99$ m

Rotation of Right Angle Triangle to Form a Cone
एक शंकु बनाने के लिए समकोण त्रिभुज का घूर्णन

- If ΔABC is a right angle triangle, then/यदि ΔABC एक समकोण त्रिभुज है, तब

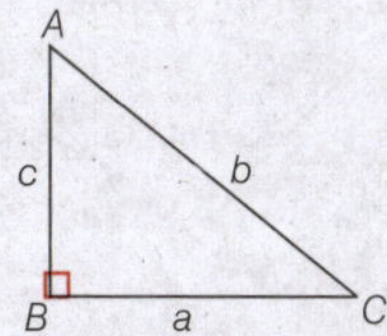

(i) Rotation along base BC/आधार BC के साथ घूमना
Here/यहाँ, $r = c, h = a, l = b$

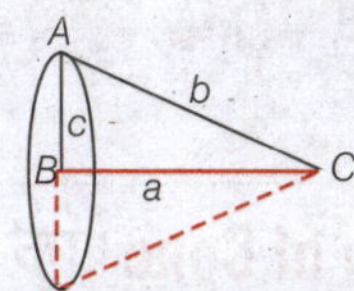

(ii) Rotation along perpendicular AB/ लम्ब AB के सापेक्ष घूमना

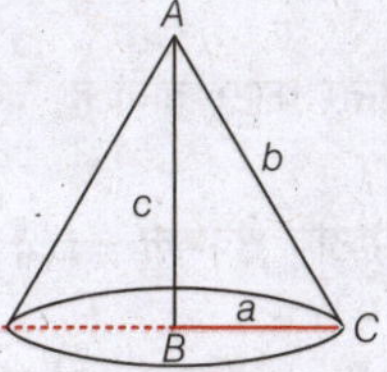

Here/यहाँ, $r = a,\ h = c,\ l = b$

(iii) Rotation along hypotenuse AC/कर्ण AC के साथ घूमना

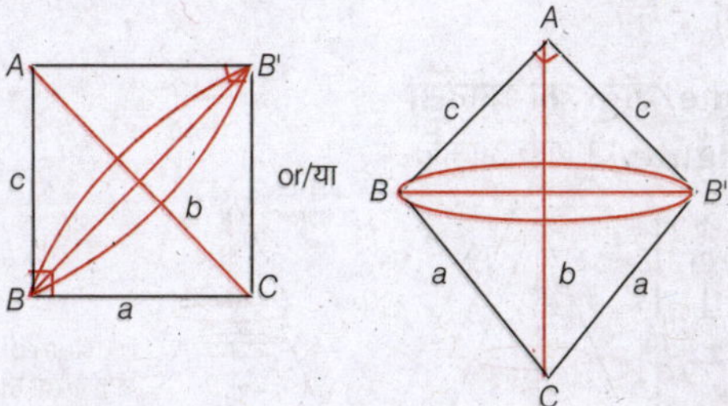

Here/यहाँ,

Sum of volume of 2 cones/द्विशंकु का आयतन $= \frac{1}{3}\pi\frac{(ac)^2}{b}$ *M. Imp*

त्रिभुज के घूर्णन से शंकु बनाने पर आधारित उदाहरण निम्न है

Example **A right angle triangle is there with perpendicular side 15 cm and 20 cm. It is rotated along the hypotenuse. Find the volume of the shape formed by this.**

एक समकोण त्रिभुज 15 सेमी और 20 सेमी के लम्बवत् भुजाओं के साथ है। इसे कर्ण के साथ घुमाया जाता है। इस प्रकार बनी आकृति का आयतन ज्ञात कीजिए।

Sol. By Pythagoras theorem,

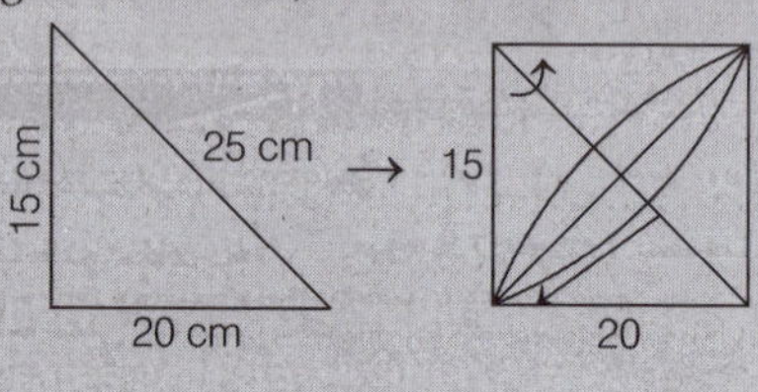

$$H^2 = P^2 + B^2$$
$$H^2 = 15^2 + 20^2$$
$$H = 25 \text{ cm}$$

$\therefore$ Volume of cone formed $= \frac{1}{3}\pi\frac{(15\times20)^2}{25} = \frac{1}{3}\pi\times\frac{300\times300}{25} = 1200\pi \text{ cm}^3$

Frustum of Cone शंकु के छिन्नक

When a cone is cut parallel to its base, then the lower portion is called **frustum of cone**.

जब एक शंकु को उसके आधार के समान्तर काटा जाता है, तब निचला भाग **शंकु का छिन्नक** कहलाता है।

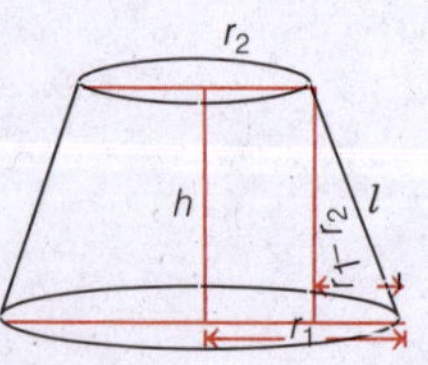

(i) Curved surface area/वक्रपृष्ठीय क्षेत्रफल $= \pi(r_1 + r_2)l$

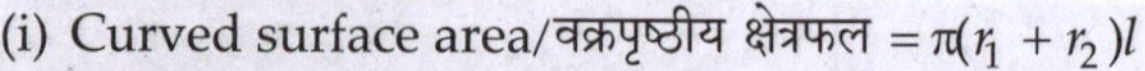

(ii) Total surface area
कुल पृष्ठीय क्षेत्रफल $= \pi(r_1 + r_2)\,l + \pi(r_1^2 + r_2^2)$

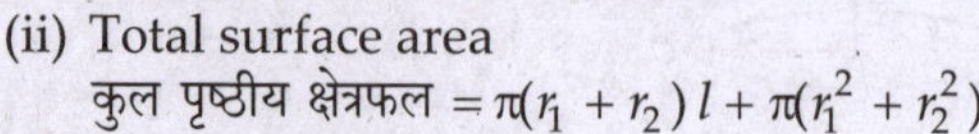

(iii) Volume/आयतन $= \frac{1}{3}\pi(r_1^2 + r_2^2 + r_1r_2)\times h$

(iv) $l = \sqrt{h^2 + (r_1 - r_2)^2}$

- **Cutting of Cone/शंकु को काटना**

In the given figure/दी गई आकृति में,

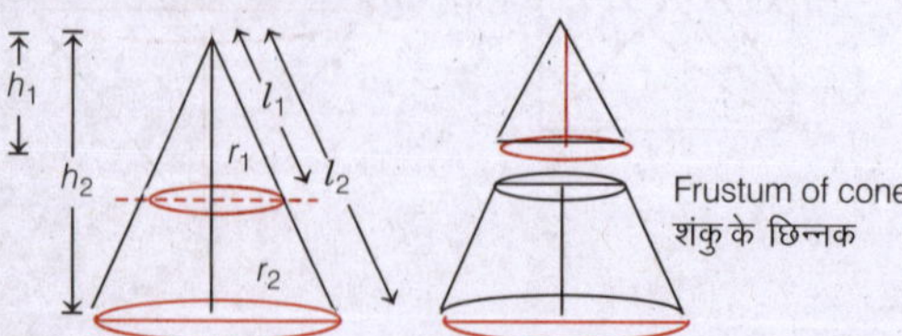

(i) $\frac{r_1}{r_2} = \frac{h_1}{h_2} = \frac{l_1}{l_2}$

(ii) $\dfrac{\text{Curved surface area of small cone}}{\text{Curved surface area of larger cone}}$

$\dfrac{\text{छोटे शंकु का वक्रपृष्ठीय क्षेत्रफल}}{\text{बड़े शंकु का वक्रपृष्ठीय क्षेत्रफल}} = \dfrac{\pi r_1 l_1}{\pi r_2 l_2} = \left(\dfrac{r_1}{r_2}\right)^2 = \left(\dfrac{l_1}{l_2}\right)^2 = \left(\dfrac{h_1}{h_2}\right)^2$

शंकु के छिन्नक पर आधारित उदाहरण निम्नलिखित है

Example **The radii of the two circular faces of the frustum of a cone of height 21 cm are 5 cm and 3 cm. What is its volume in cm^3? $\left(\pi = \dfrac{22}{7}\right)$**

21 सेमी ऊँचाई वाले एक शंकु के छिन्नक के दो वृत्ताकार सतहों की त्रिज्याएँ 5 सेमी और 3 सेमी हैं। सेमी^{3} में इसका आयतन क्या है? $\left(\pi = \dfrac{22}{7}\right)$

SSC CGL 2018

Sol. Volume of the frustum of cone $= \dfrac{1}{3}\pi [r_1^2 + r_2^2 + r_1 r_2] H$

$= \dfrac{1}{3} \times \dfrac{22}{7} \times [5^2 + 3^2 + 5 \times 3] \times 21$

$= 22 \times [25 + 9 + 15]$

$= 22 \times 49 = 1078 \text{ cm}^3$

Sphere गोला

If the radius of sphere is r, then

गोले की त्रिज्या r है, तब

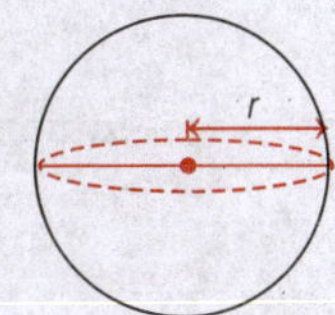

(i) Volume/आयतन $= \dfrac{4}{3}\pi r^3$

(ii) Curved surface area = Total surface area

वक्रपृष्ठीय क्षेत्रफल = कुल पृष्ठीय क्षेत्रफल $= 4\pi r^2$

Hollow Sphere खोखला गोला

If the inner and outer radius of hollow sphere are r and R respectively, then

यदि एक खोखले गोले की आन्तरिक और बाह्य त्रिज्या क्रमशः r और R हैं, तब

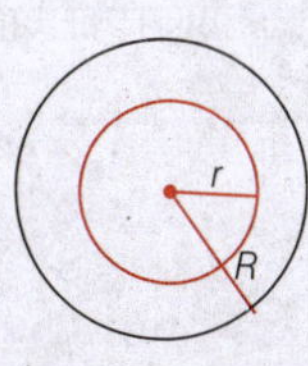

(i) Volume/आयतन $= \dfrac{4}{3}\pi(R^3 - r^3)$

(ii) Thickness of metal/ धातु की मोटाई $= R - r$

(iii) Total-surface area/कुल पृष्ठीय क्षेत्रफल $= 4\pi(R^2 + r^2)$

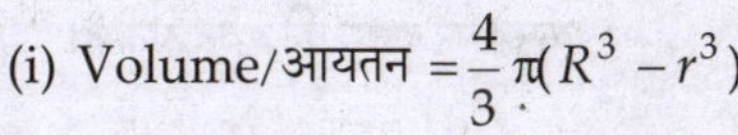

If the radius of hemi-sphere is r, then

यदि अर्द्धगोले की त्रिज्या r है, तब

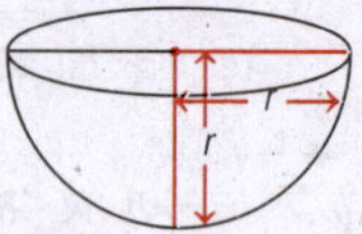

(i) Volume/आयतन $= \frac{2}{3}\pi r^3$

(ii) Curved surface area/वक्रपृष्ठीय क्षेत्रफल $= 2\pi r^2$

(iii) Total surface area/कुल पृष्ठीय क्षेत्रफल $= 3\pi r^2$

गोले और अर्द्धगोले पर आधारित उदाहरण निम्नलिखित हैं

Example **Find the surface area of a ball of diameter 14 cm.**

14 सेमी व्यास की एक गेंद का पृष्ठीय क्षेत्रफल ज्ञात कीजिए।

DP HEAD CONSTABLE 2023

Sol. Radius of the ball $= \frac{14}{2} = 7$ cm

$\therefore$ Surface area of the ball $= 4 \times \frac{22}{7} \times 7^2 = 616 \text{ cm}^2$

Example **If the radius of a sphere is thrice of the radius of a hemisphere, then what will be the respective ratio of their volumes?**

यदि एक गोले की त्रिज्या एक गोलार्द्ध की त्रिज्या की तीन गुनी है, तो उनके आयतन का सम्बन्धित अनुपात क्या होगा?

SSC (10+2) 2023

Sol. Radius of sphere (R) = Radius of hemisphere $(r) \times 3$

$$R = 3r$$

$$\therefore \text{ Ratio of the volume} = \frac{\frac{4}{3}\pi R^3}{\frac{2}{3}\pi r^3} = \frac{R^3}{r^3} = \frac{2(3r)^3}{r^3} = \frac{54}{1} = 54 : 1$$

(i) Volume of prism = Area of base × Height

प्रिज्म का आयतन = आधार का क्षेत्रफल × ऊँचाई

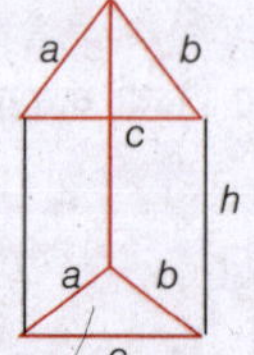

प्रिज्म टॉपिक के सूत्र M. Imp हैं। इसके प्रश्न को सूत्र से हल करने पर समय की बचत होती है।

(ii) Lateral surface area = Perimeter of base × Height

पार्श्व पृष्ठीय क्षेत्रफल = आधार का परिमाप × ऊँचाई

(iii) Total surface area = Lateral surface area + 2 × Base area
कुल पृष्ठीय क्षेत्रफल = पार्श्व पृष्ठीय क्षेत्रफल + 2 × आधार का क्षेत्रफल

Equilateral Triangular Prism समबाहु त्रिभुजाकार प्रिज्म

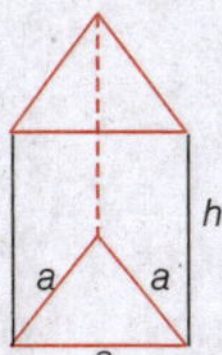

(i) Lateral surface area/पार्श्व पृष्ठीय क्षेत्रफल $= 3ah$

(ii) Total surface area/कुल पृष्ठीय क्षेत्रफल $= 3ah + \frac{\sqrt{3}}{2} a^2$

(iii) Volume/आयतन $= \frac{\sqrt{3}}{4} a^2 h$

Square Prism वर्गाकार प्रिज्म

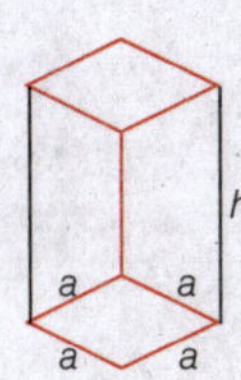

(i) Lateral surface area / पार्श्व पृष्ठीय क्षेत्रफल $= 4ah$

(ii) Total surface area कुल पृष्ठीय क्षेत्रफल $= 4ah + 2a^2$

(iii) Volume/आयतन $= a^2 h$

Hexagonal Prism षट्भुज प्रिज्म

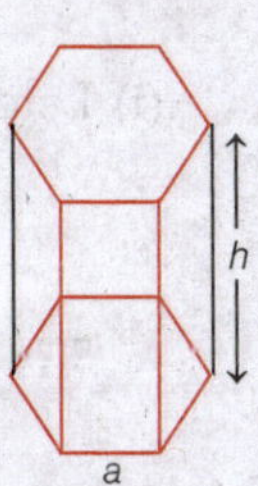

(i) Surface area of hexagon/ षट्भुज का सतह क्षेत्रफल $= \frac{3\sqrt{3}}{2} a^2$

(ii) Lateral surface area/पार्श्व पृष्ठीय क्षेत्रफल $= 6a \times h = 6ah$

(iii) Total surface area/कुल पृष्ठीय क्षेत्रफल $= 6ah + 3\sqrt{3}a^2$

(iv) Volume/आयतन $= \frac{3\sqrt{3}}{2} a^2 h$

General Formulae for a Prism प्रिज्म के लिए सामान्य सूत्र

(i) Curved surface area/वक्रपृष्ठीय क्षेत्रफल $= na \times h$

(ii) Total surface area/ कुल पृष्ठीय क्षेत्रफल $= nah + 2 \times \frac{na^2}{4} \cot \frac{\pi}{n}$

(iii) Volume/आयतन $= \frac{na^2}{4} \cot \frac{\pi}{n} \times h$

Here/यहाँ, $n \rightarrow$ Number of sides of regular polygon
नियमित बहुभुज की भुजाओं की संख्या

$a \rightarrow$ Side length of regular polygon
नियमित बहुभुज की पार्श्व लम्बाई

$h \rightarrow$ Height of prism/प्रिज्म की ऊँचाई

प्रिज्म पर आधारित उदाहरण निम्न है

Example **The base of a right prism is an equilateral triangle, each side of which is 4 cm. If the lateral surface area is 120 cm^2. Find the volume (in cm) of the prism.**

एक लम्ब प्रिज्म का आधार का समबाहु त्रिभुज है, जिसकी प्रत्येक भुजा 4 सेमी की है। यदि पार्श्व पृष्ठीय क्षेत्रफल 120 सेमी^{2} है, तो प्रिज्म का आयतन (सेमी में) ज्ञात कीजिए। *SSC (10+2) 2024*

Sol. Perimeter of base of prism $= 4 \times 3 = 12$ cm

$\therefore$ Height of prism $= \dfrac{120}{12} = 10$ cm

$\therefore$ Volume of prism $= \dfrac{\sqrt{3}}{4} \times 4 \times 4 \times 10 = 40\sqrt{3}$ cm^3

Pyramid पिरामिड

(i) Lateral surface area

$$= \frac{1}{2} \times \text{Perimeter of base} \times \text{Slant height}$$

पार्श्व पृष्ठीय क्षेत्रफल

$$= \frac{1}{2} \times \text{आधार का परिमाप} \times \text{तिर्यक ऊँचाई}$$

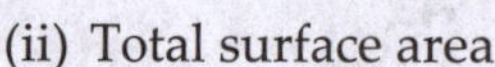

(ii) Total surface area

$$= \text{Lateral surface area} + \text{Base area}$$

कुल पृष्ठीय क्षेत्रफल = पार्श्व पृष्ठीय क्षेत्रफल + आधार का क्षेत्रफल

(iii) Volume $= \dfrac{1}{3} \times \text{Base area} \times \text{Height}$

आयतन $= \dfrac{1}{3} \times$ आधार का क्षेत्रफल $\times$ ऊँचाई

Equilateral Triangular Pyramid समबाहु त्रिभुजाकार पिरामिड

(i) Lateral surface area/पार्श्व पृष्ठीय क्षेत्रफल $= \dfrac{1}{2} \times 3a \times l$

(ii) Total surface area/कुल पृष्ठीय क्षेत्रफल $= \dfrac{1}{2} \times 3al + \dfrac{\sqrt{3}}{4} a^2$

(iii) Volume/आयतन $= \dfrac{1}{3} \times \dfrac{\sqrt{3}}{4} a^2 h$

(iv) Slant height/तिरछी ऊँचाई $l = \sqrt{h^2 + r^2}$ [r = Inradius/अन्तःवृत्त की त्रिज्या]

$$= \sqrt{h^2 + \left(\frac{a}{2\sqrt{3}}\right)^2}$$

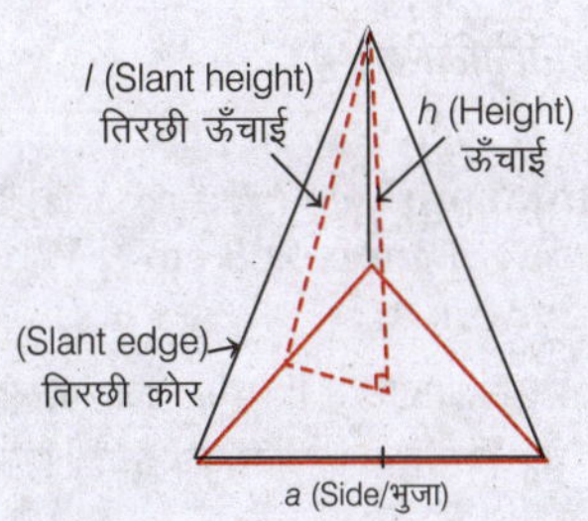

(v) Slant edge $l = \sqrt{h^2 + R^2}$. [Here, R = Circum radius/परित्रिज्या]

$$= \sqrt{h^2 + \left(\frac{a}{\sqrt{3}}\right)^2}$$

Square Pyramid वर्गाकार पिरामिड

(i) Lateral surface area/पार्श्व पृष्ठीय क्षेत्रफल $= \frac{1}{2} \times 4a \times l = 2al$

(ii) Total surface area/कुल पृष्ठीय क्षेत्रफल $= \frac{1}{2} \times 4al + a^2 = 2al + a^2$

(iii) Volume/आयतन $= \frac{1}{3} a^2 h$

(iv) Slant height/तिरछी ऊँचाई,

$$l = \sqrt{h^2 + \left(\frac{a}{2}\right)^2}$$

(v) Slant edge/तिरछी कोर $= \sqrt{h^2 + \left(\frac{a}{\sqrt{2}}\right)^2}$

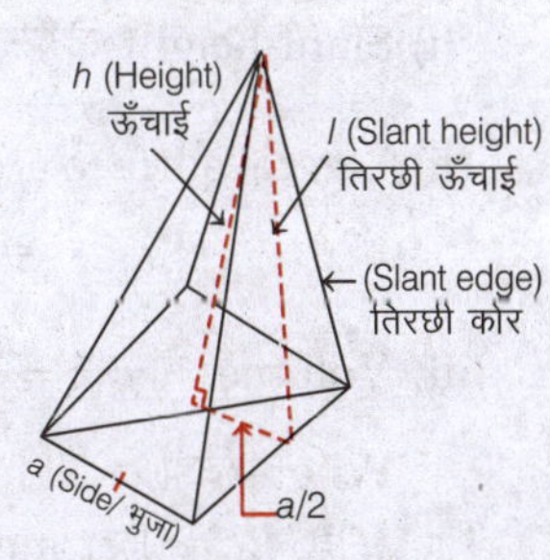

Hexagonal Pyramid षट्भुज पिरामिड

(i) Lateral surface area
पार्श्व पृष्ठीय क्षेत्रफल $= \frac{1}{2} \times 6a \times s = 3al$

(ii) Total surface area
कुल पृष्ठीय क्षेत्रफल $= 3al + \frac{3\sqrt{3}}{2} a^2$

(iii) Volume/आयतन $= \frac{\sqrt{3}}{2} a^2 h$

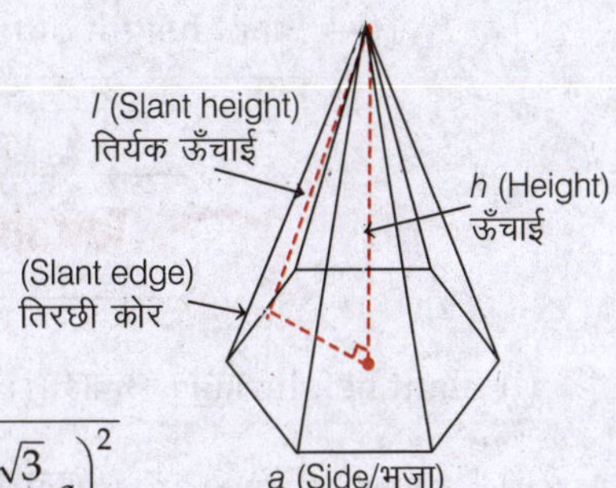

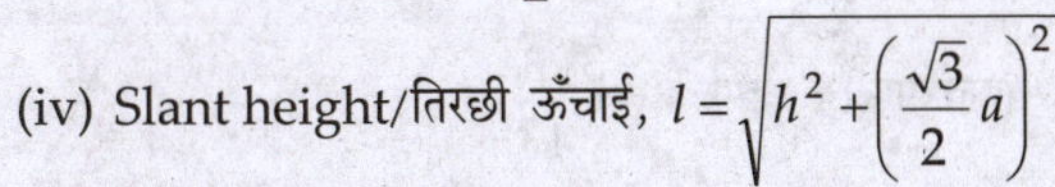
(iv) Slant height/तिरछी ऊँचाई, $l = \sqrt{h^2 + \left(\frac{\sqrt{3}}{2} a\right)^2}$

(v) Slant edge/तिरछी कोर $= \sqrt{h^2 + a^2}$

पिरामिड पर आधारित उदाहरण निम्नलिखित है

Example **The total surface area of a right pyramid with base as a square of side 8 cm is 208 cm^2. What is the slant height (in cm) of the pyramid?**

एक लम्ब पिरामिड, जिसका आधार 8 सेमी भुजा वाला एक वर्ग है, का सम्पूर्ण पृष्ठीय क्षेत्रफल 208 सेमी^{2} है। पिरामिड की तिर्यक ऊँचाई (सेमी में) क्या है? *SSC CGL 2021*

Sol. According to the question,

$$\text{Area of base} = 8^2 = 64 \text{ cm}^2$$

$$\text{Lateral surface area} + 64 = 208$$

$$\therefore \quad \text{Lateral surface area} = 208 - 64 = 144 \text{ cm}^2$$

$$\therefore \quad 4 \times \frac{1}{2} \times 8 \times \text{Slant height} = 144 \Rightarrow \text{Slant height} = 9 \text{ cm}$$

Frustum of Pyramid पिरामिड का छिन्नक

(i) Slant-height/तिर्यक ऊँचाई, $l = \sqrt{h^2 + \left(\frac{a-b}{2}\right)^2}$

(ii) Lateral surface area/पार्श्व पृष्ठीय क्षेत्रफल

$$= \frac{1}{2}(P_1 + P_2) \times l$$

(iii) Volume/आयतन $= \frac{1}{3}(A_1 + A_2 + \sqrt{A_1 \times A_2}) \times h$

Where/जहाँ,

$P_1, P_2 \rightarrow$ Perimeter of bases/आधारों का परिमाप

$A_1, A_2 \rightarrow$ Area of bases/आधारों का क्षेत्रफल

$h \rightarrow$ Height of frustum of pyramid/पिरामिड के छिन्नक की ऊँचाई

$l \rightarrow$ Slant height of frustum of pyramid
पिरामिड के छिन्नक की तिर्यक ऊँचाई

Tetrahedron समचतुष्फलक

(i) Slant height/तिर्यक ऊँचाई $(l) = \frac{\sqrt{3}}{2}a$

(ii) Lateral surface area/पार्श्व पृष्ठीय क्षेत्रफल $= \frac{3\sqrt{3}}{4}a^2$

(iii) Total surface area/कुल पृष्ठीय क्षेत्रफल $= \sqrt{3}a^2$

(iv) Height/ऊँचाई $= \sqrt{\frac{2}{3}}a$

(v) Volume/आयतन $= \frac{\sqrt{2}}{12}a^3$

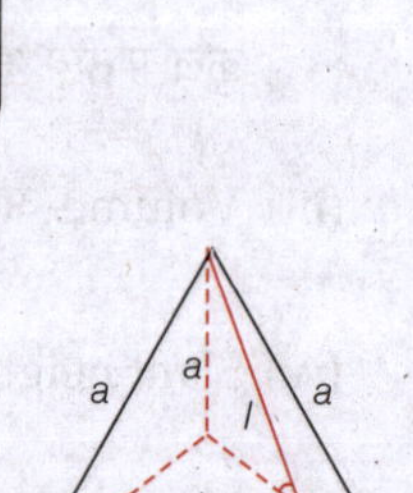

समचतुष्फलक पर आधारित उदाहरण निम्नलिखित है

Example **A box in the form of a tetrahedron having side $2\sqrt{2}$ m. What would be the volume of the box?**

एक बॉक्स चतुष्फलक के रूप में है जिसकी भुजा $2\sqrt{2}$ मी है। बॉक्स का आयतन क्या होगा?

Sol. Volume of tetrahedron $= \frac{\sqrt{2}}{12} \times (2\sqrt{2})^3$

$$= \frac{\sqrt{2}}{12} \times 16\sqrt{2} = \frac{8}{3} \text{ cm}^3$$

Combined Figures संयुक्त आकृतियाँ

1. When a cube of maximum size is put inside a hemisphere, then

   जब एक अर्द्धगोले के अन्दर अधिकतम आकार का घन रखा जाता है, तब

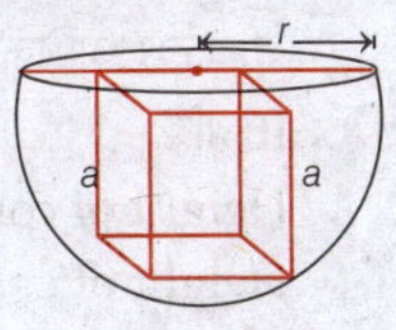

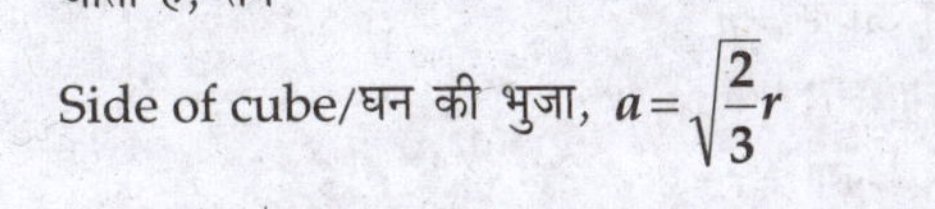

   Side of cube/घन की भुजा, $a = \sqrt{\frac{2}{3}}r$

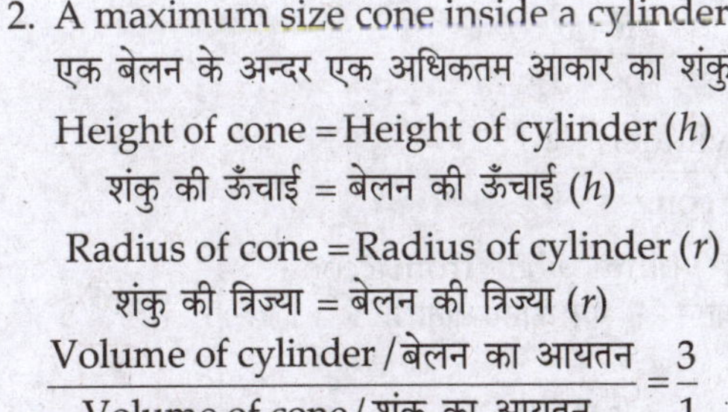

2. A maximum size cone inside a cylinder

   एक बेलन के अन्दर एक अधिकतम आकार का शंकु

   Height of cone = Height of cylinder (h)

   शंकु की ऊँचाई = बेलन की ऊँचाई (h)

   Radius of cone = Radius of cylinder (r)

   शंकु की त्रिज्या = बेलन की त्रिज्या (r)

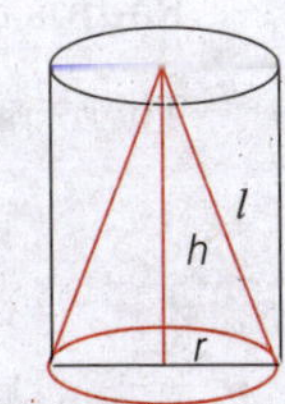

$$\frac{\text{Volume of cylinder / बेलन का आयतन}}{\text{Volume of cone / शंकु का आयतन}} = \frac{3}{1}$$

$$r^2 = \frac{l^2}{2h}$$

3. A cylinder encloses a sphere

   एक बेलन, एक गोले को घेरता है।

   Then/तब,

   (i) Height of cylinder = Diameter of sphere

   बेलन की ऊँचाई = गोले का व्यास

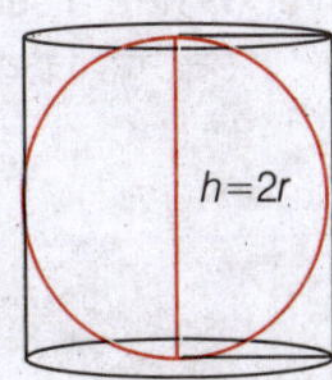

   (ii) $\frac{\text{Volume of cylinder / बेलन का आयतन}}{\text{Volume of sphere / गोले का आयतन}} = \frac{3}{2}$

4. A cylinder of maximum size inside a cube, then
एक घन के अन्दर अधिकतम आकार का एक बेलन हो, तब

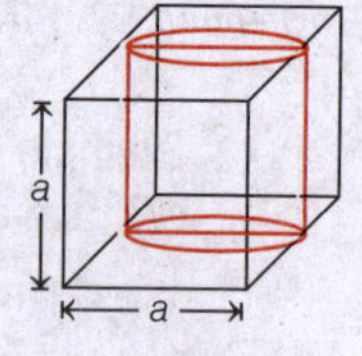

(i) Radius of cylinder/बेलन की त्रिज्या $=\frac{a}{2}$

(ii) Height of cylinder/बेलन की ऊँचाई $=a$

(iii) $\frac{\text{Volume of cube / घन का आयतन}}{\text{Volume of cylinder / बेलन का आयतन}}=\frac{14}{11}$

5. When a hemisphere of maximum size is cut from a cube, then
जब एक घन के अन्दर से अधिकतम आकार का अर्द्धगोला काटा जाता है, तब

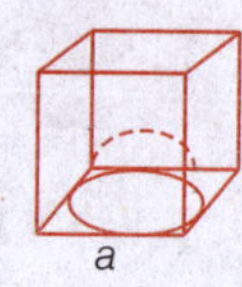

Total surface area/कुल पृष्ठीय क्षेत्रफल $=6a^2+\frac{\pi a^2}{4}$

6. A cylinder of maximum size inside a cone/एक शंकु के अन्दर अधिकतम आकार का एक बेलन

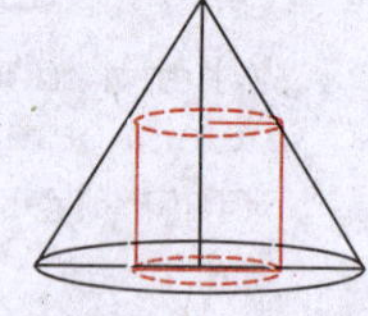

If/यदि,
Height of cone/शंकु की ऊँचाई $=H$
Height of cylinder/बेलन की ऊँचाई $=h$
Radius of cone/शंकु की त्रिज्या $=R$
Radius of cylinder/बेलन की त्रिज्या $=r$

Then/तब, $$\frac{R}{r}=\frac{H}{H-h}$$

$$\frac{\text{Volume of cylinder}}{\text{Volume of cone}}=\frac{4}{9}$$

7. When a cube of maximum volume is cut from a cone
जब एक शंकु से अधिकतम आयतन का घन काटा जाता है
Then/तब,

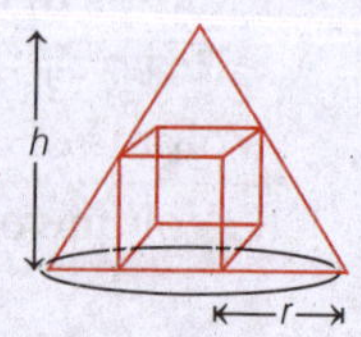

Side of cube/घन की भुजा $=\frac{\sqrt{2}rh}{\sqrt{2}r+h}$

8. When a sector is folded to make a cone
जब एक त्रिज्यखण्ड को मोड़कर एक शंकु बनाया जाता है

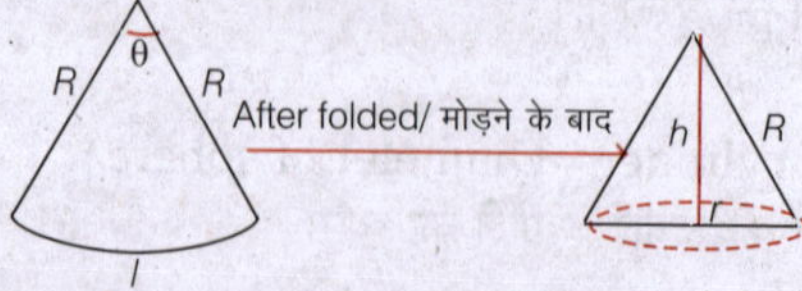

(i) Circumference of base of cone = arc l
शंकु के आधार की परिधि = चाप l

(ii) Slant height of cone = Radius of sector, $h=\sqrt{R^2-r^2}$
शंकु की तिर्यक ऊँचाई = त्रिज्यखण्ड की त्रिज्या, $h=\sqrt{R^2-r^2}$

9. A sphere of maximum size inside a cube
एक घन के अन्दर एक अधिकतम आकार का गोला

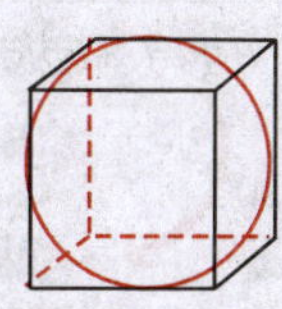

Then/तब,

(i) Diameter of sphere/गोले का व्यास $= a$

or Radius of sphere/गोले की त्रिज्या $= \frac{a}{2}$

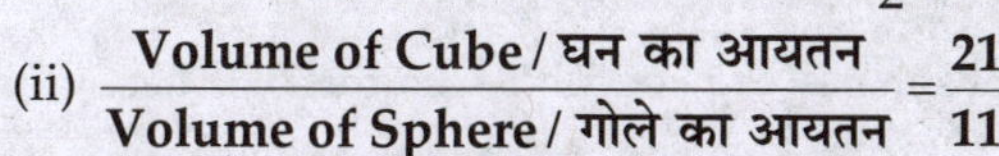

(ii) $\frac{\textbf{Volume of Cube / घन का आयतन}}{\textbf{Volume of Sphere / गोले का आयतन}} = \frac{\mathbf{21}}{\mathbf{11}}$

10. A cube of maximum size inside a sphere
एक गोले के अन्दर एक अधिकतम आकार का घन

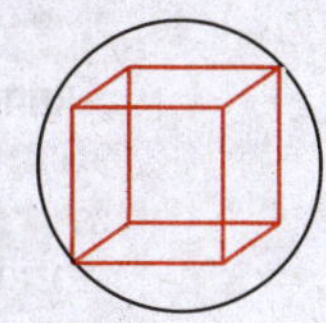

Then/तब,

(i) Diagonal of cube = Diameter of sphere.

घन का विकर्ण = गोले का व्यास

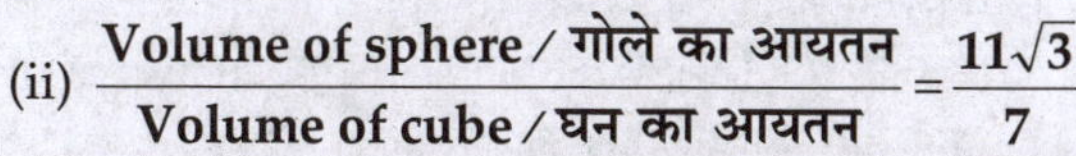

(ii) $\frac{\textbf{Volume of sphere / गोले का आयतन}}{\textbf{Volume of cube / घन का आयतन}} = \frac{\mathbf{11\sqrt{3}}}{\mathbf{7}}$

11. (i) When two spheres of same radius inside cube diagonally

जब घन के अन्दर दो गोले विकर्णवत् हों

$$\frac{a}{r} = \frac{2(\sqrt{3}+1)}{\sqrt{3}}$$

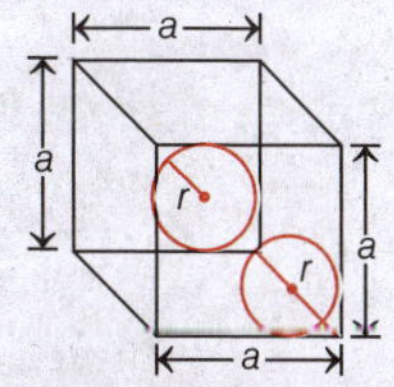

(ii) When one maximum size and one small size sphere inside cube

जब घन के अन्दर एक अधिकतम आकार का और एक छोटे आकार का गोला हो

$$\frac{r}{R} = \frac{\sqrt{3}-1}{\sqrt{3}+1}$$

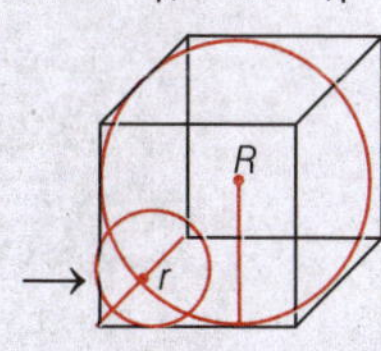

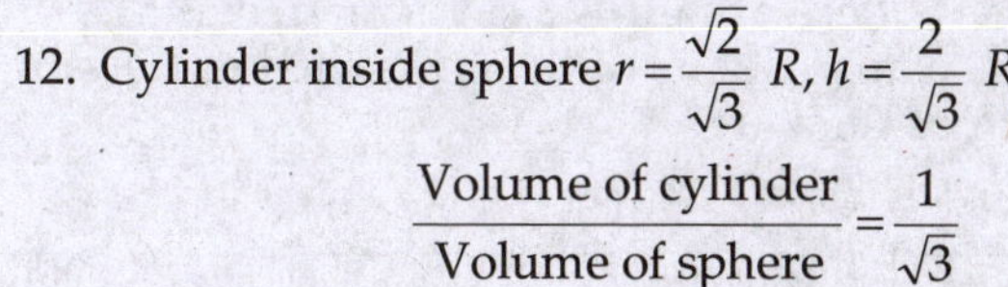

12. Cylinder inside sphere $r = \frac{\sqrt{2}}{\sqrt{3}} R, h = \frac{2}{\sqrt{3}} R$

$$\frac{\text{Volume of cylinder}}{\text{Volume of sphere}} = \frac{1}{\sqrt{3}}$$

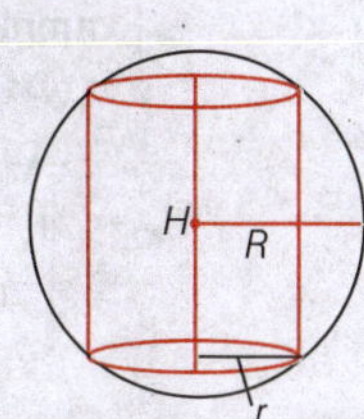

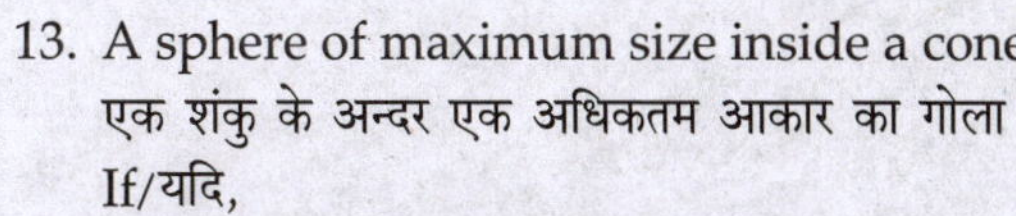

13. A sphere of maximum size inside a cone
एक शंकु के अन्दर एक अधिकतम आकार का गोला

If/यदि,

h = Height of cone/शंकु की ऊँचाई

l = Slant height of cone/शंकु की तिर्यक ऊँचाई

r = Radius of base of cone/शंकु के आधार की त्रिज्या

Then/तब,

Radius of sphere/गोले की त्रिज्या, $R = \frac{hr}{l+r}$

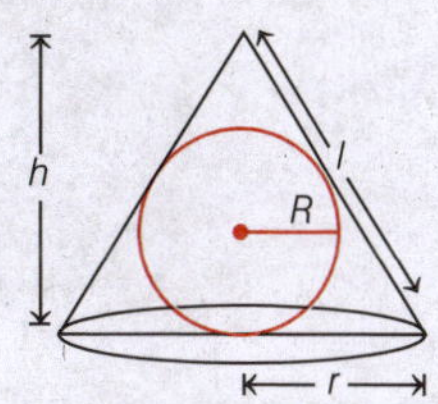

संयुक्त आकृतियों पर आधारित उदाहरण निम्नलिखित हैं

Example **Find the volume of the largest sphere that can be carved out of a cube of side 21 cm.**

सबसे बड़े गोले का आयतन ज्ञात कीजिए जिसे 21 सेमी भुजा वाले घन से काटकर बनाया जा सकता है।

Sol. Diameter of sphere = Side of cube ∴ Radius of sphere = $\frac{21}{2}$ cm

Volume of sphere = $\frac{4}{3} \times \frac{22}{7} \times \frac{21}{2} \times \frac{21}{2} \times \frac{21}{2} = 11 \times 21 \times 21 = 4851 \text{ cm}^3$

Example **From a solid cylindrical wooden block of height 18 cm and radius 7.5 cm, a conical cavity of the same height and same radius is taken out. What is total surface area of the remaining solid?**

ठोस बेलनाकार लकड़ी के ब्लॉक की ऊँचाई 18 सेमी और त्रिज्या 7.5 सेमी है जिसमें समान ऊँचाई और समान त्रिज्या की शंक्वाकार खोल (cavity) बाहर निकाली जाती है। शेष ठोस का कुल क्षेत्रफल (सेमी में) क्या है?

SSC CGL (TIER II) 2020

Sol. Slant height of the cone

$$l = \sqrt{18^2 + 7.5^2} = \sqrt{324 + 56.25}$$

$$= \sqrt{380.25} = 19.5 \text{ cm}$$

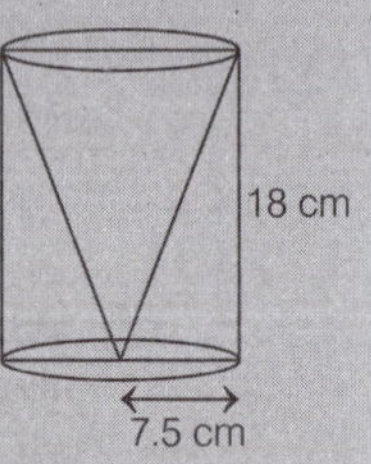

Curved surface area of cone

$$= \pi \times 7.5 \times 19.5 = 146.25\pi \text{ cm}^2$$

Curved surface area of cylinder

$$= 2\pi \times 7.5 \times 18 = 270\pi \text{ cm}^2$$

Area of the base of cylinder = $\pi \times (7.5)^2 = 56.25\pi \text{ cm}^2$

∴ Total surface area of the remaining solid

$$= 146.25\pi + 270\pi + 56.25\pi = 472.5\pi$$

Example **A metalic sphere of radius 6 cm is melted and drawn into a wire, whose radius of the cross-section is 8 cm. What is the length of the wire?**

6 सेमी त्रिज्या के एक धातु के गोले को पिघलाया जाता है और उसे एक तार के रूप में बदला जाता है जिसकी अनुप्रस्थ काट की त्रिज्या 8 सेमी है। तार की लम्बाई क्या है?

RRB JE 2019

Sol. Volume of sphere = Volume of wire

$$\Rightarrow \quad \pi r^2 h = \frac{4}{3}\pi R^3 \Rightarrow 8 \times 8 \times h = \frac{4}{3} \times 6 \times 6 \times 6$$

$$\Rightarrow \quad h = \frac{4 \times 6 \times 6 \times 6}{3 \times 8 \times 8} \Rightarrow h = 4.5 \text{ cm}$$

∴ Length of wire = 4.5 cm

अध्याय 08

Coordinate Geometry

निर्देशांक ज्यामिति

Coordinate geometry is the branch of geometry, in which we study the position of any point on a plane with the help of its coordinates.

निर्देशांक ज्यामिति, ज्यामिति की वह शाखा है, जिसमें हम किसी तल पर किसी भी बिन्दु की स्थिति का अध्ययन, उसके निर्देशांकों की सहायता से करते हैं।

A cartesian system in two dimensions is defined by an ordered pair of perpendicular lines called axes. In cartesian system as shown in given figure, the lines XOX' and YOY' are called the X-axis and Y-axis, respectively and the two lines taken together are called the coordinate axes or the axes of coordinate.

दो आयामों (2-D) में एक कार्तीय पद्धति को लम्बवत् रेखाओं के क्रमबद्ध युग्म द्वारा परिभाषित किया जाता है, जिन्हें अक्ष कहा जाता है। जैसा कि दिए गए चित्र में दिखाया गया है, कार्तीय पद्धति में रेखाओं XOX' और YOY' को क्रमशः X-अक्ष और Y-अक्ष कहा जाता है और एकसाथ ली गई दोनों रेखाओं को निर्देशांक अक्ष कहा जाता है।

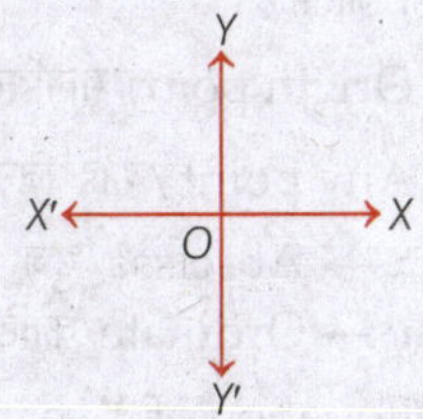

These two lines XOX' and YOY' mutually intersect each other at the point O, which is known as origin.

ये दो रेखाएँ XOX' और YOY' परस्पर बिन्दु O पर एक-दूसरे को प्रतिच्छेद करती हैं, जिसे मूल-बिन्दु कहा जाता है।

Cartesian Coordinates of a Point

एक बिन्दु के कार्तीय निर्देशांक

- Consider a point P at the distances of x and y from Y-axis and X-axis respectively, as shown in figure.
  माना एक बिन्दु P है, जिसकी Y-अक्ष और X-अक्ष से दूरी क्रमशः x और y है, जैसा कि चित्र में दर्शाया गया है।

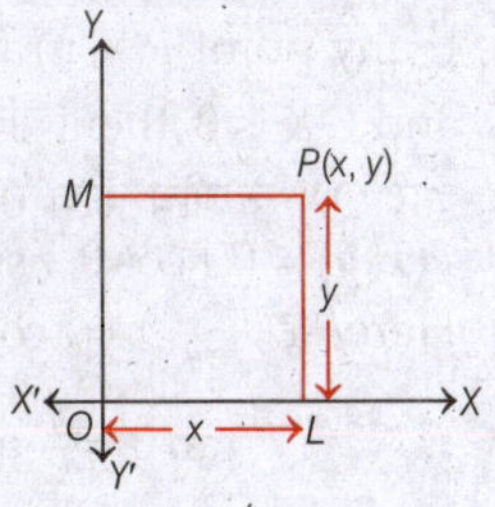

- Here, x is called the **x-coordinate** or **abscissa** of point P and y is called the **y-coordinate** or **ordinate** of point P.
  यहाँ, x को बिन्दु P का **x-निर्देशांक** या **भुज** कहा जाता है और y को बिन्दु P का **y-निर्देशांक** या **कोटि** कहा जाता है।
- This pair of real numbers (x, y) is called **cartesian coordinates** of **point P.**
  वास्तविक संख्याओं (x, y) के युग्म को **कार्तीय निर्देशांक** या **बिन्दु** P कहा जाता है।
- The coordinates of point O (origin) are taken as (0, 0).
  मूल-बिन्दु O के निर्देशांक (0, 0) के रूप में लिए जाते हैं।
- The coordinates of any point on X-axis are of the form $(x, 0)$.
  X-अक्ष पर किसी बिन्दु के निर्देशांक $(x, 0)$ रूप में होते हैं।
- The coordinates of any point on Y-axis are of the form $(0, y)$.
  Y-अक्ष पर किसी बिन्दु के निर्देशांक $(0, y)$ रूप में होते हैं।

Rectangular Coordinate System
समकोणीय निर्देशांक पद्धति

If, there are two mutually perpendicular lines, then they are called **coordinate axes**.

यदि समकोणीय निर्देशांक पद्धति में दो परस्पर लम्बवत् रेखाएँ होती हैं, तो इन्हें **निर्देशांक अक्ष** कहा जाता है।

- Origin point/मूलबिन्दु O (0,0)
- Any point /कोई बिन्दु $P(x, y)$
  $x \rightarrow$ Abscissa/भुज
  $y \rightarrow$ Ordinate/कोटि
- Equation of X-axis
  X-अक्ष का समीकरण $y = 0$
- Equation of Y-axis
  Y-अक्ष का समीकरण $x = 0$

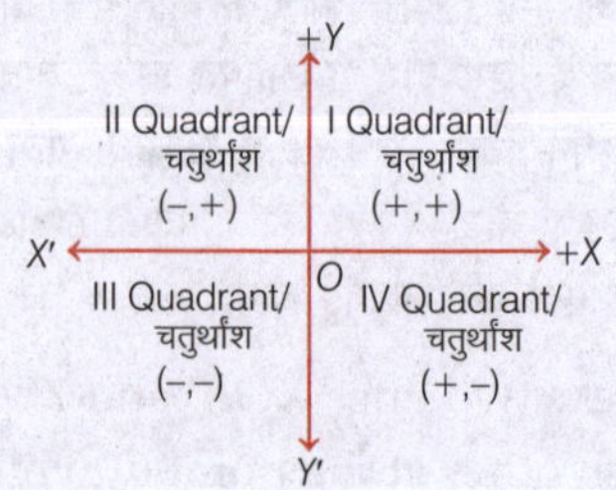

Polar Coordinates of a Point एक बिन्दु के ध्रुवीय निर्देशांक

- If any point $P(x, y)$ and angle between OP and OX is θ, then यदि कोई बिन्दु $P(x,y)$ तथा OP और OX के बीच कोण θ है, तब
  $P(x, y) = P(r\cos\theta, r\sin\theta)$
  Here/यहाँ, $x = r\cos\theta$ and/और $y = r\sin\theta$
  $\therefore \quad \tan\theta = \dfrac{y}{x}$ and/और $x^2 + y^2 = r^2$

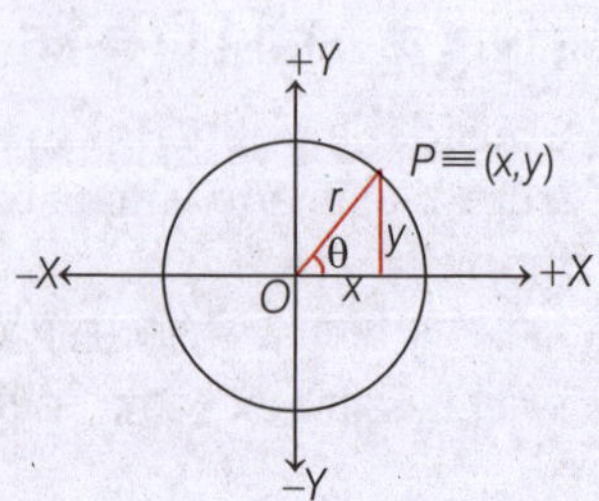

Distance Formulae दूरी सूत्र

- Distance between two points $A(x_1, y_1)$ and $B(x_2, y_2)$,
  दो बिन्दुओं $A(x_1, y_1)$ और $B(x_2, y_2)$ के बीच की दूरी,

$A(x_1, y_1)$ ——— $B(x_2, y_2)$

$$AB = \sqrt{(x_2 - x_1)^2 + (y_2 - y_1)^2}$$

यह सूत्र RRBs परीक्षा की दृष्टि से बहुत महत्त्वपूर्ण है इस पर 1 प्रश्न परीक्षा में अवश्य पूछा जाता है।

- **Distance from the Origin :** Distance between the point (x, y) and origin *i.e.*, $(0, 0)$
  मूल-बिन्दु से दूरी : बिन्दु (x, y) और मूल-बिन्दु अर्थात् $(0, 0)$ के बीच की दूरी,

$$d = \sqrt{x^2 + y^2}$$

दूरी सूत्र पर आधारित उदाहरण निम्न है

Example What is the distance between the points (–2, 4) and (4, 4)?

बिन्दु (–2, 4) और (4, 4) के बीच की दूरी क्या है? *RRB ALP 2019*

Sol. Required distance $= \sqrt{(4+2)^2 + (4-4)^2} = \sqrt{36} = 6$ units

Section Formulae विभाजन सूत्र

- The coordinate of a point which internally divides the line PQ joined by the point $P(x_1, y_1)$ and $Q(x_2, y_2)$ in the ratio $m : n$, is
  उस बिन्दु के निर्देशांक, जो बिन्दु $P(x_1, y_1)$ तथा $Q(x_2, y_2)$ को मिलाने वाले रेखाखण्ड PQ को $m : n$ के अनुपात में आन्तरिक विभाजित करता है।

$P(x_1, y_1)$ —— m —— $R(x, y)$ —— n —— $Q(x_2, y_2)$

$$x = \frac{mx_2 + nx_1}{m + n}, \; y = \frac{my_2 + ny_1}{m + n}$$

- If $R(x, y)$ divides the line segment PQ externally in the ratio $m : n$, then
  यदि $R(x, y)$ रेखाखण्ड PQ को $m : n$ के अनुपात में बाह्य विभाजित करता है, तब

$P(x_1, y_1)$ —— $Q(x_2, y_2)$ —— $R(x, y)$; $PR = m$, $QR = n$

$$x = \frac{mx_2 - nx_1}{m - n}, \; y = \frac{my_2 - ny_1}{m - n}$$

Mid-Point Formula मध्य-बिन्दु सूत्र

If $m = n = 1$, then R is the mid-point of PQ.

यदि $m = n = 1$, तब R, PQ का मध्य-बिन्दु है।

Then/तब,

P (x_1, y_1) R (x, y) Q (x_2, y_2)

$m = 1$ $n = 1$

$$x = \frac{x_1 + x_2}{2}, \quad y = \frac{y_1 + y_2}{2}$$

विभाजन सूत्र पर आधारित उदाहरण निम्न हैं

Example **Find the coordinates of the point which divides the line segment joining the points (4, −3) and (8, 5) internally in the ratio 1 : 3.**

उस बिन्दु के निर्देशांक ज्ञात कीजिए, जो बिन्दु (4, −3) और (8, 5) को मिलाने वाले रेखाखण्ड को 1 : 3 के अनुपात में अन्तःविभाजित करता है।

RRB ALP 2019

Sol. Let $P(x, y)$ is the point which divides the line segment internally.

Given, $m : n = 1 : 3$, $x_1 = 4$, $y_1 = -3$, $x_2 = 8$ and $y_2 = 5$

By section formula,

$$x = \frac{mx_2 + nx_1}{m + n} = \frac{1 \times 8 + 3 \times 4}{1 + 3} = \frac{20}{4} = 5$$

$$y = \frac{my_2 + ny_1}{m + n} = \frac{1 \times 5 - 3 \times 3}{1 + 3} = \frac{-4}{4} = -1$$

So, the required point is (5, − 1).

Example **Point P is the mid-point of segment AB. Coordinates of P are (5, −1) and A are (2, −4). What are the coordinates of point B?**

बिन्दु P, रेखाखण्ड AB का मध्य-बिन्दु है। P के निर्देशांक (5, −1) हैं और A के (2, −4) हैं, तब बिन्दु B के निर्देशांक क्या है?

SSC CGL (TIER-I) 2017

Sol.

A (2, −4) P (5, −1) B (x, y)

By using mid-point formula,

$$\frac{x + 2}{2} = 5 \Rightarrow x = 8$$

and

$$\frac{y - 4}{2} = -1 \Rightarrow y - 4 = -2$$

$$y = 2$$

$\therefore$ Coordinates of point B is (8, 2).

Reflection Point प्रतिबिम्ब बिन्दु

- Mirror image of (x, y) with respect to X-axis is $(x, -y)$.
X-अक्ष के परितः (x, y) का दर्पण प्रतिबिम्ब $(x, -y)$ है।

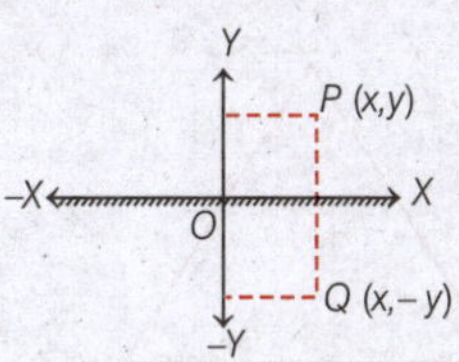

अतिसम्भावित! दर्पण प्रतिबिम्ब SSC एवं RRBs परीक्षाओं के लिए महत्त्वपूर्ण टॉपिक है। इससे 1 प्रश्न अवश्य पूछा जाता है।

- Mirror image of (x, y) with respect to Y-axis is $(-x, y)$.
Y-अक्ष के परितः बिन्दु (x, y) का दर्पण प्रतिबिम्ब $(-x, y)$ है।

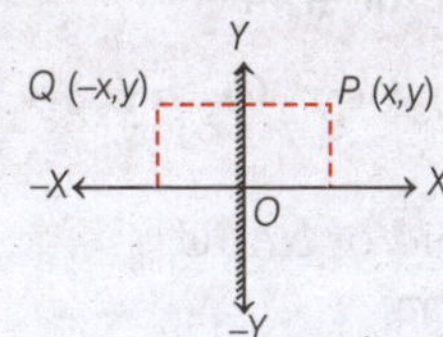

- Mirror image of (x, y) with respect to origin is $(-x, -y)$.
मूलबिन्दु के परितः (x, y) का दर्पण प्रतिबिम्ब $(-x, -y)$ है।

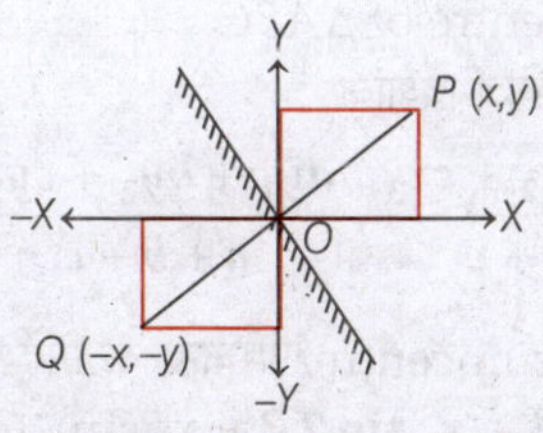

दर्पण प्रतिबिम्ब पर आधारित प्रश्नों को निम्न उदाहरण की सहायता से समझें

Example **What is the reflection of the point (5, –3) in the line $y = 3$?**
रेखा $y = 3$ पर बिन्दु $(5, -3)$ का प्रतिबिम्ब क्या है? ***SSC CGL 2020***

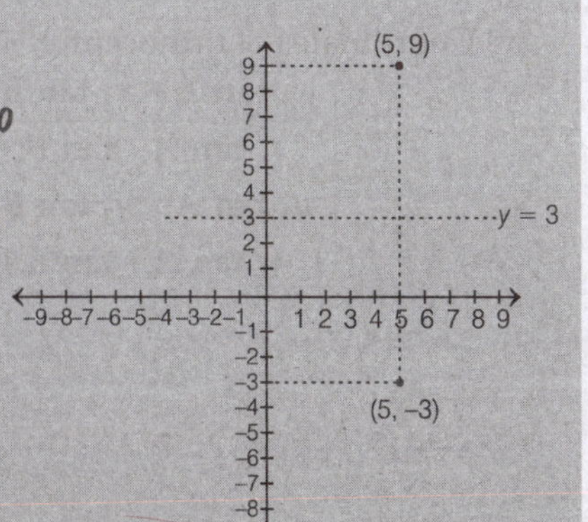

Sol. In the above figure, we can see the reflection of point $(5, -3)$ in the line $y = 3$ is $(5, 9)$ as the distance between the point and the line is 6 units.

So, the mirror image will also be 6 units above the line.

Some Formulae Related to Triangle त्रिभुज से सम्बन्धित कुछ सूत्र

- If the vertex of a triangle ABC are $A(x_1, y_1)$, $B(x_2, y_2)$ and $C(x_3, y_3)$, then

  यदि एक त्रिभुज ABC के शीर्ष $A(x_1, y_1)$, $B(x_2, y_2)$ और $C(x_3, y_3)$ हैं, तब

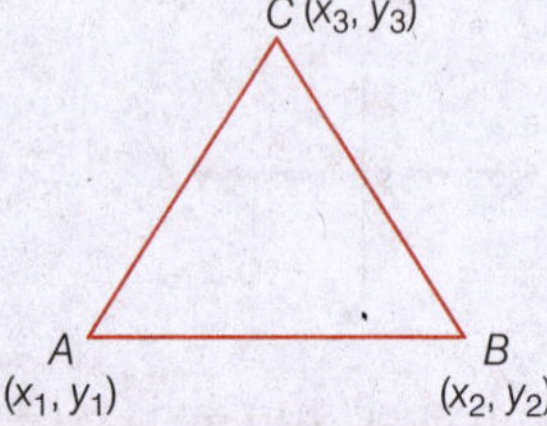

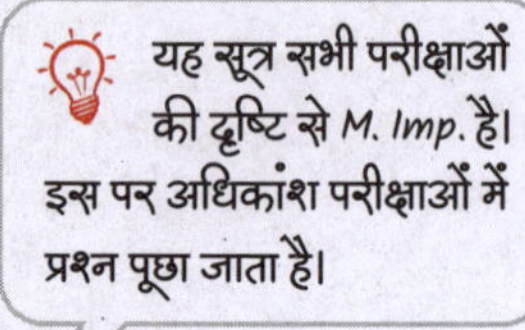

(i) **Area of ΔABC / ΔABC का क्षेत्रफल**

$$= \frac{1}{2} | x_1 (y_2 - y_3) + x_2 (y_3 - y_1) + x_3 (y_1 - y_2) |$$

(ii) Coordinates of centroid of ΔABC

ΔABC के केन्द्रक के निर्देशांक

$$= \left(\frac{x_1 + x_2 + x_3}{3}, \frac{y_1 + y_2 + y_3}{3} \right)$$

(iii) Coordinates of incentre of ΔABC

ΔABC के अन्तः केन्द्र के निर्देशांक

$$= \left(\frac{ax_1 + bx_2 + cx_3}{a + b + c}, \frac{ay_1 + by_2 + cy_3}{a + b + c} \right)$$

(iv) Coordinates of circumcentre/परिकेन्द्र के निर्देशांक

$$= \left(\frac{x_1 \sin 2A + x_2 \sin 2B + x_3 \sin 2C}{\sin 2A + \sin 2B + \sin 2C}, \frac{y_1 \sin 2A + y_2 \sin 2B + y_3 \sin 2C}{\sin 2A + \sin 2B + \sin 2C} \right)$$

(v) Coordinates of orthocentre/लम्ब केन्द्र के निर्देशांक

$$= \left(\frac{x_1 \tan A + x_2 \tan B + x_3 \tan C}{\tan A + \tan B + \tan C}, \frac{y_1 \tan A + y_2 \tan B + y_3 \tan C}{\tan A + \tan B + \tan C} \right)$$

त्रिभुज से सम्बन्धित कुछ सूत्र पर आधारित उदाहरण निम्न हैं

Example **If $A(-5, -2)$, $B(1, 3)$ and $C(2, -2)$ are the vertices of a triangle, then what is the area (in sq. units) of the ΔABC?**

यदि $A(-5,-2)$, $B(1,3)$ और $C(2,-2)$ एक त्रिभुज के शीर्ष हैं, तब ΔABC का क्षेत्रफल (वर्ग इकाई में) क्या होगा? *CRPF HEAD CONSTABLE 2023*

Sol. Area of ΔABC

$$= \frac{1}{2}|-5(3+2) + (-2+2) + 2(-2-3)|$$

$$= \frac{1}{2}|-25-10| = \frac{1}{2} \times 35$$

$= 17.5$ sq. units [∵ area can't be negative]

Example **Find the centroid of a triangle formed by the points (2, 5), (3, 9) and (4, 1).**

बिन्दुओं (2, 5), (3, 9) और (4, 1) द्वारा बने त्रिभुज का केन्द्रक ज्ञात कीजिए। *RRB NTPC 2021*

Sol. Let centroid of the triangle is (x, y).

$$x = \frac{x_1 + x_2 + x_3}{3} = \frac{2+3+4}{3} = \frac{9}{3} = 3$$

$$y = \frac{y_1 + y_2 + y_3}{3} = \frac{5+9+1}{3} = \frac{15}{3} = 5$$

Straight Line सरल रेखा

Slope of a Line रेखा की प्रवणता (ढलान)

The slope of a line is a measure of its steepness.

किसी रेखा की प्रवणता (ढलान), उसकी ढाल का माप होता है।

Slope of a line/रेखा का ढलान, $m = \tan\theta = \dfrac{y_2 - y_1}{x_2 - x_1}$

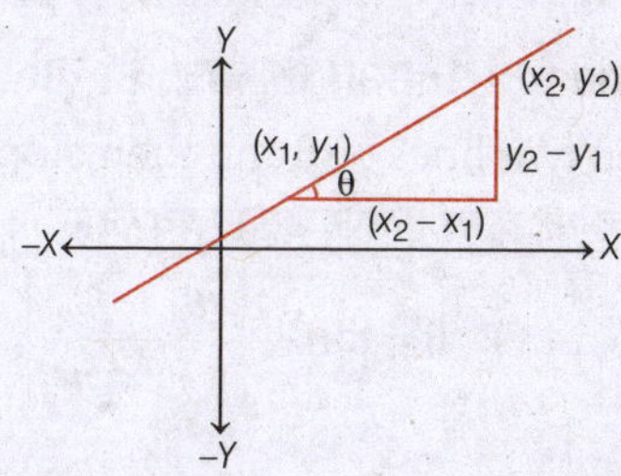

If standard equation of line is $ax + by + c = 0$, then

यदि रेखा का मानक समीकरण $ax + by + c = 0$ है, तब

$$\text{Slope of line/रेखा की प्रवणता} = \frac{-\text{Coefficient of } x / -x \text{ का गुणांक}}{\text{Coefficient of } y / y \text{ का गुणांक}} = \frac{-a}{b}$$

- If two lines are parallel to each other, then their slopes will be equal.
  यदि दो रेखाएँ एक-दूसरे के समान्तर हैं, तो उनकी प्रवणता बराबर होती है, $m_1 = m_2$
- If two lines are perpendicular to each other, then the product of their slopes be equal to -1, i.e. $m_1 m_2 = -1$
  यदि दो रेखाएँ एक-दूसरे के लम्बवत् हैं, तब इनकी प्रवणताओं का गुणनफल -1 के बराबर होता है अर्थात् $m_1 m_2 = -1$

सरल रेखा की प्रवणता पर आधारित उदाहरण निम्न हैं

Example **What is the slope of the line $2x+3y=12$?**
रेखा $2x+3y=12$ की प्रवणता क्या है?
SSC CGL (TIER-I) 2017

Sol. Given line, $2x + 3y = 12 \Rightarrow y = -\frac{2x}{3} + \frac{12}{3}$

Compare it with equation of line $y = mx + c$

$$\therefore \quad m = \text{Slope} = \frac{-2}{3}$$

Example **What is the slope of the line perpendicular to the line passing through the points (–2, 3) and (2, 0)?**
बिन्दु (–2,3) और (2, 0) से गुजरने वाली रेखा के लम्बवत् रेखा की प्रवणता क्या होगी?
SSC CGL (TIER-I) 2017

Sol. Slope of line that passes through (x_1, y_1) and (x_2, y_2) is $m_1 = \frac{y_2 - y_1}{x_2 - x_1}$.

Given, points are (–2, 3) and (2, 0).

$$\therefore \quad \text{Slope } (m_1) = \frac{0-3}{2+2} = \frac{-3}{4}$$

Condition for two perpendicular lines is

$$m_1 \times m_2 = -1 \quad \Rightarrow -\frac{3}{4} \times m_2 = -1 \Rightarrow m_2 = \frac{4}{3}$$

$\therefore$ Required slope $= m_2 = \frac{4}{3}$

Angle Between Two Lines दो रेखाओं के बीच का कोण

Let angle between two lines be θ and their slopes are m_1 and m_2, then
माना दो रेखाओं के बीच का कोण θ और उनकी प्रवणताएँ m_1 और m_2 हैं, तब

$$\tan\theta = \left|\frac{m_1 - m_2}{1 + m_1 m_2}\right| \quad \text{or/या} \quad \theta = \tan^{-1}\left|\frac{m_1 - m_2}{1 + m_1 m_2}\right|$$

Equation of Straight Line in Various Form
विभिन्न रूप में सरल रेखा के समीकरण

- Intercept on Y-axis/Y-अक्ष पर अवरोधन (अन्तःखण्ड) $= c$
  Equation of line/रेखा का समीकरण, $y = mx + c$

- Intercept Form of a Line/एक रेखा का अन्तःखण्ड रूप
  Intercept at X-axis $= a$ and Intercept at Y-axis $= b$
  X-अक्ष पर अन्तःखण्ड $= a$ तथा Y-अक्ष पर अन्तःखण्ड $= b$
  $\therefore$ Equation of line/रेखा का समीकरण,
  $$\frac{x}{a}+\frac{y}{b}=1$$

- The equation of a line passing through point (x_1, y_1) and (x_2, y_2)
  बिन्दुओं (x_1, y_1) और (x_2, y_2) से होकर जाने वाली सरल रेखा का समीकरण,
  $$y-y_1=\frac{y_2-y_1}{x_2-x_1}(x-x_1)$$
- The equation of a straight line passing through (x_1, y_1) and its slope is m
  बिन्दु (x_1, y_1) से होकर जाने वाली उस रेखा का समीकरण, जिसकी प्रवणता m है।
  $$y-y_1=m(x-x_1)$$
- Equation of line parallel to line $ax+by+c=0$,
  रेखा $ax+by+c=0$ के समान्तर रेखा का समीकरण, $ax+by=k$
- Equation of line perpendicular to line $ax+by+c=0$,
  रेखा $ax+by+c=0$ के लम्बवत् रेखा का समीकरण,
  $$bx-ay=k$$

सरल रेखा के समीकरण पर आधारित उदाहरण निम्न हैं

Example **What is the equation of the line which passes through the points (2, 3) and (–4, 1)?**

बिन्दु (2, 3) और (–4, 1) से गुजरने वाली रेखा का समीकरण क्या है?

SSC (10+2) 2017

Sol. $x_1=2, x_2=-4, y_1=3$ and $y_2=1$

Equation of the line is $(y-y_1)=\left(\frac{y_2-y_1}{x_2-x_1}\right)(x-x_1)$

$\Rightarrow \quad (y-3)=\frac{(1-3)}{(-4-2)}(x-2) \Rightarrow y-3=\frac{-2}{-6}(x-2)$

$\Rightarrow \quad y-3=\frac{1}{3}(x-2) \Rightarrow 3y-9=x-2 \Rightarrow 3y-x=7$

$\therefore \quad x-3y=-7$

Example **The equation of a line which passes through a point (1, 5) and perpendicular to the line $x-y=5$.**

एक रेखा का समीकरण, जो एक बिन्दु (1, 5) से होकर गुजरती है और रेखा $x-y=5$ के लम्बवत् है।

RRB NTPC 2021

Sol. Given line, $x - y = 5 \Rightarrow y = x - 5$

By comparing with $y = mx + c$, slope $(m) = 1$

Slope of the perpendicular line $= -\frac{1}{m} = -1$

Equation of the line passes through the point (1, 5) with slope −1 is

$$y - y_1 = m(x - x_1) \Rightarrow y - 5 = -1(x - 1)$$

$$\Rightarrow \quad y + x = 1 + 5 \Rightarrow y + x = 6$$

(iv) Perpendicular distance of a line $ax + by + c = 0$ from a point (x, y)

रेखा $ax + by + c = 0$ की बिन्दु (x, y) से लम्बवत् दूरी $= \left|\frac{ax_1 + by_1 + c}{\sqrt{a^2 + b^2}}\right|$

(v) The perpendicular distance between two parallel lines

$ax + by + c_1 = 0$ and $ax + by + c_2 = 0$ is $\left|\frac{c_1 - c_2}{\sqrt{a^2 + b^2}}\right|$

समान्तर रेखा $ax + by + c_1 = 0$ तथा $ax + by + c_2 = 0$ के बीच लम्बवत् दूरी

$$= \left|\frac{c_1 - c_2}{\sqrt{a^2 + b^2}}\right|$$

दो समान्तर रेखाओं के बीच की दूरी पर आधारित उदाहरण निम्न है

Example **What is the distance between two parallel lines $5x + 12y + 2 = 0$ and $5x - 12y - 3 = 0$?**

दो समान्तर रेखाओं $5x - 12y + 2 = 0$ और $5x - 12y - 3 = 0$ के बीच की दूरी क्या है?

Sol. $d = \frac{|c_1 - c_2|}{\sqrt{a^2 + b^2}} = \frac{|2 - (-3)|}{\sqrt{5^2 + (-12)^2}} = \frac{|5|}{\sqrt{169}} = \frac{5}{13}$

Equation of Circle वृत्त का समीकरण

The coordinates of centre of circle are (a, b) and r is the radius of circle, then

वृत्त के केन्द्र के निर्देशांक (a, b) और त्रिज्या r है, तब

Equation of circle/वृत्त का समीकरण,

$$(x - a)^2 + (y - b)^2 = r^2$$

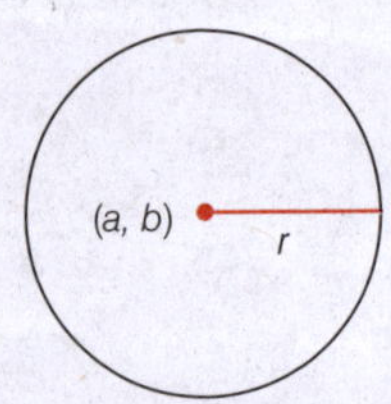

If centre is origin, then/यदि केन्द्र मूलबिन्दु है, तब

Equation of circle/ वृत्त का समीकरण, $x^2 + y^2 = r^2$

Standard Equation of Circle वृत्त का मानक समीकरण

Equation of circle having centre $(-g, -f)$ is

केन्द्र $(-g, -f)$ वाले वृत्त का समीकरण, $x^2 + y^2 + 2gx + 2fy + c = 0$

$$\text{Radius/त्रिज्या} = \sqrt{g^2 + f^2 - c}$$

अध्याय 09

Algebra बीजगणित

Algebraic Formulae
बीजगणितीय सूत्र

Square Formulae वर्ग सूत्र

1. $a^2 - b^2 = (a+b)(a-b)$
2. $(a+b)^2 = a^2 + b^2 + 2ab$
3. $(a-b)^2 = a^2 + b^2 - 2ab$
4. $(a+b)^2 = (a-b)^2 + 4ab$
5. $(a+b)^2 + (a-b)^2 = 2(a^2 + b^2)$
6. $(a+b)^2 - (a-b)^2 = 4ab$ M. Imp.
7. $(a-b)^2 = (a+b)^2 - 4ab$
8. $a^2 + b^2 = (a+b)^2 - 2ab = (a-b)^2 + 2ab$
9. $a^4 + b^4 = (a^2 + b^2)^2 - 2a^2b^2$
10. $ab = \left(\frac{a+b}{2}\right)^2 - \left(\frac{a-b}{2}\right)^2$
11. $a^4 + b^4 = (a^2 - b^2)^2 + 2a^2b^2$
12. $\frac{(a+b)^2 + (a-b)^2}{a^2 + b^2} = 2$
13. $a^4 + a^2b^2 + b^4 = (a^2 - ab + b^2)(a^2 + ab + b^2)$
14. $a^4 - b^4 = (a^2)^2 - (b^2)^2 = (a^2 + b^2)(a^2 - b^2)$
15. $(a+b+c)^2 = a^2 + b^2 + c^2 + 2(ab + bc + ca)$ M. Imp.
16. $(a-b-c)^2 = a^2 + b^2 + c^2 - 2(ab - bc + ca)$
17. $2(ab + bc + ca) = (a+b+c)^2 - (a^2 + b^2 + c^2)$
18. $a^2 + b^2 + c^2 = (a+b+c)^2 - 2(ab + bc + ca)$ M. Imp.
19. $(ax + by)^2 + (ay - bx)^2 = (a^2 + b^2)(x^2 + y^2)$
20. $\frac{a+b}{a-b} + \frac{a-b}{a+b} = \frac{2(a^2 + b^2)}{(a^2 - b^2)}$
21. $\frac{a+b}{a-b} - \frac{a-b}{a+b} = \frac{4ab}{a^2 - b^2}$

परीक्षा ब्रह्मास्त्र

ये सूत्र बहुत प्रश्नों को हल करने में सहायक हैं। इन पर आधारित 1-2 प्रश्न सभी परीक्षाओं में पूछे जाते हैं।

बीजगणितीय सूत्रों पर आधारित उदाहरण निम्न हैं

Example **If $x-y=25$ and $xy=78$, then what is the value of x^2+y^2?**

यदि $x-y=25$ और $xy=78$, तो x^2+y^2 का मान क्या है?

SSC (10+2) 2022

Sol. $\because (x-y)^2 = x^2+y^2-2xy$

According to the question,

$$(25)^2 = x^2+y^2-2\times 78$$

$\Rightarrow \quad 625+156 = x^2+y^2$

$\Rightarrow \quad x^2+y^2 = 781$

Example **If $(a+b+c)=17$ and $(a^2+b^2+c^2)=115$, then what is the value of $(a+b)^2+(b+c)^2+(c+a)^2$?**

यदि $(a+b+c)=17$ और $(a^2+b^2+c^2)=115$, तो $(a+b)^2+(b+c)^2+(c+a)^2$ का मान क्या है? *SSC SELECTION POST 2024*

Sol. $\because (a+b)^2+(b+c)^2+(c+a)^2 = a^2+b^2+c^2+(a+b+c)^2$

$$= 115+(17)^2 = 115+289 = 404$$

Example **Find the value of $\dfrac{(a+b)^2-(a-b)^2}{ab}$.**

$\dfrac{(a+b)^2-(a-b)^2}{ab}$ का मान ज्ञात कीजिए। *DSSSB LAB. ATTENDANT 2024*

Sol. $\because (a+b)^2-(a-b)^2 = 4ab$

$\therefore \quad \dfrac{(a+b)^2-(a-b)^2}{ab} = \dfrac{4ab}{ab} = 4$

Example **If $a+b+c=9$ and $ab+bc+ca=23$, then what is the value of $a^2+b^2+c^2$?**

यदि $a+b+c=9$ और $ab+bc+ca=23$, तो $a^2+b^2+c^2$ का मान क्या है?

Sol. $\because (a+b+c)^2 = a^2+b^2+c^2+2(ab+bc+ca)$

$\Rightarrow \quad (9)^2 = a^2+b^2+c^2+2(23)$

$81-46 = a^2+b^2+c^2$

$\therefore \quad a^2+b^2+c^2 = 35$

Cube Formulae घन सूत्र

1. $(a+b)^3 = a^3+3a^2b+3ab^2+b^3 = a^3+b^3+3ab(a+b)$
2. $a^3+b^3 = (a+b)^3-3ab(a+b)$ M. Imp.
3. $a^3+b^3 = (a+b)[(a+b)^2-3ab]$

4. $a^3 + b^3 = (a+b)(a^2 + b^2 - ab)$

5. $a^2 + b^2 - ab = \dfrac{a^3 + b^3}{a+b}$

6. $(a+b) = \dfrac{a^3 + b^3}{a^2 + b^2 - ab}$

7. $(a-b)^3 = a^3 - 3a^2b + 3ab^2 - b^3 = a^3 - b^3 - 3ab(a-b)$

8. $a^3 - b^3 = (a-b)^3 + 3ab(a-b)$

9. $a^3 - b^3 = (a-b)[(a-b)^2 + 3ab]$ M. Imp.

10. $a^3 - b^3 = (a-b)(a^2 + b^2 + ab)$

11. $a^2 + b^2 + ab = \dfrac{a^3 - b^3}{a-b}$

12. $a - b = \dfrac{a^3 - b^3}{a^2 + b^2 + ab}$

13. $(a+b)^3 - (a-b)^3 = 2b^3 + 6a^2b = 2b(b^2 + 3a^2)$

14. $(a+b)^3 + (a-b)^3 = 2a^3 + 6ab^2 = 2a(a^2 + 3b^2)$

15. $(a+b+c)^3 = a^3 + b^3 + c^3 + 3a^2(b+c) + 3b^2(c+a) + 3c^2(a+b) + 6abc$

अतिसम्भावित!
घन के सूत्रों पर 1 प्रश्न सभी परीक्षाओं में पूछे जाने की सम्भावना रहती है। इनका निरन्तर अभ्यास कीजिए।

घन सूत्रों पर आधारित उदाहरण निम्न हैं

Example **If $a^3 + b^3 = 432$ and $a + b = 12$, then the value of $(a+b)^2 - 3ab$ is equal.**

यदि $a^3 + b^3 = 432$ और $a + b = 12$, तो $(a+b)^2 - 3ab$ का मान बराबर है।

Sol. $(a+b)^2 - 3ab = \dfrac{a^3 + b^3}{a+b} = \dfrac{432}{12} = 36$

Example **If $(a-b) = 9$ and $a^3 - b^3 = 4401$, then what is the value of ab?**

यदि $(a-b) = 9$ और $a^3 - b^3 = 4401$ है, तो ab का मान क्या है?

SSC (10+2) 2023

Sol. $a^3 - b^3 = (a-b)\{(a-b)^2 + 3ab\}$

$\Rightarrow \quad 4401 = 9(9^2 + 3ab)$

$\Rightarrow \quad 489 = 81 + 3ab \quad \Rightarrow 489 - 81 = 3ab$

$\Rightarrow \quad 408 = 3ab$

$\therefore \quad ab = 136$

Example **Simplify the expression.**

दिए गए व्यंजक को सरल कीजिए।

$$\frac{(x^3 - y^3)(x + y)}{x^2 + xy + y^2}$$ SSC CPO 2023

Sol. $\because x^3 - y^3 = (x - y)(x^2 + xy + y^2)$

$$\therefore \frac{(x^3 - y^3)(x + y)}{x^2 + xy + y^2} = \frac{(x - y)(x^2 + xy + y^2)(x + y)}{(x^2 + xy + y)}$$

$$= (x - y)(x + y) = x^2 - y^2$$

Special Cases विशेष स्थितियाँ

1. If $a^2 - ab + b^2 = 0$, then $a^3 + b^3 = 0$

   यदि $a^2 - ab + b^2 = 0$ है, तो $a^3 + b^3 = 0$ होगा।

2. If $a^2 + a + 1 = 0$, then $a^3 - 1 = 0$ or $a^3 = 1$

   यदि $a^2 + a + 1 = 0$ है, तो $a^3 - 1 = 0$ या $a^3 = 1$ होगा।

3. If $\frac{a}{b} + \frac{b}{a} = 1$, then $a^3 + b^3 = 0$

   यदि $\frac{a}{b} + \frac{b}{a} = 1$ है, तो $a^3 + b^3 = 0$ होगा।

4. If $\frac{1}{a} - \frac{1}{b} = \frac{1}{a - b}$, then $a^3 + b^3 = 0$

   यदि $\frac{1}{a} - \frac{1}{b} = \frac{1}{a - b}$ है, तो $a^3 + b^3 = 0$ होगा।

5. If $\frac{a}{b} + \frac{b}{a} = -1$, then $a^3 - b^3 = 0$

   यदि $\frac{a}{b} - \frac{b}{a} = -1$ है, तो $a^3 - b^3 = 0$ होगा।

6. If $\frac{1}{b} + \frac{1}{a} = \frac{1}{a + b}$, then $a^3 - b^3 = 0$

   यदि $\frac{1}{b} + \frac{1}{a} = \frac{1}{a + b}$ है, तो $a^3 - b^3 = 0$ होगा।

7. If $ab(a + b) = 1$, then $\frac{1}{a^3 b^3} - a^3 - b^3 = 3$

   यदि $ab(a + b) = 1$ है, तो $\frac{1}{a^3 b^3} - a^3 - b^3 = 3$ होगा।

2 Exponential Formulae / 2 घातीय सूत्र

1. If/यदि $x + \frac{1}{x} = k$, then/तब $x^2 - kx + 1 = 0$

2. If/यदि $x + \frac{1}{x} = k$, then/तब $x^2 + \frac{1}{x^2} = k^2 - 2$ M. Imp.

3. If/यदि $x-\frac{1}{x}=k$, then/तब $x^2-kx-1=0$

4. If/यदि $x-\frac{1}{x}=k$, then/तब $x^2+\frac{1}{x^2}=k^2+2$ *M. Imp.*

5. If/यदि $x+\frac{1}{x}=\pm\sqrt{2}$, then/तब $x^2+\frac{1}{x^2}=0$

or/या $x^4+1=0$

6. If/यदि $x+\frac{1}{x}=k$, then/तब $\sqrt{x}+\frac{1}{\sqrt{x}}=\pm\sqrt{k+2}$

7. If/यदि $x^2+\frac{1}{x^2}=k$, then/तब $x+\frac{1}{x}=\pm\sqrt{k+2}$ *M. Imp.*

8. If/यदि $x^2+\frac{1}{x^2}=k$, then/तब $x-\frac{1}{x}=\pm\sqrt{k-2}$

Same Exponential Formulae समान घातीय सूत्र

1. If/यदि $x+\frac{1}{x}=k$, then/तब $x-\frac{1}{x}=\pm\sqrt{k^2-4}$
2. If/यदि $x-\frac{1}{x}=k$, then/तब $x+\frac{1}{x}=\pm\sqrt{k^2+4}$
3. If/यदि $\frac{a+b}{\sqrt{ab}}=k$, then/तब $\frac{a-b}{\sqrt{ab}}=\pm\sqrt{k^2-4}$
4. If/यदि $\sqrt{\frac{a}{b}}+\sqrt{\frac{b}{a}}=k$, then/तब $\sqrt{\frac{a}{b}}-\sqrt{\frac{b}{a}}=\sqrt{k^2-4}$

2 घातीय सूत्र तथा समघातीय सूत्र बहुत महत्त्वपूर्ण हैं। इन पर आधारित उदाहरण निम्न हैं

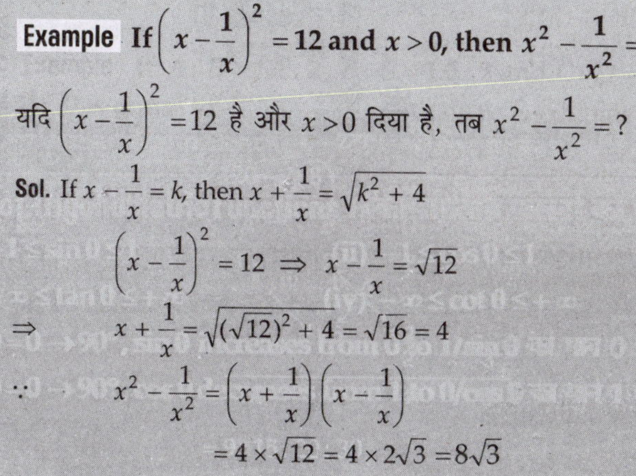

Example **If $\left(x-\frac{1}{x}\right)^2=12$ and $x>0$, then $x^2-\frac{1}{x^2}=?$**

यदि $\left(x-\frac{1}{x}\right)^2=12$ है और $x>0$ दिया है, तब $x^2-\frac{1}{x^2}=?$

Sol. If $x-\frac{1}{x}=k$, then $x+\frac{1}{x}=\sqrt{k^2+4}$

$$\left(x-\frac{1}{x}\right)^2=12 \Rightarrow x-\frac{1}{x}=\sqrt{12}$$

$$\Rightarrow \quad x+\frac{1}{x}=\sqrt{(\sqrt{12})^2+4}=\sqrt{16}=4$$

$$\because \quad x^2-\frac{1}{x^2}=\left(x+\frac{1}{x}\right)\left(x-\frac{1}{x}\right)$$

$$=4\times\sqrt{12}=4\times2\sqrt{3}=8\sqrt{3}$$

Example If $x^2 - \frac{1}{x^2} = 4\sqrt{2}$, then what is the value of $x^4 - \frac{1}{x^4}$?

यदि $x^2 - \frac{1}{x^2} = 4\sqrt{2}$ है, तब $x^4 - \frac{1}{x^4}$ का मान क्या है? *SSC (10+2) 2023*

Sol. $\because x - \frac{1}{x} = k$, then $x + \frac{1}{x} = \sqrt{k^2 + 4}$

$$\Rightarrow \quad x^2 - \frac{1}{x^2} = 4\sqrt{2},$$

$$\Rightarrow \quad x^2 + \frac{1}{x^2} = \sqrt{(4\sqrt{2})^2 + 4} = \sqrt{36} = 6$$

$$\because \quad x^4 - \frac{1}{x^4} = \left(x^2 + \frac{1}{x^2}\right)\left(x^2 - \frac{1}{x^2}\right) = 6 \times 4\sqrt{2} = 24\sqrt{2}$$

Example If $a + \frac{1}{a} = 6$, then what is the value of $\frac{3}{4}\left(a^2 + \frac{1}{a^2}\right)$?

यदि $a + \frac{1}{a} = 6$ है, तो $\frac{3}{4}\left(a^2 + \frac{1}{a^2}\right)$ का मान क्या है? *SSC CPO 2023*

Sol. $\because a + \frac{1}{a} = k$, then $a^2 + \frac{1}{a^2} = k^2 - 2$

$$\because \quad a + \frac{1}{a} = 6 \Rightarrow a^2 + \frac{1}{a^2} = (6)^2 - 2 = 36 - 2 = 34$$

$$\therefore \quad \frac{3}{4}\left(a^2 + \frac{1}{a^2}\right) = \frac{3}{4} \times 34 = \frac{102}{4} = \frac{51}{2} = 25.5$$

Example If $w + \frac{1}{w} = 6$, then what is the value of $w - \frac{1}{w}$?

यदि $w + \frac{1}{w} = 6$ है, तो $w - \frac{1}{w}$ का मान क्या है? *SSC (10+2) 2023*

Sol. $\because x + \frac{1}{x} = k$, then $x - \frac{1}{x} = \sqrt{k^2 - 4}$

$$\therefore \quad w - \frac{1}{w} = \sqrt{(6)^2 - 4} = \sqrt{36 - 4} = \sqrt{32} = 4\sqrt{2}$$

3 Exponential Formulae/3 घातीय सूत्र

1. If/यदि $x - \frac{1}{x} = k$, then/तब $x^3 - \frac{1}{x^3} = k^3 + 3k$ *M. Imp.*
2. If/यदि $x + \frac{1}{x} = k$, then/तब $x^3 + \frac{1}{x^3} = k^3 - 3k$
3. If/यदि $x - \frac{1}{x} = \sqrt{t}$, then/तब $x^3 - \frac{1}{x^3} = (t + 3)\sqrt{t}$
4. If/यदि $x + \frac{1}{x} = \sqrt{t}$, then/तब $x^3 + \frac{1}{x^3} = (t - 3)\sqrt{t}$
5. If/यदि $\sqrt{x} + \frac{1}{\sqrt{x}} = a$, then/तब $x\sqrt{x} + \frac{1}{x\sqrt{x}} = a^3 - 3a$
6. If/यदि $\sqrt{x} - \frac{1}{\sqrt{x}} = a$, then/तब $x\sqrt{x} - \frac{1}{x\sqrt{x}} = a^3 + 3a$

7. If/यदि $\frac{a-b}{\sqrt{ab}}=k$, then/तब $\left(\frac{a}{b}\right)^{3/2}-\left(\frac{b}{a}\right)^{3/2}=k^3+3k$

8. If/यदि $\sqrt{\frac{a}{b}}-\sqrt{\frac{b}{a}}=k$, then/तब $\left(\frac{a}{b}\right)^{3/2}-\left(\frac{b}{a}\right)^{3/2}=k^3+3k$

9. If/यदि $\frac{a+b}{\sqrt{ab}}=k$, then/तब $\left(\frac{a}{b}\right)^{3/2}+\left(\frac{b}{a}\right)^{3/2}=k^3-3k$

10. If/यदि $\sqrt{\frac{a}{b}}+\sqrt{\frac{b}{a}}=k$, then/तब $\left(\frac{a}{b}\right)^{3/2}+\left(\frac{b}{a}\right)^{3/2}=k^3-3k$

11. If/यदि $x+\frac{1}{x}=\pm\sqrt{3}$, then/तब $x^3+\frac{1}{x^3}=0$ or/या $x^6+1=0$

3 घातीय सूत्रों पर आधारित उदाहरण निम्न हैं

Example **If $y+\frac{1}{y}=11$, then what is the value of $y^3-\frac{1}{y^3}$?**

यदि $y+\frac{1}{y}=11$ है, तो $y^3-\frac{1}{y^3}$ का मान क्या है? *SSC CGL 2022*

Sol. $x+\frac{1}{x}=N$, then $x-\frac{1}{x}=\sqrt{N^2-4}$

$$y+\frac{1}{y}=11,\text{ then } y-\frac{1}{y}=\sqrt{(11)^2-4}=\sqrt{117}$$

$$y-\frac{1}{y}=N,\text{ then } y^3-\frac{1}{y^3}=N^3+3N$$

$$y-\frac{1}{y}=\sqrt{117},\text{ then } y^3-\frac{1}{y^3}=(\sqrt{117})^3+3\times\sqrt{117}$$

$$=117\sqrt{117}+3\sqrt{117}=\sqrt{117}\,(117+3)$$

$$=120\sqrt{117}=120\times3\sqrt{13}=360\sqrt{13}$$

Example **If $a^2+\frac{1}{a^2}=\frac{7}{3}$, then what is the value of $a^3-\frac{1}{a^3}$?**

यदि $a^2+\frac{1}{a^2}=\frac{7}{3}$ है, तो $a^3-\frac{1}{a^3}$ का मान क्या है? *SSC (10+2) 2022*

Sol. $x^2+\frac{1}{x^2}=k$, then $x-\frac{1}{x}=\pm\sqrt{k-2}$

$$a^2+\frac{1}{a^2}=\frac{7}{3},\text{ then } a-\frac{1}{a}=\sqrt{\frac{7}{3}-\frac{2}{1}}=\sqrt{\frac{7-6}{3}}=\frac{1}{\sqrt{3}}$$

If $a-\frac{1}{a}=k$, then $a^3-\frac{1}{a^3}=k^3+3k$

$$a-\frac{1}{a}=\frac{1}{\sqrt{3}},\text{ then } a^3-\frac{1}{a^3}=\left(\frac{1}{\sqrt{3}}\right)^3+3\times\frac{1}{\sqrt{3}}$$

$$=\frac{1}{3\sqrt{3}}+\frac{3}{\sqrt{3}}=\frac{1+9}{3\sqrt{3}}=\frac{10}{3\sqrt{3}}$$

Example **If $2x+\frac{2}{x}=5$, then what is the value of $x^3+\frac{1}{x^3}+2$?**

यदि $2x+\frac{2}{x}=5$, तो $\left(x^3+\frac{1}{x^3}+2\right)$ का मान क्या है? *SSC CGL 2022*

Sol. $2x+\frac{2}{x}=5$

$$2\left(x+\frac{1}{x}\right)=5$$

$$x+\frac{1}{x}=\frac{5}{2}$$

If $x+\frac{1}{x}=k$, then $x^3+\frac{1}{x^3}=k^3-3k$

$x+\frac{1}{x}=\frac{5}{2}$, then $x^3+\frac{1}{x^3}=\left(\frac{5}{2}\right)^3-3\times\frac{5}{2}=\frac{125}{8}-\frac{15}{2}=\frac{125-60}{8}=\frac{65}{8}$

$\therefore \quad x^3+\frac{1}{x^3}+2=\frac{65}{8}+\frac{2}{1}=\frac{65+16}{8}=\frac{81}{8}$

Example **If $x^2+\frac{1}{x^2}=18$ and $x>1$, then what is the value of $x^3-\frac{1}{x^3}$?**

यदि $x^2+\frac{1}{x^2}=18$ और $x>1$ है, तो $x^3-\frac{1}{x^3}$ का मान क्या है?

SSC SELECTION POST 2023

Sol. If $x^2+\frac{1}{x^2}=k$, then $x+\frac{1}{x}=\sqrt{k+2}$

$$x^2+\frac{1}{x^2}=18, \text{ then } x+\frac{1}{x}=\sqrt{18+2}=\sqrt{20}$$

If $x+\frac{1}{x}=k$, then $x-\frac{1}{x}=\sqrt{k^2-4}$

$$x+\frac{1}{x}=\sqrt{20}, \text{ then } x-\frac{1}{x}=\sqrt{(\sqrt{20})^2-4}=\sqrt{20-4}=\sqrt{16}=4$$

If $x-\frac{1}{x}=k$, then $x^3-\frac{1}{x^3}=k^3+3k$

$x-\frac{1}{x}=4$, then $x^3-\frac{1}{x^3}=(4)^3+3\times 4=64+12$

$$=76$$

Higher Exponential Formulae उच्च घातीय सूत्र

1. $(a^5-b^5)=(a-b)(a^4+a^3b+a^2b^2+ab^3+b^4)$
2. $(a^5+b^5)=(a+b)(a^4-a^3b+a^2b^2-ab^3+b^4)$
3. $x^4-\frac{1}{x^4}=\left(x^2+\frac{1}{x^2}\right)\left(x+\frac{1}{x}\right)\left(x-\frac{1}{x}\right)$
4. $x^5+\frac{1}{x^5}=\left(x^2+\frac{1}{x^2}\right)\left(x^3+\frac{1}{x^3}\right)-\left(x+\frac{1}{x}\right)$ *M. Imp.*

5. $\left(x^5 - \frac{1}{x^5}\right) = \left(x^2 + \frac{1}{x^2}\right)\left(x^3 - \frac{1}{x^3}\right) - \left(x - \frac{1}{x}\right)$ *M. Imp.*

6. $\left(x^7 + \frac{1}{x^7}\right) = \left(x^4 + \frac{1}{x^4}\right)\left(x^3 + \frac{1}{x^3}\right) - \left(x + \frac{1}{x}\right)$

7. $\left(x^7 - \frac{1}{x^7}\right) = \left(x^4 + \frac{1}{x^4}\right)\left(x^3 - \frac{1}{x^3}\right) + \left(x - \frac{1}{x}\right)$

8. $\left(x^8 - \frac{1}{x^8}\right) = \left(x^4 + \frac{1}{x^4}\right)\left(x^2 + \frac{1}{x^2}\right)\left(x + \frac{1}{x}\right)\left(x - \frac{1}{x}\right)$ *M. Imp.*

उच्च घातीय सूत्रों पर आधारित उदाहरण निम्न हैं

Example **If $x + \frac{1}{x} = 5$ and $x > 1$, then what will be the value of $x^8 - \frac{1}{x^8}$?**

यदि $x + \frac{1}{x} = 5$ और $x > 1$ है, तो $x^8 - \frac{1}{x^8}$ का मान क्या होगा? ***SSC CGL 2023***

Sol. $x^8 - \frac{1}{x^8} = \left(x^4 + \frac{1}{x^4}\right)\left(x^2 + \frac{1}{x^2}\right)\left(x + \frac{1}{x}\right)\left(x - \frac{1}{x}\right)$

$\because \quad x + \frac{1}{x} = k$, then $x - \frac{1}{x} = \sqrt{k^2 + 4}$

$\because \quad x + \frac{1}{x} = k$, then $x^2 + \frac{1}{x^2} = k^2 - 2$

$x + \frac{1}{x} = 5$, then $x^2 + \frac{1}{x^2} = (5)^2 - 2 = 23$

$x^2 + \frac{1}{x^2} = 23$, then $x^4 + \frac{1}{x^2} = (23)^2 - 2 = 527$

$x + \frac{1}{x} = 5$, then $x - \frac{1}{x} = \sqrt{(5)^2 - 4} = \sqrt{21}$

$\therefore \quad x^8 - \frac{1}{x^8} = 527 \times 23 \times 5 \times \sqrt{21} = 60605\sqrt{21}$

Example **If $x - \frac{1}{x} = -6$, then what is the value of $x^5 - \frac{1}{x^5}$?**

यदि $x - \frac{1}{x} = -6$ है, तो $x^5 - \frac{1}{x^5}$ का मान क्या है? ***SSC CGL 2023***

Sol. $x^5 - \frac{1}{x^5} = \left(x^2 + \frac{1}{x^2}\right)\left(x^3 - \frac{1}{x^3}\right) - \left(x - \frac{1}{x}\right)$

If $x - \frac{1}{x} = k$, then $x^2 + \frac{1}{x^2} = k^2 + 2$

If $x - \frac{1}{x} = k$, then $x^3 - \frac{1}{x^3} = k^3 + 3k$

$x - \frac{1}{x} = -6$, then $x^2 + \frac{1}{x^2} = (-6)^2 + 2 = 38$

$x - \frac{1}{x} = -6$, then $x^3 - \frac{1}{x^3} = (-6)^3 + 3 \times (-6) = -216 - 18 = -234$

$\therefore x^5 - \frac{1}{x^5} = 38 \times (-234) - (-6) = 38 \times (-234) + 6 = -8892 + 6 = -8886$

Formulae Based on $x+\frac{1}{x}$ / $x+\frac{1}{x}$ पर आधारित सूत्र

1. If/यदि $x+\frac{1}{x}=2$, then/तब $x=1$
2. If/यदि $x+\frac{1}{x}=-2$, then/तब $x=-1$
3. If/यदि $x+\frac{1}{x}=1$, then/तब $x^3=-1$, or/या $x^2-x+1=0$
4. If/यदि $x+\frac{1}{x}=-1$, then/तब $x^3=1$ or/या $x^2+x+1=0$
5. If/यदि $x+\frac{1}{x}=\sqrt{3}$, then/तब $x^3+\frac{1}{x^3}=0$ or/या $x^6=-1$.
6. If/यदि $x+\frac{1}{x}=\sqrt{2}$, then/तब $x^2+\frac{1}{x^2}=0$ or/या $x^4=-1$.

Relation Between 2 Exponent and 4 Exponent
2 घात और 4 घात के बीच सम्बन्ध

1. $(x^2+y^2)^2=x^4+y^4+2x^2y^2$
2. $x^4+y^4+x^2y^2=(x^2+y^2+xy)(x^2+y^2-xy)$
3. $\frac{x^6-y^6}{x^2-y^2}=x^4+y^4+x^2y^2$
4. $a^2+b^2+ab=(a+b+\sqrt{ab})(a+b-\sqrt{ab})$
5. $x^4+x^2+1=(x^2+x+1)(x^2-x+1)$

Variable Formulae चर सूत्र

1. If/यदि $x^2+y^2+z^2=0$, then/तब $x=0, y=0, z=0$
2. $a^2+b^2+c^2-ab-bc-ca=\frac{1}{2}[(a-b)^2+(b-c)^2+(c-a)^2]$
3. If a, b and c are in AP with common difference d, then
   यदि a, b तथा c, AP (समान्तर श्रेढ़ी) में हैं, जिसका सार्व अन्तर d है, तब
   $a^2+b^2+c^2-ab-bc-ca=3d^2$
4. If/यदि $a^2+b^2+c^2-ab-bc-ca=0$,
   then/तब $a=b=c$
5. $a^3+b^3+c^3-3abc=(a+b+c)(a^2+b^2+c^2-ab-bc-ca)$
   $=\frac{1}{2}[(a+b+c)\{(a-b)^2+(b-c)^2+(c-a)^2\}]$
6. $a^3+b^3+c^3-3abc=\frac{1}{2}(a+b+c)[3(a^2+b^2+c^2)-(a+b+c)^2]$

7. If/यदि $a^3 + b^3 + c^3 = 3abc$, then/तब $a + b + c = 0$ or $a = b = c$ *M. Imp.*

8. If a, b and c are in AP, then $a^3 + b^3 + c^3 - 3abc = 9bd^2$

   यदि a, b तथा c समान्तर श्रेढ़ी (AP) में हैं, तब

$$a^3 + b^3 + c^3 - 3abc = 9bd^2$$

   Where, d is common difference/जहाँ, d सार्व अन्तर है।

9. If/यदि $x + \frac{1}{y} = a, y + \frac{1}{z} = b, z + \frac{1}{x} = c$, then/तब

$$xyz + \frac{1}{xyz} = abc - (a + b + c)$$

10. If/यदि $x - \frac{1}{y} = a, y - \frac{1}{z} = b, z - \frac{1}{x} = c$, then/तब

$$xyz - \frac{1}{xyz} = abc + (a + b + c)$$

11. If/यदि $a + b + c = x$ and/और $\frac{1}{a} + \frac{1}{b} + \frac{1}{c} = y$, then/तब

$$\frac{a}{b} + \frac{b}{a} + \frac{c}{a} + \frac{a}{c} + \frac{b}{c} + \frac{c}{b} = xy - 3$$

चर सूत्र पर आधारित महत्त्वपूर्ण उदाहरण निम्नलिखित है

Example **If $a + b + c = 6$ and $a^2 + b^2 + c^2 = 40$, then what is the value of $a^3 + b^3 + c^3 - 3abc$?**

यदि $a + b + c = 6$ और $a^2 + b^2 + c^2 = 40$, तो $a^3 + b^3 + c^3 - 3abc$ का मान क्या है?

Sol. $(a + b + c)^2 = a^2 + b^2 + c^2 + 2(ab + bc + ca)$

$\Rightarrow \quad 6^2 = 40 + 2(ab + bc + ca) \Rightarrow 36 - 40 = 2(ab + bc + ca)$

$\Rightarrow \quad -4 = 2(ab + bc + ca)$

$\Rightarrow \quad (ab + bc + ca) = -2$

Now, $a^3 + b^3 + c^3 - 3abc = (a + b + c)\{(a^2 + b^2 + c^2 - (ab + bc + ca)\}$

$= (6)\{(40) - (-2)\} = 6 \times 42 = 252$

Polynomials बहुपद

Degree or Order of a Polynomial बहुपद की घात (कोटि)

The maximum power of variable in an algebraic polynomial, is called degree or order of that polynomial.

एक बीजगणितीय बहुपद में चर की अधिकतम घात, उस बहुपद की कोटि या घात कहलाती है।

- If $a_0x^n + a_1x^{n-1} + a_2x^{n-2} + ... + a_{n-1} = 0$ is a n degree polynomial, then

  Sum of all zeroes $= (-1)^1 \dfrac{\text{Coefficient of } x^{n-1}}{\text{Coefficient of } x^n}$

  Sum of products of two-two zeroes $= (-1)^2 \dfrac{\text{Coefficient of } x^{n-2}}{\text{Coefficient of } x^n}$

  ....

  ....

  यदि $a_0x^n + a_1x^{n-1} + a_2x^{n-2} + ... + a_{n-1} = 0$, एक n घातीय बहुपद है, तब सभी

  शून्यकों का योग $= \dfrac{(-1)^1 x^{n-1} \text{ का गुणांक}}{x^n \text{ का गुणांक}}$

  दो-दो शून्यकों के गुणनफलों का योग $= \dfrac{(-1)^2 x^{n-2} \text{ का गुणांक}}{x^n \text{ का गुणांक}}$

  ....

  ....

  Product of all n zeroes $= (-1)^n \dfrac{\text{Constant term}}{\text{Coefficient of } x^n}$

  सभी n शून्यकों का गुणनफल $= \dfrac{(-1)^n \text{ अचर पद}}{x^n \text{ का गुणांक}}$

Some Results कुछ परिणाम

1. $(x^2 + ax + bx + ab) = (x + a)(x + b)$
2. $1 + a + b + ab = (1 + a)(1 + b)$
3. $(1 + a)(1 + b)(1 + c) = 1 + a + b + c + ab + bc + ca + abc$
4. If/यदि $xy = 1$ or/या $x = \dfrac{1}{y}$, then/तब $\dfrac{1}{1 + x^n} + \dfrac{1}{1 + y^n} = 1$
5. If/यदि $a + b + c = 2s$, then/तब

   $$\frac{(s-a)^2 + (s-b)^2 + (s-c)^2 + s^2}{a^2 + b^2 + c^2} = 1$$

6. If/यदि $xy + yz + zx = 0$, then/तब

   $$\left(\frac{1}{x^2 - yz} + \frac{1}{y^2 - zx} + \frac{1}{z^2 - xy}\right) = 0$$

   (Where/जहाँ, $x, y, z \neq 0$)

7. If/यदि $pq+qr+rp=0$, then/तब

$$\left(\frac{p^2}{p^2-qr}+\frac{q^2}{q^2-rp}+\frac{r^2}{r^2-pq}\right)=1$$

8. If/यदि $[\sqrt{(a^2+b^2+ab)}+\sqrt{(a^2+b^2-ab)}]=1$, then/तब $(1-a^2)(1-b^2)=\frac{3}{4}$

9. If/यदि $(x+\sqrt{1+x^2})(y+\sqrt{1+y^2})=1$,
   Where x and y are real numbers/जहाँ, x और y वास्तविक संख्याएँ हैं।
   Then/तब, $(x+y)^2=0$

10. If/यदि $a+b+c=abc$, then/तब

$$\frac{(1-a^2)(1-b^2)}{ab}+\frac{(1-b^2)(1-c^2)}{bc}+\frac{(1-c^2)(1-a^2)}{ca}=4$$

11. If/यदि $bc+ca+ab=abc$, then/तब

$$\frac{b+c}{bc(a-1)}+\frac{c+a}{ca(b-1)}+\frac{a+b}{ab(c-1)}=1$$

12. If/यदि $xy+yz+xz=1$, then/तब

$$\frac{x+y}{1-xy}+\frac{y+z}{1-yz}+\frac{z+x}{1-xz}=\frac{1}{xyz}$$

13. If/यदि $x=\frac{a-b}{a+b}, y=\frac{b-c}{b+c}, z-\frac{c-a}{c+a}$, then/तब

$$\frac{(1-x)(1-y)(1-z)}{(1+x)(1+y)(1+z)}=1$$

14. If/यदि $x^2+y^2=z+1, y^2+z^2=x+1, z^2+x^2=y+1$,
    then/तब $xyz=1$ or/या $-\frac{1}{8}$

15. If/यदि $a+\frac{1}{b}=b+\frac{1}{c}=c+\frac{1}{a}$ (Where/जहाँ $a\neq b\neq c$), then/तब $abc=\pm 1$

Symmetric Polynomials/Expressions समघातीय बहुपद/व्यंजक

If degree of each term is same on both sides, then it is called symmetric polynomial or expression.

यदि दोनों पक्षों में प्रत्येक पद की घात समान हो, तब वह समघातीय बहुपद या व्यंजक कहलाता है।

Some Results Based on Symmetric Polynomials/Expressions
समघातीय बहुपदों/व्यंजकों पर आधारित कुछ परिणाम

1. $ab(a-b)+bc(b-c)+ca(c-a)=-(a-b)(b-c)(c-a)$
2. $(a+b+c)(ab+bc+ca)-abc=(a+b)(b+c)(c+a)$ *M. Imp.*

3. $(a+b+c)(ab+bc+ca)-3abc=a^2(b+c)+b^2(c+a)+c^2(a+b)$

4. $\dfrac{a^3(b^2-c^2)+b^3(c^2-a^2)+c^3(a^2-b^2)}{a^2(b-c)+b^2(c-a)+c^2(a-b)}=ab+bc+ca$ M. Imp.

5. $\dfrac{a^3(b+c)}{(a-b)(a-c)}+\dfrac{b^3(c+a)}{(b-c)(b-a)}+\dfrac{c^3(a+b)}{(c-a)(c-b)}=ab+bc+ca$

6. $\dfrac{a(b-c)^2}{(c-a)(a-b)}+\dfrac{b(c-a)^2}{(a-b)(b-c)}+\dfrac{c(a-b)^2}{(b-c)(c-a)}=a+b+c$

7. $\dfrac{x^4}{(x^2-y^2)(x^2-z^2)}+\dfrac{y^4}{(y^2-x^2)(y^2-z^2)}+\dfrac{z^4}{(z^2-x^2)(z^2-y^2)}=1$

8. If/यदि $x+y=2z$, then/तब $\dfrac{z}{x-z}+\dfrac{y}{y-z}=0$

9. If/यदि $x+y=2z$, then/तब $\dfrac{x}{x-z}+\dfrac{y}{y-z}=2$

10. If/यदि $a+b+c=0$, then/तब

(i) $\dfrac{1}{(a+b)(b+c)}+\dfrac{1}{(a+c)(b+a)}+\dfrac{1}{(c+a)(c+b)}=0$

(ii) $\dfrac{a^2}{a^2-bc}+\dfrac{b^2}{b^2-ca}+\dfrac{c^2}{c^2-ab}=2$ M. Imp.

(iii) $\dfrac{a^2}{2a^2+bc}+\dfrac{b^2}{2b^2+ca}+\dfrac{c^2}{2c^2+ab}=1$

(iv) $\dfrac{a^2+b^2+c^2}{(a-b)^2+(b-c)^2+(c-a)^2}=\dfrac{1}{3}$

(v) $\dfrac{1}{a^2+b^2-c^2}+\dfrac{1}{b^2+c^2-a^2}+\dfrac{1}{c^2+a^2-b^2}=0$

(vi) $\dfrac{2(a^4+b^4+c^4)}{a^2b^2+b^2c^2+c^2a^2}=4$

(vii) $\dfrac{a+b}{ab}(a^2+b^2-c^2)+\dfrac{b+c}{bc}(b^2+c^2-a^2)+\dfrac{c+a}{ca}(c^2+a^2-b^2)=0$

(viii) $\dfrac{2a^2}{b^2+c^2-a^2}+\dfrac{2b^2}{a^2+c^2-b^2}+\dfrac{2c^2}{a^2+b^2-c^2}=-3$

समघातीय बहुपद पर आधारित उदाहरण निम्नलिखित है

Example If $\frac{x^2}{by+cz}=\frac{y^2}{ax+cz}=\frac{z^2}{ax+by}=1$, then find the value of $\frac{x}{x+a}=\frac{y}{y+b}+\frac{z}{z+a}$. /यदि $\frac{x^2}{by+cz}=\frac{y^2}{ax+cz}=\frac{z^2}{ax+by}=1$ हों, तो $\left(\frac{x}{x+a}+\frac{y}{y+b}+\frac{z}{z+a}\right)$ का मान ज्ञात कीजिए।

Sol. Put $x=y=z=2$ and $a=b=c=1$

$$\frac{x}{x+a}+\frac{y}{y+b}+\frac{z}{z+a}=\frac{2}{2+1}+\frac{2}{2+1}+\frac{2}{2+1}=\frac{2}{3}+\frac{2}{3}+\frac{2}{3}=\frac{2}{3}\times3=3$$

Linear Equation रैखिक समीकरण

The equation which has degree 1, is called a linear equation.
वह समीकरण जिसकी घात 1 हो, रैखिक समीकरण कहलाती है।

Conditions for Solution of Pair of Equations समीकरण युग्म के हल के प्रतिबन्ध

A system of linear equations is/रैखिक समीकरणों का एक निकाय है

$a_1x+b_1y+c_1=0$ and/तथा $a_2x+b_2y+c_2=0$

(i) If/यदि $\frac{a_1}{a_2}\neq\frac{b_1}{b_2}$, then system has unique solution.
तब निकाय का एक अद्वितीय हल होगा।

(ii) If/यदि $\frac{a_1}{a_2}=\frac{b_1}{b_2}=\frac{c_1}{c_2}$, then system has infinite solutions.
तब निकाय के अनेक हल होंगे।

(iii) If/यदि $\frac{a_1}{a_2}=\frac{b_1}{b_2}\neq\frac{c_1}{c_2}$, then system has no solution.
तब निकाय का कोई हल नहीं होगा।

अतिसम्भावित! इन सूत्रों पर 1 प्रश्न परीक्षाओं में पूछे जाने की सम्भावना होती है।

समीकरण युग्म के हल के प्रतिबन्ध पर आधारित उदाहरण निम्न हैं

Example For what value of m, will the system of equations $18x-72y+13=0$ and $7x-my-17=0$, have no solution?
m के किस मान के लिए समीकरणों $18x-72y+13=0$ तथा $7x-my-17=0$ के कोई भी हल नहीं होंगे? *SSC CGL (TIER-II) 2023*

Sol. $a_1=18, b_1=-72, c_1=13$ and $a_2=7, b_2=-m, c_2=-17$

For equations having no solution,

$$\frac{a_1}{a_2}=\frac{b_1}{b_2}\neq\frac{c_1}{c_2}\Rightarrow\frac{18}{7}=\frac{72}{m}\neq\frac{13}{17}\Rightarrow\frac{18}{7}=\frac{72}{m}\Rightarrow m=\frac{72\times7}{18}=28$$

Example **Determine the value of *a* and *b* for which the following system of equations has infinite solutions.**

a और b के मान की गणना कीजिए, जिसके लिए निम्नलिखित समीकरणों के निकाय के अनन्त हल हैं।

$2x-(a-4)y=2b+1, 4x-(a-1)y=5b-1$ **RRB NTPC 2022**

Sol. Given, $2x-(a-4)y-(2b+1)=0$

$$4x-(a-1)y-(5b-1)=0$$

$$a_1=2, b_1=-(a-4), c_1=-(2b+1)$$

$$a_2=4, b_2=-(a-1), c_2=-(5b-1)$$

For infinite solution,

$$\frac{a_1}{a_2}=\frac{b_1}{b_2}=\frac{c_1}{c_2} \Rightarrow \frac{2}{4}=\frac{a-4}{a-1}=\frac{2b+1}{5b-1}$$

When, $\frac{a-4}{a-1}=\frac{1}{2} \Rightarrow 2a-8=a-1 \quad \Rightarrow a=7$

Again, $\frac{2b+1}{5b-1}=\frac{1}{2} \Rightarrow 5b-4b=3$

$\therefore \quad b=3$

Hence, $a=7$ and $b=3$

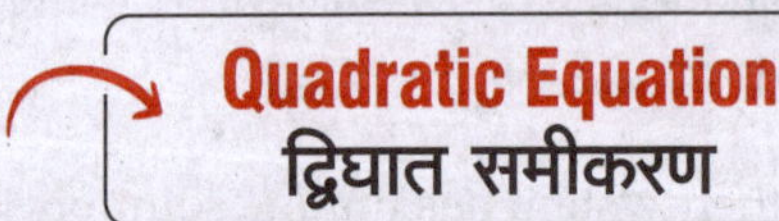

Quadratic Equation
द्विघात समीकरण

Quadratic Equation : An equation in which the highest power of the variable appearing is two, is called quadratic equation. It is written as $ax^2+bx+c=0$

द्विघात समीकरण : एक ऐसा समीकरण जिसमें चर राशि की अधिकतम घात दो हो, द्विघात समीकरण कहलाता है। इसका व्यापक रूप $ax^2+bx+c=0$ होता है।

Solution of Quadratic Equation द्विघात समीकरण का हल

(i) **Factorisation Method/गुणनखण्ड विधि**

Let $ax^2+bx+c=(x-\alpha)(x-\beta)=0$, then $x=\alpha$ and $x=\beta$ will satisfy the given equation.

मान लीजिए $ax^2+bx+c=(x-\alpha)(x-\beta)=0$, तब $x=\alpha$ और $x=\beta$ दिए गए समीकरण को सन्तुष्ट करेंगे।

(ii) **Quadratic Formula/द्विघाती सूत्र**

Quadratic equation $ax^2+bx+c=0 (a\neq 0)$ has two roots, given by

द्विघात समीकरण $ax^2+bx+c=0 (a\neq 0)$ के दो मूल हैं, जो निम्नलिखित हैं

$$\alpha = \frac{-b + \sqrt{b^2 - 4ac}}{2a}$$

$$\beta = \frac{-b - \sqrt{b^2 - 4ac}}{2a}$$

it is also called Sridharacharya formula./इसे श्रीधराचार्य सूत्र भी कहते हैं।

(iii) Discriminate of Quadratic Equation $ax^2 + bx + c = 0$/ द्विघात समीकरण $ax^2 + bx + c = 0$ का विविक्तकर, $D = b^2 - 4ac$

Nature of Roots मूलों की प्रकृति

If D is discriminate of the quadratic equation $ax^2 + bx + c = 0 (a \neq 0)$, then
यदि D, द्विघात समीकरण $ax^2 + bx + c = 0 (a \neq 0)$ का विविक्तकर है, तब

(i) If $D = 0$, then roots are real and equal.
यदि $D = 0$, तब मूल वास्तविक और समान हैं।

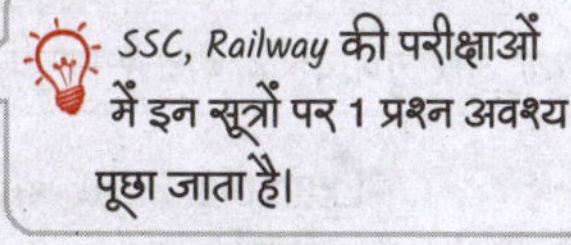

(ii) If $D > 0$ and perfect square, then roots are real, unequal and rational.
यदि $D > 0$ और पूर्ण वर्ग है, तब मूल वास्तविक, असमान और परिमेय है।

(iii) If $D > 0$ and is not perfect square, then roots are real, unequal and irrational.
यदि $D > 0$ और पूर्ण वर्ग नहीं है, तब मूल वास्तविक, असमान और अपरिमेय है।

(iv) If $D < 0$, then roots are imaginary or complex.
यदि $D < 0$, तब मूल अधिकल्पित (काल्पनिक) या सम्मिश्र हैं।

मूलों की प्रकृति पर आधारित महत्त्वपूर्ण उदाहरण निम्न है

Example **What kind of roots does the quadratic equation $5x^2 + 7x + 4 = 0$ have?**

द्विघात समीकरण $5x^2 + 7x + 4 = 0$ के मूलों की प्रकृति क्या है?

Sol. Given quadratic equation is, $5x^2 + 7x + 4 = 0$

Here, $a = 5$, $b = 7$ and $c = 4$

$$D = b^2 - 4ac = (7)^2 - 4 \times 5 \times 4 = 49 - 80 = -31 < 0$$

$\because \quad D < 0$

Hence, roots are imaginary.

(v) **Conjugate Roots/संयुग्मी मूल**

- If one root be $\alpha + i\beta$, then other root will be $\alpha - i\beta$.
  यदि एक मूल $\alpha + i\beta$ है, तब दूसरा मूल $\alpha - i\beta$ होगा।
- If one root be $\alpha + \sqrt{\beta}$, then other root will be $\alpha - \sqrt{\beta}$.
  यदि एक मूल $\alpha + \sqrt{\beta}$ है, तो दूसरा मूल $\alpha - \sqrt{\beta}$ होगा।

Relation Between Roots and Coefficients मूल और गुणांक के बीच सम्बन्ध

(i) **Quadratic Equation/द्विघात समीकरण**

If roots of quadratic equation $ax^2 + bx + c = 0$, are α, β, then

यदि द्विघात समीकरण $ax^2 + bx + c = 0$ के मूल α और β हैं, तो

- Sum of roots/मूलों का योग $= \alpha + \beta = \dfrac{-b}{a}$

$$= \frac{-\text{ Coefficient of } x / -x \text{ का गुणांक}}{\text{Coefficient of } x^2 / x^2 \text{ का गुणांक}}$$

- Product of roots/मूलों का गुणनफल $= \alpha\beta = \dfrac{c}{a}$, also, $|\alpha - \beta| = \dfrac{\sqrt{D}}{|a|}$

$$= \frac{\text{Constant term / अचर पद}}{\text{Coefficient of } x^2 / x^2 \text{ का गुणांक}}$$

मूल और गुणांक के बीच सम्बन्ध पर आधारित उदाहरण निम्न है

Example **The sum of the reciprocals of the roots of the equation, $abx^2 = (a^2 + b^2 + 2ab)(x - 1)$ is**

दिए गए समीकरण $abx^2 = (a^2 + b^2 + 2ab)(x - 1)$ के मूलों के व्युत्क्रम का योग है

Sol. Given, $abx^2 = (a^2 + b^2 + 2ab)(x - 1)$

$\Rightarrow abx^2 - (a^2 + b^2 + 2ab)x + (a^2 + b^2 + 2ab) = 0$

Sum of roots $= \dfrac{(a^2 + b^2 + 2ab)}{ab} \quad \Rightarrow \alpha + \beta = \dfrac{(a^2 + b^2 + 2ab)}{ab}$

and product of roots, $\alpha\beta = \dfrac{a^2 + b^2 + 2ab}{ab}$

Now, $\quad \dfrac{1}{\alpha} + \dfrac{1}{\beta} = \dfrac{\alpha + \beta}{\alpha\beta} = 1$

(ii) **Formation of Quadratic Equation/द्विघात समीकरण की संरचना**

If α and β be two roots of a quadratic equation, then quadratic equation is

यदि किसी द्विघात समीकरण के मूल α और β हैं, तो द्विघात समीकरण $= x^2 -$ (Sum of roots/मूलों का योग) $x +$ (Product of roots/मूलों का गुणनफल) $= 0$

Or/या $\quad x^2 - (\alpha + \beta)x + \alpha\beta = 0$

द्विघात समीकरण की संरचना पर आधारित उदाहरण निम्न है

Example **If a and b are the roots of $x^2 + x - 2 = 0$, then the quadratic equation is in x, whose roots are $\left(\dfrac{1}{a} + \dfrac{1}{b}\right)$ and ab.**

यदि a और b समीकरण $x^2 + x - 2 = 0$ के मूल हैं, तब x के रूप में द्विघात समीकरण है, जिसका मूल $\left(\dfrac{1}{a} + \dfrac{1}{b}\right)$ तथा ab हैं।

Sol. Given quadratic equation is, $x^2 + x - 2 = 0$

Sum of roots $= \dfrac{-\text{Coefficient of } x}{\text{Coefficient of } x^2} = \dfrac{-1}{1} = -1$

$\therefore \quad a + b = -1$

Product of roots $= \dfrac{\text{Constant term}}{\text{Coefficient of } x^2} = \dfrac{-2}{1}$

$\therefore \quad ab = -2$

Let $\alpha = \dfrac{1}{a} + \dfrac{1}{b} = \dfrac{a+b}{ab} = \dfrac{-1}{-2} = \dfrac{1}{2}$ and $\beta = -2$

Required quadratic equation, $x^2 - (\alpha + \beta)x + \alpha\beta = 0$

$\Rightarrow \quad x^2 - \left(\dfrac{1}{2} - 2\right)x + \dfrac{1}{2} \times (-2) = 0$

$\Rightarrow \quad x^2 + \dfrac{3x}{2} - 1 = 0$

$\therefore \quad 2x^2 + 3x - 2 = 0$

(iii) **Cubic Equation/घन समीकरण**

If α, β and γ are the roots of cubic equation $ax^3 + bx^2 + cx + d = 0$, then

यदि α, β और γ, घन समीकरण $ax^3 + bx^2 + cx + d = 0$ के मूल हैं, तब

- Sum of all roots/मूलों का योग $= \alpha + \beta + \gamma$

  $= \Sigma\alpha = \dfrac{-b}{a} = \dfrac{-\text{Coefficient of } x^2 / -x^2 \text{ का गुणांक}}{-\text{Coefficient of } x^3 / x^3 \text{ का गुणांक}}$

- $\Sigma\alpha\beta = \alpha\beta + \beta\gamma + \gamma\alpha = \dfrac{c}{a} = \dfrac{-\text{Coefficient of } x / x \text{ का गुणांक}}{-\text{Coefficient of } x^2 / x^2 \text{ का गुणांक}}$

- $\alpha\beta\gamma = -\dfrac{d}{a} = \dfrac{-\text{Constant term} / -\text{अचर पद}}{-\text{Coefficient of } x^2 / x^2 \text{ का गुणांक}}$

- Cubic equation/घन समीकरण,
  $x^3 - (\alpha + \beta + \gamma)x^2 + (\alpha\beta + \beta\gamma + \gamma\alpha)x - \alpha\beta\gamma = 0$

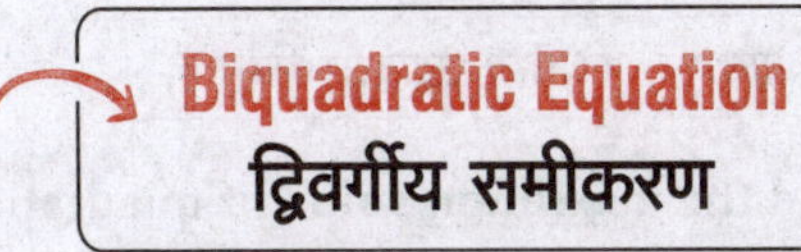

If α, β, γ and δ are the roots of the biquadratic equation $ax^4 + bx^3 + cx^2 + dx + e = 0$, then

यदि α, β, γ और δ द्विवर्गीय समीकरण $ax^4 + bx^3 + cx^2 + dx + e = 0$ के मूल हैं, तब

- $\alpha + \beta + \gamma + \delta = -\dfrac{b}{a}$
- $\alpha\beta + \alpha\gamma + \alpha\delta + \beta\gamma + \beta\delta + \gamma\delta = \dfrac{c}{a}$
- $\alpha\beta\gamma + \beta\gamma\delta + \gamma\delta\alpha + \alpha\beta\delta = -\dfrac{d}{a}$
- $\alpha\beta\gamma\delta = \dfrac{e}{a}$

- Biquadratic equation/द्विवर्गीय समीकरण

$$x^4 - (\alpha+\beta+\gamma+\delta)x^3 + (\alpha\beta+\alpha\gamma+\alpha\delta+\beta\gamma+\beta\delta+\gamma\alpha)x^3 - (\alpha\beta\gamma+\beta\gamma\delta+\gamma\delta\alpha+\alpha\beta\delta)x^2 + \alpha\beta\gamma\delta = 0$$

Maximum and Minimum Values of Quadratic Expression
द्विघात व्यंजक का अधिकतम और न्यूनतम मान

If $ax^2 + bx + c$ is a quadratic expression, then

यदि $ax^2 + bx + c$ एक द्विघात व्यंजक है, तब

(i) If $a > 0$, quadratic expression has least value at $x = -\frac{b}{2a}$.

This least value is given by $\frac{4ac - b^2}{4a} = -\frac{D}{4a}$, But there is no greatest value.

यदि $a > 0, x = -\frac{b}{2a}$ पर द्विघात व्यंजक का मान न्यूनतम है, तो न्यूनतम मान $= \frac{4ac - b^2}{4a} = -\frac{D}{4a}$, परन्तु कोई अधिकतम मान नहीं है।

(ii) If $a < 0$, quadratic expression has greatest value at $x = -\frac{b}{2a}$. This greatest value is given by $\frac{4ac - b^2}{4a} = -\frac{D}{4a}$, but there is no least value.

यदि $a < 0$, द्विघात व्यंजक का $x = -\frac{b}{2a}$ पर सबसे बड़ा मान है, तो सबसे बड़ा मान $= \frac{4ac - b^2}{4a} = -\frac{D}{4a}$, परन्तु कोई न्यूनतम मान नहीं है।

द्विघात व्यंजक के अधिकतम मान पर आधारित उदाहरण निम्न है

Example **Find the maximum value of quadratic expression $5x^2 - 3x - 6$.**

द्विघात व्यंजक $5x^2 - 3x - 6$ का अधिकतम मान ज्ञात कीजिए।

Sol. Let $f(x) = 5x^2 - 3x - 6$

Here, $a = 5$. So, $a > 0$

$$\therefore F(x)_{max} = \frac{4ac - b^2}{4a} = \frac{4 \times 5 \times (-6) - (-3)^2}{4 \times 5} = \frac{-120 - 9}{20} = -\frac{129}{20}$$

Linear Inequality रैखिक असमिका

A statement involving the symbols >, <, ≤ or ≥, is called an inequality.
>, <, ≤ या ≥ प्रतीकों को सम्मिलित करने वाले एक कथन को असमिका कहते हैं।

Some Facts Related to Linear Inequality रैखिक असमिका से सम्बन्धित कुछ तथ्य

(i) Equal numbers may be added or subtracted from both sides of an inequality without changing the sign of inequality.
एक असमिका के दोनों पक्षों में समान राशि जोड़ने अथवा घटाने से असमिका में कोई परिवर्तन नहीं होता है।

(ii) If both sides of an inequality are multiplied or divided by the same positive number, then the sign of inequality remains the same.
यदि एक असमिका के दोनों पक्षों में समान धनात्मक राशि से गुणा या भाग दिया जाए तब असमिका में कोई परिवर्तन नहीं होता है।

(iii) If both sides of an inequality are multiplied or divided by the same negative number, then the sign of inequality is reversed.
यदि एक असमिका के दोनों पक्षों में समान ऋणात्मक संख्या से गुणा या भाग की जाए, तब असमिका का चिन्ह बदलकर पहले का विपरीत हो जाता है।

Laws of Surds करणी के नियम

1. $\sqrt[n]{a} = a^{1/n}$

2. $\sqrt[n]{ab} = \sqrt[n]{a} \times \sqrt[n]{b} = a^{1/n} \times b^{1/n} = (ab^{1/n})$

3. $\sqrt[n]{\frac{a}{b}} = \frac{\sqrt[n]{a}}{\sqrt[n]{b}} = \left(\frac{a}{b}\right)^{\frac{1}{n}}$

4. $(\sqrt[n]{a})^m = a^{\frac{m}{n}} = \sqrt[n]{a^m}$

5. $(\sqrt[n]{a})^n = a$

6. $\sqrt[n]{\sqrt[m]{a}} = (a)^{\frac{1}{mn}}$

7. $\left(\sqrt[z]{(\sqrt[y]{(\sqrt[x]{a})^m})^n}\right)^p = (a)^{\frac{mnp}{xyz}}$

8. Find Square Root/वर्गमूल ज्ञात करना

(i) $7 + 4\sqrt{3} = \sqrt{2^2 + (\sqrt{3})^2 + 2 \times 2 \times \sqrt{3}} = \sqrt{(2+\sqrt{3})^2} = 2 + \sqrt{3}$

(ii) $\sqrt{99 - 70\sqrt{2}} = \sqrt{50 + 49 - 2 \times 5\sqrt{2} \times 7} = \sqrt{(5\sqrt{2} - 7)^2} = 5\sqrt{2} - 7$

9. $\sqrt{a\sqrt{a\sqrt{a \ldots\ldots\ldots\ldots \infty}}} = a$ M. Imp.

10. $\sqrt{a\sqrt{a\sqrt{a \ldots\ldots n \text{ times} / n \text{ बार}}}} = (a)^{\frac{2^n - 1}{2^n}}$

11. $\sqrt[3]{a\sqrt[3]{a\sqrt[3]{a \ldots\ldots\ldots \infty}}} = \sqrt{a}$

12. $\sqrt[n]{a\sqrt[n]{a\sqrt[n]{a \ldots\ldots \infty}}} = \sqrt[n-1]{a}$

13. $\sqrt{a \div \sqrt{a \div \sqrt{a......\infty}}} = \sqrt[3]{a}$

14. $\sqrt[n]{a \div \sqrt[n]{a \div \sqrt[n]{a \div \sqrt[n]{a......\infty}}}} = \sqrt[n+1]{a}$

15. $\sqrt{a + \sqrt{a + \sqrt{a + ...\infty}}} = \dfrac{\sqrt{4a+1}+1}{2} = x$ *M. Imp.*

$\sqrt{a - \sqrt{a - \sqrt{a ......\infty}}} = \dfrac{\sqrt{4a+1}-1}{2} = y$

$x - y = 1$ and/तथा $xy = a$

16. $x = \sqrt{a + \sqrt[b]{a + \sqrt[b]{a + ...\infty}}} = \dfrac{\sqrt{4a+b^2}+b}{2}$

$y = \sqrt{a - \sqrt[b]{a - \sqrt[b]{a - ....\infty}}} = \dfrac{\sqrt{4a+b^2}-b}{2}$

$x - y = b, xy = a$

$x + y = \sqrt{4a + b^2}$

17. $x = \sqrt{a + \sqrt{a - \sqrt{a + ...\infty}}} = \dfrac{\sqrt{4a-3}+1}{2}$

$y = \sqrt{a - \sqrt{a + \sqrt{a - ...\infty}}} = \dfrac{\sqrt{4a-3}-1}{2}$

$x - y = 1, xy = 2(a-1)$

18. $\sqrt{a + \sqrt[b]{a - \sqrt[b]{a + ...\infty}}} = \dfrac{\sqrt{4a-3b^2}+b}{2}$

$\sqrt{a - \sqrt[b]{a + \sqrt[b]{a - ....\infty}}} = \dfrac{\sqrt{4a-3b^2}-b}{2}$

करणी पर आधारित उदाहरण निम्नलिखित है

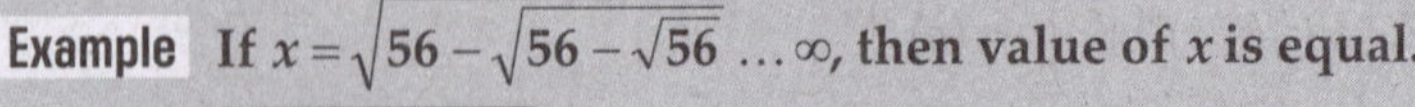

Example **If $x = \sqrt{56 - \sqrt{56 - \sqrt{56}}} \ldots \infty$, then value of x is equal.**

यदि $x = \sqrt{56 - \sqrt{56 - \sqrt{56}}} \ldots \infty$, तब x का मान बराबर है।

Sol. $x = \sqrt{56 - \sqrt{56 - \sqrt{56}}} \ldots \infty$

Let, $x = \sqrt{56 - x}$

Squaring on both sides, we get

$$x^2 = 56 - x \Rightarrow x^2 + x - 56 = 0$$

$$\Rightarrow x^2 + 8x - 7x - 56 = 0 \Rightarrow (x+8)(x-7) = 0 \Rightarrow x = 7$$

Alternate Method

By formula, $x = \dfrac{\sqrt{49+1}-1}{2}$

$$\therefore \quad x = \frac{\sqrt{4 \times 56 + 1} - 1}{2} = \frac{\sqrt{225}-1}{2} = \frac{15-1}{2} = \frac{14}{2} = 7$$

Indices घातांक

Law of Indices घातांक के नियम

1. $a \times a \times a \times \ldots . n$ times/n बार $= a^n$
2. $a^m \times a^n = (a)^{m+n}$
3. $a^m \times a^n \times a^p \times \ldots = (a)^{m+n+p+\ldots}$
4. $\frac{a^m}{a^n} = (a)^{m-n}$
5. $(a^m)^n = (a)^{mn}$
6. $(abc)^n = a^n \times b^n \times c^n$
7. $\left(\frac{a}{b}\right)^n = \frac{a^n}{b^n}$
8. $a^0 = 1$
9. $a^{-1} = \frac{1}{a}$
10. $\left(\frac{a}{b}\right)^m = \left(\frac{b}{a}\right)^{-m}$
11. $(-1)^n = +1$ ($n \to$ even/सम); $(-1)^n = -1$ ($n \to$ odd/विषम)
12. If/यदि $a^m = a^n$, then/तब $m = n$ and/तथा *M. Imp.*
    If/यदि $a^m = b^m$, then/तब $a = b$

परीक्षा ब्रह्मास्त्र
ये सूत्र सभी परीक्षाओं की दृष्टि से *M. Imp* हैं। इन पर आधारित 1 प्रश्न अवश्य पूछा जाता है।

घातांक पर आधारित उदाहरण निम्न है

Example **If $3^{x+2} = 243$, then find the value of x.**

यदि $3^{x+2} = 243$ है, तो x का मान ज्ञात कीजिए। *MP PATWARI 2017*

Sol. $3^{x+2} = 243 \Rightarrow 3^{x+2} = 3^5$

$\therefore x + 2 = 5 \Rightarrow x = 3$

Logarithm लघुगणक

Laws of Logarithm लघुगणक के नियम

If a, m, n are real positive numbers and $a \neq 0$, then
यदि a, m, n धनात्मक वास्तविक संख्याएँ हैं और $a \neq 0$, तब

1. $\log_a mn = \log_a m + \log_a n$
2. $\log_a (m_1, m_2, m_3 \ldots..) = \log_a m_1 + \log_a m_2 + \log_a m_3 + \ldots$
3. $\log_a \left(\frac{m}{n}\right) = \log_a m - \log_a n$
4. $\log_a (m^n) = n \log_a m$
5. $\log_{a^b} m^n = \frac{n}{b} \log_a m$
6. $\log_a m = \frac{\log m}{\log a}$
7. $\log_a m = \frac{1}{\log_m a}$
8. $\log_a m \cdot \log_m a = 1$
9. $a^{\log_a m} = m$
10. $\log_a a = 1$
11. $\log_a 1 = 0$

Progressions श्रेढ़ियाँ

1. Arithmetic Progression (A P) समान्तर श्रेढ़ी

Sequence $a, a+d, a+2d, ...$ is an AP.

अनुक्रम $a, a+d, a+2d, ...$ एक समान्तर श्रेढ़ी है।

(i) n th term of AP/AP का nवाँ पद, $a_n = a+(n-1)d.$

(ii) n th term of AP from last/AP का अन्त से nवाँ पद $= l-(n-1)d$

समान्तर श्रेढ़ी के nवें पद पर आधारित उदाहरण निम्न है

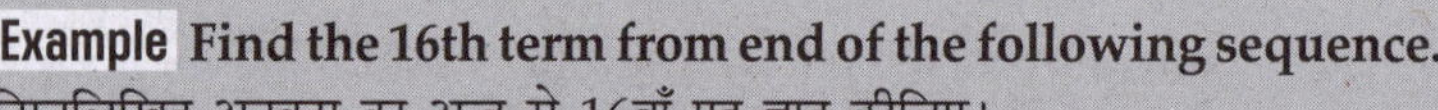

Example **Find the 16th term from end of the following sequence.**

निम्नलिखित अनुक्रम का अन्त से 16वाँ पद ज्ञात कीजिए।

7, 11, 15, ..., 215

Sol. Here, $l = 215, a = 7, d = 4$

16th term from end $= l-(n-1)d = 215-(16-1)\times 4 = 215-60 = 155$

(iv) Sum of first n terms of AP/AP के n पदों का योग, *M. Imp.*

$$S_n = \frac{n}{2}[2a+(n-1)d] = \frac{n}{2}[a+l]$$

Where/जहाँ, $a \rightarrow$ First term/प्रथम पद

$d \rightarrow$ Common ratio/सार्व अन्तर

$l \rightarrow$ Last term/अन्तिम पद

Example **Find the sum of 12 term of AP 51, 53, 55.........**

समान्तर श्रेढ़ी 51, 53, 55......... के 12 पदों का योग ज्ञात कीजिए।

Sol. Here, $a = 51, d = 2, n = 12$

$\therefore\ S_{12} = \frac{12}{2}[2\times 51+(12-1)2] = 6\{102+11\times 2\} = 6\{102+22\} = 6\times 124 = 744$

Some Points Related to AP / AP से सम्बन्धित कुछ बिन्दु

(i) If a constant is added or subtracted from each term of an AP, then the resulting progression is also an AP with same common difference.
यदि किसी AP के प्रत्येक पद में एक अचर राशि जोड़ी या घटाई जाए, तब परिणामी श्रेढ़ी समान सार्व अन्तर वाली AP होती है।

(ii) If each term of an AP is multiplied or divided by a non-zero constant k, then the resulting progression is also an AP.
यदि किसी AP के प्रत्येक पद में एक अचर राशि k से गुणा या भाग किया जाए, तब परिणामी श्रेढ़ी AP ही होती है।

(iii) Consecutive terms in AP can be taken as
AP में क्रमागत पद इस रूप में लिए जा सकते हैं
- 3 terms/3 पद $\rightarrow a-d, a, a+d$
- 4 terms/4 पद $\rightarrow a-3d, a-d, a+d, a+3d$
- 5 terms/5 पद $\rightarrow a-2d, a-d, a, a+d, a+2d$

(iv) $a_n = S_n - S_{n-1}$

Note $\Sigma n = 1+2+3+\ldots+n = \frac{n(n+1)}{2}; \Sigma n^2 = 1^2+2^2+3^2+\ldots+n^2 = \frac{n(n+1)(2n+1)}{6}$

$\Sigma n^3 = 1^3+2^3+3^3+\ldots+n^3 = \left[\frac{n(n+1)}{2}\right]^2$

Arithmetic Mean समान्तर माध्य

If a, b, c are in AP, then b is called arithmetic mean of a and c. *i.e.*
यदि a, b, c, AP में हैं, तब b, a और c का समान्तर माध्य कहलाता है, अर्थात् $\mathbf{b = \frac{a+c}{2}}$

Note Arithmetic mean of a and b/a और b का समान्तर माध्य, $A = \frac{a+b}{2}$

Some Results on AP / AP पर कुछ परिणाम

(i) If/यदि $a_p = q$ and/और $a_q = p$, then/तब $a_{p+q} = 0, a_r = p+q-r$

(ii) If/यदि $a_p = \frac{1}{q}$ and/और $a_q = \frac{1}{p}$, then/तब $a_{pq} = 1$

(iii) If/यदि $S_p = q$ and/और $S_q = p$, then/तब $S_{p+q} = -(p+q)$

(iv) If/यदि $S_p = S_q$, then/तब $S_{p+q} = 0$

(v) If a^2, b^2 and c^2 are in AP, then $\frac{1}{b+c}, \frac{1}{c+a}, \frac{1}{a+b}$ and $\frac{a}{b+c}, \frac{b}{c+a}, \frac{c}{a+b}$ both are also in AP.
यदि a^2, b^2 और c^2 AP में हैं, तब $\frac{1}{b+c}, \frac{1}{c+a}, \frac{1}{a+b}$ और $\frac{a}{b+c}, \frac{b}{c+a}, \frac{c}{a+b}$ दोनों भी AP में होंगे।

2. Geometric Progression (GP) गुणोत्तर श्रेढ़ी

Sequence $a, ar, ar^2, \ldots\ldots$ is a GP. / अनुक्रम $a, ar, ar^2, \ldots\ldots$ एक गुणोत्तर श्रेढ़ी है।

(i) nth term of a GP/गुणोत्तर श्रेढ़ी का nवाँ पद, $a_n = ar^{n-1}$

(ii) The sum of the first n terms of a GP/गुणोत्तर श्रेढ़ी के प्रथम n पदों का योग

$$S_n = \frac{a(1-r^n)}{1-r}; r<1 \text{ or/या } S_n = \frac{a(r^n-1)}{r-1}; r>1$$

Where/जहाँ, $a \rightarrow$ First term/प्रथम पद, $r \rightarrow$ Common ratio/सार्व अनुपात

(iii) The sum of infinite terms of a GP

GP के अनन्त पदों का योग, $S_\infty = \frac{a}{1-r}; |r| < 1$

गुणोत्तर श्रेढ़ी पर आधारित उदाहरण निम्न हैं

Example **Find the 9th terms in the given series.**

दी गई श्रृंखला 9वाँ पद ज्ञात कीजिए।

3, 6, 12, ...

Sol. $a = 3, r = \frac{6}{3} = 2$ $\quad \because$ nth term $(a_n) = ar^{n-1}$

$\therefore$ 9th term $(a_9) = 3 \times (2)^{9-1} = 3 \times 2^8 = 3 \times 256$

$= 768$

Example **Find the sum of given series**

दी गई श्रेणी का योग ज्ञात कीजिए

$3 + 3^2 + 3^3 + \ldots + 3^8$ *SSC CGL 2023*

Sol. $a = 3, r = \frac{3^2}{3} = 3, n = 8$

$\because \quad S_n = \frac{a(r^n - 1)}{r-1} = \frac{3 \times (3^8 - 1)}{3-1} = \frac{3}{2} \times 6560 = 9840$

Some Points Based on GP / GP पर आधारित कुछ बिन्दु

(i) If all terms of GP are multiplied or divided by same non-zero constant, then the resulting progression is also a GP with the same common ratio.

यदि GP के सभी पदों को एक ही अशून्य अचर संख्या से गुणा या भाग किया जाए, तब परिणामी श्रेढ़ी भी समान सार्व अनुपात वाली GP होती है।

(ii) The reciprocal of terms of a given GP also form a GP.

किसी दी गई GP के पदों के व्युत्क्रम से प्राप्त श्रेढ़ी GP ही होती है।

(iii) Consecutive term in a GP can be taken as

GP में क्रमागत पद निम्न रूपों में लिए जा सकते हैं

- 3 terms/ तीन पद $\rightarrow \frac{a}{r}, a, ar$
- 4 terms/4 पद $\rightarrow \frac{a}{r^3}, \frac{a}{r}, ar, ar^3$
- 5 terms/5 पद $\rightarrow \frac{a}{r^2}, \frac{a}{r}, a, ar, ar^2$

(iv) If a, b and c are in GP, then b is called Geometric Mean of a and c. *i.e.*, $\mathbf{b = \sqrt{ac}}$

यदि a, b और c GP में हैं, तब b, a और c का गुणोत्तर माध्य कहलाता है, अर्थात् $\mathbf{b = \sqrt{ac}}$

Note Geometric mean of a and b/ a और b का गुणोत्तर माध्य $G = \sqrt{ab}$

3. Harmonic Progression हरात्मक श्रेढ़ी

A series $a_1, a_2, a_3 \ldots a_n$ is called Harmonic Progression, if $\frac{1}{a_1}, \frac{1}{a_2}, \frac{1}{a_3} \ldots \frac{1}{a_n}$ is AP.

एक श्रेणी $a_1, a_2, a_3 \ldots a_n$ HP कहलाती है, यदि $\frac{1}{a_1}, \frac{1}{a_2}, \frac{1}{a_3} \ldots \frac{1}{a_n}$ एक AP है।

Harmonic mean of a and b/a और b का हरात्मक माध्य $H = \frac{2ab}{a+b}$

Relation Between AP, GP and HP / AP, GP तथा HP में सम्बन्ध

If a and b are two numbers, then/यदि a और b दो संख्याएँ हैं, तब

(i) $A = \frac{a+b}{2}, G = \sqrt{ab}$ and/और $H = \frac{2ab}{a+b}$

(ii) $A \geq G \geq H$

(iii) $G^2 = AH$

Exponential Series चरघातांकी श्रेणी

- $e = 1 + \frac{1}{1!} + \frac{1}{2!} + \frac{1}{3!} + \ldots$
- $e^x = 1 + \frac{x}{1!} + \frac{x^2}{2!} + \frac{x^3}{3!} + \ldots$
- $e^{-x} = 1 - \frac{x}{1!} + \frac{x^2}{2!} - \frac{x^3}{3!} + \frac{x^4}{4!} - \ldots$
- $a^x = 1 + x(\log_e a) + \frac{x^2}{2!}(\log_e a)^2 + \frac{x^3}{3!}(\log_e a)^3 + \ldots \infty, x \in R.$

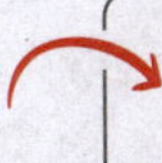

Permutation and Combination
क्रमचय तथा संचय

Fundamental Principles of Counting गणना के मौलिक सिद्धान्त

(i) **Multiplication Principle** If first operation can be performed in m ways and second operation can be performed in n ways. Then, the two operations taken together can be performed in mn ways.

गुणन सिद्धान्त यदि पहली संक्रिया m तरीकों से की जा सकती है और दूसरी संक्रिया को n तरीकों से किया जा सकता है। फिर एकसाथ ली गई दोनों संक्रियाओं को $m \times n$ विधियों से निष्पादित किया जा सकता है।

(ii) **Addition Principle** If an operation can be performed in m ways and another operation, which is independent of the first can be performed in n ways. Then, either of the two operations can be performed in $m + n$ ways.

जोड़ सिद्धान्त यदि एक संक्रिया को m तरीकों से किया जा सकता है और दूसरी संक्रिया, जो पहले से स्वतन्त्र है, n विधियों से की जा सकती है। फिर दोनों में से कोई भी संक्रिया $m + n$ तरीकों से की जा सकती है।

Factorial क्रम गुणित

For any natural number n, we defined factorial as/

किसी भी प्राकृतिक संख्या n के लिए, हम क्रम गुणित को इस प्रकार परिभाषित करते हैं,

$n!$ or/या $n! = n(n-1)(n-2)..3 \cdot 2 \cdot 1$

(i) $0! = 1! = 1$

(ii) $n! = n(n-1)! = n(n-1)(n-2)! = n(n-1)(n-2)...3.2.1$

(iii) $E_p(n!) = \left[\frac{n}{p}\right] + \left[\frac{n}{p^2}\right] + \left[\frac{n}{p^3}\right] + ...$

Permutation क्रमचय

The systematic arrangement of r objects by selecting it from 'n' different object is called as permutation. It is denoted by '${}^nP_r$'.

दी गई 'n' असमान वस्तुओं में से 'r' वस्तुओं का चयन करके उन्हें क्रम में व्यवस्थित करना क्रमचय कहलाता है। इसे '${}^nP_r$' प्रदर्शित करते हैं।

$$^nP_r = \frac{n!}{(n-r)!}$$

Properties of Permutation क्रमचय के गुण

(i) ${}^nP_n = n(n-1)(n-2)...1 = n!$

(ii) ${}^nP_0 = 1$

(iii) ${}^nP_1 = n$

(iv) ${}^nP_{n-1} = n!$

(v) ${}^nP_r = n({}^{n-1}P_{r-1})$

(vi) ${}^{n-1}P_r + r_1\, {}^{n-1}P_{r-1} = {}^nP_r$

(vii) $\frac{{}^nP_r}{{}^nP_{r-1}} = n - r + 1$

क्रमचय के गुण पर आधारित उदाहरण निम्नलिखित है

Example **If ${}^{10}P_r = 5040$, then find the value of r.**

यदि ${}^{10}P_r = 5040$, तब r का मान ज्ञात कीजिए।

Sol. Given, ${}^{10}P_r = 5040$

$\because \quad {}^nP_r = \frac{n!}{(n-r)!}$

$\Rightarrow \quad {}^{10}P_r = \frac{10!}{(10-r)!} = 5040 \Rightarrow (10-r)! = \frac{10!}{5040}$

$\Rightarrow \quad (10-r)! = \frac{10 \times 9 \times 8 \times 7 \times 6!}{5040} \Rightarrow (10-r)! = 6!$

$\therefore \quad r = 10 - 6 = 4$

Number of Permutation in Special Cases
विशेष स्थितियों में क्रमचयों की संख्या

1. The number of permutation of n different objects taken 'r' at a time where repetition is allowed $= n^r$

   एक समय में r वस्तुएँ लेकर, n विभिन्न वस्तुओं के क्रमचयों की संख्या, जहाँ पुनरावृत्ति की अनुमति है $= n^r$

2. The number of permutation of n objects, where P objects are of same kind, and rest are all different $\frac{n!}{P!}$

   n वस्तुओं के क्रमचयों की संख्या, जहाँ P वस्तुएँ एक ही प्रकार की हैं और शेष सभी भिन्न हैं $= \frac{n!}{P!}$

3. The number of permutation of n objects, where P_1 objects are of one kind, P_2 are of second kind, ...P_k are of kth kind and the rest if any are of different kind $= \frac{n!}{P_1!\,P_2!\ldots.P_k!}$

   n वस्तुओं के क्रमचयों की संख्या, जहाँ P_1 वस्तुएँ एक प्रकार की, P_2 वस्तुएँ दूसरे प्रकार की....P_k वस्तुएँ k प्रकार की हैं और शेष वस्तुएँ भिन्न प्रकार की हैं $= \frac{n!}{P_1!\,P_2!\ldots.P_k!}$

Combination संचय

The number of combinations of n different things taken r at a times

एक समय में r वस्तुएँ लेकर, n विभिन्न वस्तुओं के संचयों की संख्या

$$= C(n, r) \text{ or/या } \left(\frac{n}{r}\right)$$

or/या $${}^nC_r = \frac{n!}{r!(n-r)!}, 0 \le r \le n$$

Properties of Combination संचय के गुण

(i) ${}^nC_0 = {}^nC_n = 1$ (ii) ${}^nC_1 = {}^nC_{n-1} = n$ (iii) ${}^nC_r = {}^nC_{n-r}$

(iv) If/यदि ${}^nC_r = {}^nC_p$, then/तब $r = p$ or/या $r + p = n$

(v) ${}^nC_r = \frac{{}^nP_r}{r!}$

(vi) ${}^nC_r + {}^nC_{r-1} = {}^{n+1}C_r$ (Pascal Rule/पास्कल का नियम)

(vii) $n\,{}^{n-1}C_{r-1} = (n-r+1)\,{}^nC_{r-1}$

(viii) ${}^nC_0 + {}^nC_1 + {}^nC_2 + \ldots + {}^nC_n = 2^n$

Number of Combinations in Special Cases
विशेष स्थितियों में संचयों की संख्या

1. Total numbers of handshakes to each other by n persons

$$= {}^nC_2 = \frac{n(n-1)}{2}$$

n व्यक्तियों द्वारा एक-दूसरे से हस्तमिलापों की संख्या $= {}^nC_2 = \frac{n(n-1)}{2}$

2. If in n persons, each person gives a gift another person, then total number of gifts $= n(n-1)$

यदि n व्यक्तियों में से प्रत्येक व्यक्ति दूसरे व्यक्ति को उपहार देता है, तो कुल उपहारों $= n(n-1)$

3. If there are n non-collinear points in a plane. Then,

यदि किसी समतल में n असंरेख बिन्दु हैं, तो

   (i) Number of straight lines formed $= {}^nC_2$
   बनी हुई सीधी रेखाओं की संख्या $= {}^nC_2$

   (ii) Number of triangles formed $= {}^nC_3$
   बने हुए त्रिभुजों की संख्या $= {}^nC_3$

   (iii) Number of quadrilaterals formed $= {}^nC_4$
   बने हुए चतुर्भुजों की संख्या $= {}^nC_4$

4. If there are n points in a plane out of which m are collinear, then

यदि किसी तल में n बिन्दु है। जिनमें से m संरेख है, तो

   (i) Number of straight lines formed
   बनी हुई सीधी रेखाओं की संख्या $= {}^nC_2 - {}^mC_2 + 1$

   (ii) Number of triangles formed
   बने हुए त्रिभुजों की संख्या $= {}^nC_3 - {}^nC_3 = 0$

   (iii) Number of diagonals in a polygon of n sides
   n भुजाओं वाले बहुभुज में विकर्णों की संख्या $= {}^nC_2 - n$

(i) $P(E) = \frac{\text{Number of outcomes favourable to an event } (E)}{\text{Number of all possible outcomes}}$

$P(E) = \frac{\text{एक घटना } (E) \text{ के घटित होने के अनुकूल प्रकार}}{\text{घटना के सभी सम्भव प्रकार}}$ *M. Imp.*

(ii) Probability of an event (E) + Probability of the event not $E = 1$

किसी घटना E की प्रायिकता + घटना E के नहीं होने की प्रायिकता $= 1$

or/या $P(E) + P(\overline{E}) = 1$

Some Point Related to Questions of Probability
प्रायिकता के प्रश्नों से सम्बन्धित कुछ बिन्दु

1. **Tossing a Coin** When a coin is tossed, there are two possible outcomes. So, we can say that the probability of geting Head is $\frac{1}{2}$ or the probability of getting Tail is $\frac{1}{2}$.
जब एक सिक्का उछाला जाता है, तो दो सम्भावित परिणाम आते हैं इसलिए हम कह सकते हैं, कि चित प्राप्त करने की सम्भावना $\frac{1}{2}$ है और पट प्राप्त करने की सम्भावना $\frac{1}{2}$ है।
2. **Throwing Dice** When a single dice is thrown, there are six possible outcomes 1, 2, 3, 4, 5 and 6.
जब एक पासा फेंका जाता है, तो छह सम्भावित परिणाम 1, 2, 3, 4, 5 तथा 6 प्राप्त होते हैं।
3. **In a deck of cards/एक ताश की गड्डी में,**
   (i) There are total of 52 cards/कुल 52 कार्ड होते हैं,
   (ii) There are 4 suit/4 सूट होते हैं,
   - Clubs/चिड़ियाँ
   - Diamonds/ईंट
   - Hearts/पान
   - Spades/हुकूम

   (iii) There are 13 cards in each suit/प्रत्येक सूट में 13 कार्ड होते हैं।
   (iv) There are 4 aces, 4 jacks, 4 queens and 4 kings
   4 इक्के, 4 गुलाम, 4 बेगम और 4 बादशाह होते हैं।
   (v) There are 26 red and 26 black cards.
   26 पत्ते लाल और 26 पत्ते काले होते हैं।
   (vi) There are 12 face cards; king, queen and jacks in each suit.
   12 पत्ते तस्वीर या चेहरे वाले होते हैं; प्रत्येक सूट में बादशाह, बेगम तथा गुलाम।

सिक्के और पासे पर आधारित उदाहरण निम्न हैं

Example **A coin is tossed 10 times and the outcomes are observed as,** *CTET 2018*
$H, T, H, T, T, H, H, T, H, H$ (H is Head; T is Tail)
What is the probability of getting head?
एक सिक्के को 10 बार उछाला जाता है और उसके परिणाम इस प्रकार आते हैं।
$H, T, H, T, T, H, H, T, H, H$ (H = चित है, T = पट है)
चित आने की प्रायिकता क्या है?
Sol. Given outcomes are ($H, T, H, T, T, H, H, T, H, H$)
Total number of possible outcomes $n(S) = 10$
Number of favourable (head) outcomes $n(E) = 6$
Required probability $= \frac{n(E)}{n(S)} = \frac{6}{10} = \frac{3}{5}$

Example **A dice is thrown twice and the sum of the numbers to come is 10. A number 5 is the probability of coming atleast one.**

एक पासे को दो बार फेंका जाता है और आने वाली संख्याओं का योग 10 है। एक संख्या 5, कम-से-कम एक बार आने की प्रायिकता है।

Sol. Total outcomes = 36

Favourable outcomes = (5, 5) = 1

Required probability = $\frac{1}{36}$

Example **Three cards are drawn one after another with replacement from a pack of cards. What is the probability of getting first card a Jack, second card a black card, and third card an even numbered card?**

ताश की एक गड्डी से प्रतिस्थापन के साथ एक के बाद एक तीन पत्ते निकाले जाते हैं। पहला पत्ता गुलाम, दूसरा पत्ता काला पत्ता और तीसरा पत्ता सम संख्या वाला पत्ता होने की प्रायिकता क्या है? *SSC CGL (TIER-II) 2023*

Sol. Total number of cards = 52

Number of Jack = 4

Number of black cards = 26

Cards of even number = 5 × 4 = 20

$\therefore$ Required probability = $\frac{4}{52} \times \frac{26}{52} \times \frac{20}{52} = \frac{5}{338}$

Set Theory समुच्चय सिद्धान्त

Set समुच्चय

A collection of well defined objects, is called set.

एक सुपरिभाषित वस्तुओं का संग्रह, समुच्चय कहलाता है।

Types of Set समुच्चय के प्रकार

(i) **Null Set or Empty Set** : The set which has no element, is called an empty set.

रिक्त समुच्चय : वह समुच्चय, जिसमें कोई अवयव न हो, रिक्त समुच्चय कहलाता है।

(ii) **Singleton Set** : The set which has only one element, is called a singleton set.

एकल समुच्चय : वह समुच्चय, जिसमें केवल एक अवयव हो, एकल समुच्चय कहलाता है।

(iii) **Finite Set** : The set which has finite number of elements is called a finite set.

परिमित समुच्चय : वह समुच्चय, जिसमें अवयवों की संख्या निश्चित हों, परिमित समुच्चय कहलाता है।

(iv) **Infinite Set** : The set which has infinite number of elements is called an infinite set.

अपरिमित समुच्चय : वह समुच्चय, जिसमें अवयवों की संख्या अनिश्चित हो, अपरिमित समुच्चय कहलाता है।

Operations on Sets समुच्चयों पर संक्रियाएँ

(i) Union of sets/समुच्चयों का सम्मिलन

$A \cup B$ = All elements of A and B but not repeated.

A और B के सभी अवयव जिनमें पुनरावृत्ति न हो।

(ii) Intersection of sets/समुच्चयों का प्रतिच्छेदन

$A \cap B$ = Only common elements of A and B.

A और B के केवल उभयनिष्ठ अवयव।

(iii) Difference of sets/समुच्चयों का अन्तर

$A - B$ = Only elements of A which are not in B.

A के केवल वे अवयव, जो B में न हो।

(iv) Complement of sets/पूरक समुच्चय

$A' = U - A$, Where U is universal set.

जहाँ, U एक समष्टीय समुच्चय है।

Laws of Set समुच्चय के नियम

If A, B and C are three sets, then/यदि A, B और C तीन समुच्चय हैं, तब

(i) **Idempotent laws/निष्क्रिय नियम**

(a) $A \cup A = A$ (b) $A \cap A = A$

(ii) **Complement laws/पूरक नियम**

(a) $\phi' = U$ and $U' = \phi$ (b) $(A')' = A$

(c) $A \cup A' = U$ (d) $A \cap A' = \phi$

(iii) **Identity laws/पहचान का नियम**

(a) $A \cup \phi = A$ (b) $A \cap U = A$

(iv) **Commutative laws/क्रम विनिमेय नियम**

(a) $A \cup B = B \cup A$ (b) $A \cap B = B \cap A$

(v) **Associative laws/साहचर्य नियम**

(a) $(A \cup B) \cup C = A \cup (B \cup C)$ (b) $A \cap (B \cap C) = (A \cap B) \cap C$

(vi) **Distributive laws/वितरण नियम**

(a) $A \cup (B \cap C) = (A \cup B) \cap (A \cup C)$ (b) $A \cap (B \cup C) = (A \cap B) \cup (A \cap C)$

(vii) **De-Morgan's laws/डी मॉर्गन नियम**

(a) $(A \cup B)' = A' \cap B'$ (b) $(A \cap B)' = A' \cup B'$

Important Results Based on Sets समुच्चयों पर आधारित महत्त्वपूर्ण परिणाम

Let A, B and C be finite sets in a finite universal set U. Then,
यदि A, B तथा C एक निश्चित सार्वसमुच्चय का परिमित समुच्चय है, तब

(i) $n(A \cup B) = n(A) + n(B) - n(A \cap B)$ M. Imp.

(ii) $n(A \cup B) = n(A) + n(B)$
$\Leftrightarrow$ A and B are disjoint non-void sets. / A और B असंयुक्त अरिक्त समुच्चय है।

(iii) $n(A \cup B \cup C) = n(A) + n(B) + n(C) - n(A \cap B) - n(B \cap C) - n(C \cap A) + n(A \cap B \cap C)$

(iv) $n(A - B) = n(A) - n(A \cap B)$

(v) $n(A \Delta B) = n(A) + n(B) - 2n(A \cap B)$

(vi) $n(A') = n(U) - n(A)$

(vii) $n(A' \cup B') = n(U) - n(A \cap B)$

(viii) $n(A' \cap B') = n(U) - n(A \cup B)$

(ix) If $A_1, A_2, ... A_n$ are n disjoint sets, then
यदि $A_1, A_2, ... A_n$; n असंयुक्त समुच्चय हैं, तो
$n(A_1 \cup A_2 \cup .... \cup A_n) = n(A_1) + n(A_2) + ... + n(A_n)$

समुच्चय पर आधारित महत्त्वपूर्ण परिणाम पर आधारित उदाहरण निम्न हैं

Example **If in a group of 500 students, 475 students can speak Hindi and 200 students can speak Bangali, then how many students can speak only Hindi?** *AIRFORCE GROUP X 2020*

यदि 500 विद्यार्थियों के एक समूह में 475 विद्यार्थी हिन्दी बोल सकते हैं और 200 विद्यार्थी बंगाली बोल सकते हैं, तब कितने विद्यार्थी केवल हिन्दी बोल सकते हैं?

Sol. $n(H \cup B) = n(H) + n(B) - n(H \cap B)$

$500 = 475 + 200 - n(H \cap B)$

$\Rightarrow n(H \cap B) = 475 + 200 - 500 \Rightarrow n(H \cap B) = 175$

175 students able to speak in both languages.
Number of students who only speak in Hindi = 475 – 175 = 300

Example **In an examination, 28% of the students failed in Hindi and 32% failed in English. If 20% of the students failed in both the subjects, then the percentage of students who passed in both the subjects is** *RRB NTPC 2022*

एक परीक्षा में 28% छात्र हिन्दी में और 32% अंग्रेजी में असफल रहे। यदि 20% छात्र दोनों विषयों में असफल रहे, तो उन छात्रों का प्रतिशत जो दोनों विषयों में उत्तीर्ण हुए हैं

Sol. $n(A \cup B) = n(A) + n(B) - n(A \cap B) = 28\% + 32\% - 20\% = 40\%$

Total number of fail students = 40%

The percentage of students who passed in both the subjects
= (100 – 40) = 60%

अध्याय 10

Trigonometry

त्रिकोणमिति

Measurement Systems of Angle
कोण की माप प्रणाली

(i) **Degree Measure** (Sexagesimal System) In this system, an angle is measured in degrees, minutes and seconds.

डिग्री माप (षाष्टिक पद्धति) इस पद्धति में कोणों को अंशों, मिनटों और सेकण्डों में मापा जाता है।

1 right angle/समकोण $=90^\circ$ (90 degree/अंश)

$1^\circ = 60'$ (60 min/मिनट)

$1' = 60''$ (60 sec/सेकण्ड)

(ii) **Grade Measure** (Centesimal System) In this system, an angle is measured in grade, minutes and seconds.

ग्रेड माप (शतिक पद्धति) इस पद्धति में कोणों को ग्रेड, मिनटों और सेकण्डों में मापा जाता है।

1 right angle/समकोण $=100$ grade/ ग्रेड (100 g)

$1\text{ g} = 100'$ (100 min/मिनट)

$1' = 100''$ (100 sec/सेकण्ड)

(iii) **Radian Measure** (Circular System) In this system, an angle is measured in radian. π radian $=180^\circ$ (180 degree)

रेडियन माप (वृत्तीय पद्धति) इस पद्धति में, कोणों को रेडियन में मापा जाता है।

π रेडियन $=180^\circ$ (180 अंश)

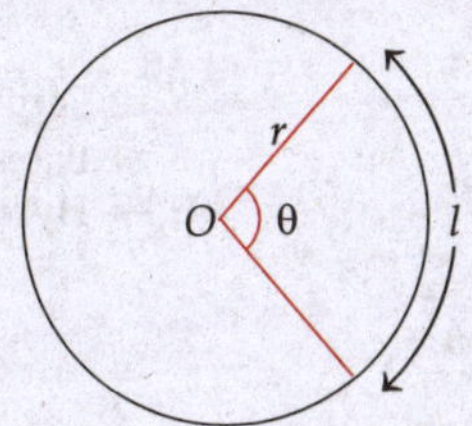

$$r = \frac{360^\circ}{2\pi} = \frac{180^\circ}{\pi} = 1^\circ \text{ and/और } l = \frac{\pi r \theta}{180^\circ}$$

Pythagoras Theorem पाइथागोरस प्रमेय

In a right angle triangle, the square of hypotenuse is equal to the sum of squares of perpendicular and base.

एक समकोण त्रिभुज में, कर्ण का वर्ग, त्रिभुज के लम्ब और आधार के वर्गों के योग के बराबर होता है।

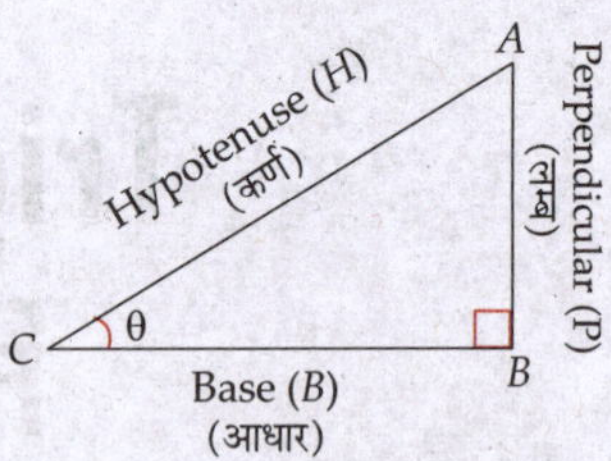

By Pythagoras theorem/पाइथागोरस प्रमेय से,

$$H^2 = P^2 + B^2$$
$$(AC)^2 = (AB)^2 + (BC)^2$$

Pythagorean Triplets पाइथागोरस त्रिक

A set of three integers *a, b* and *c* are called **Pythagorean triplets**, if they satisfy the pythagoras theorem i.e $a^2 + b^2 = c^2$.

तीन पूर्णांकों *a, b* तथा *c* के समुच्चय को **पाइथागोरस त्रिक** कहा जाता है, यदि ये पाइथागोरस प्रमेय को सन्तुष्ट करते हैं अर्थात् $a^2 + b^2 = c^2$.

> ***Note*** *If a, b and c is a pythagorean triplet, then operations of multiplication, division on triplets will also be a triplet. i.e. (ax, bx, cx) and* $\left(\frac{a}{x}, \frac{b}{x}, \frac{c}{x}\right)$ *are also triplets.*
>
> *यदि a, b व c एक पाइथागोरस त्रिक है तब इन पर गुणा व भाग की संक्रिया से प्राप्त संख्या भी त्रिक होगी। अर्थात् (ax, bx, cx) और* $\left(\frac{a}{x}, \frac{b}{x}, \frac{c}{x}\right)$ *भी त्रिक हैं।*

- The smallest triplet is (3, 4, 5)/सबसे छोटा पाइथागोरस त्रिक (3, 4, 5) है।
- Some frequently used triplets are/कुछ मुख्यतः प्रयुक्त होने वाले त्रिक हैं (3, 4, 5), (5, 12, 13), (6, 8,10), (9, 12, 15), (12, 16, 20), (8, 15, 17), (15, 20, 25), (9, 40, 41), (1, 1, $\sqrt{2}$)

In a right angle triangle/एक समकोण त्रिभुज में,

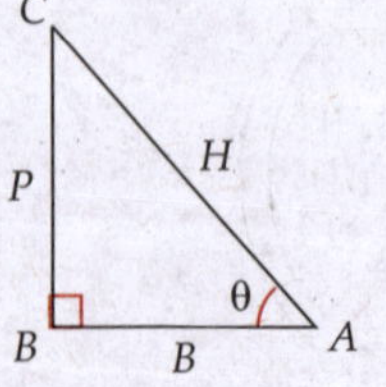

$P \rightarrow$ Perpendicular/लम्ब
$B \rightarrow$ Base/ आधार
$H \rightarrow$ Hypotenuse/ कर्ण

(i) $\sin\theta = \frac{P}{H}$ (ii) $\cos\theta = \frac{B}{H}$ (iii) $\tan\theta = \frac{P}{B}$

(iv) $\cot\theta = \frac{B}{P}$ (v) $\sec\theta = \frac{H}{B}$ (vi) $\operatorname{cosec}\theta = \frac{H}{P}$

Trigonometry Ratio Table त्रिकोणमिति अनुपात तालिका

Angle/कोण → T-Ratio/T-अनुपात	0°	30°/$\frac{\pi}{6}$	45°/$\frac{\pi}{4}$	60°/$\frac{\pi}{3}$	90°/$\frac{\pi}{2}$	180°/π	270°/$\frac{3\pi}{2}$	360°/2π
$\sin\theta$	0	$\frac{1}{2}$	$\frac{1}{\sqrt{2}}$	$\frac{\sqrt{3}}{2}$	1	0	−1	0
$\cos\theta$	1	$\frac{\sqrt{3}}{2}$	$\frac{1}{\sqrt{2}}$	$\frac{1}{2}$	0	−1	0	1
$\tan\theta$	0	$\frac{1}{\sqrt{3}}$	1	$\sqrt{3}$	∞	0	∞	0
$\cot\theta$	∞	$\sqrt{3}$	1	$\frac{1}{\sqrt{3}}$	0	∞	0	∞
$\sec\theta$	1	$\frac{2}{\sqrt{3}}$	$\sqrt{2}$	2	∞	−1	∞	1
$\operatorname{cosec}\theta$	∞	2	$\sqrt{2}$	$\frac{2}{\sqrt{3}}$	1	∞	−1	∞

Sign Change of Trigonometric-Ratios in Quadrants चतुर्थांशों में त्रिकोणमिति अनुपातों का चिह्न परिवर्तन

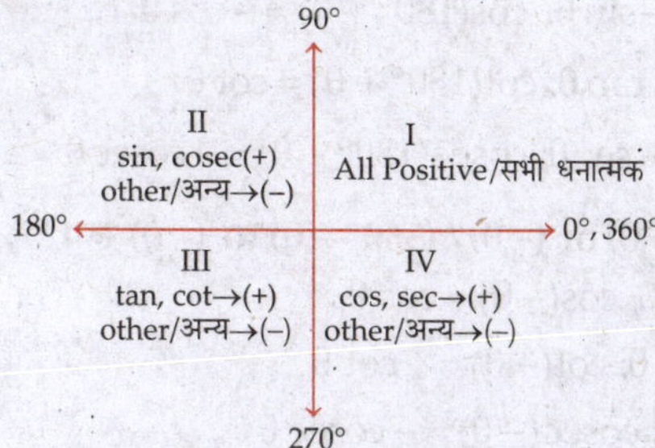

Range of Trigonometric Functions त्रिकोणमितीय फलनों का परिसर

(i) $-1 \le \sin\theta \le 1$ (ii) $-1 \le \cos\theta \le 1$

(iii) $-\infty \le \tan\theta \le +\infty$ (iv) $-\infty \le \cot\theta \le +\infty$

(v) $\theta = 0 \to 90°$, $\sin\theta$ increases from 0 to 1/$\sin\theta$ का मान 0 से 1 तक बढ़ता है।

(vi) $\theta = 0 \to 90°$, $\cos\theta$ decreases from 1 to 0/$\cos\theta$ का मान 1 से 0 तक घटता है।

Trigonometric Ratios of Complementary Angles

पूरक कोणों के त्रिकोणमितीय अनुपात

(i) **In form $(90° - \theta)$ / $(90° - \theta)$ रूप में**

$\sin(90° - \theta) = \cos\theta$; $\cos(90° - \theta) = \sin\theta$

$\tan(90° - \theta) = \cot\theta$; $\cot(90° - \theta) = \tan\theta$

$\sec(90° - \theta) = \text{cosec}\,\theta$;

$\text{cosec}(90° - \theta) = \sec\theta$

अतिसम्भावित! पूरक कोणों और सम्पूरक कोणों के त्रिकोणमितीय अनुपात के सूत्र मुख्यत: सभी परीक्षा की दृष्टि से *M. Imp* हैं। इन पर 1 प्रश्न अवश्य पूछा जाता है।

(ii) **In form $(90° + \theta)$ / $(90° + \theta)$ रूप में**

$\sin(90° + \theta) = \cos\theta$; $\cos(90° + \theta) = -\sin\theta$

$\tan(90° + \theta) = -\cot\theta$; $\cot(90° + \theta) = -\tan\theta$

$\sec(90° + \theta) = -\text{cosec}\,\theta$; $\text{cosec}(90° + \theta) = \sec\theta$

Trigonometric Ratios of Supplementary Angles

सम्पूरक कोणों के त्रिकोणमितीय अनुपात

(i) **In form $(180° - \theta)$ / $(180° - \theta)$ रूप में**

$\sin(180° - \theta) = \sin\theta$; $\cos(180° - \theta) = -\cos\theta$

$\tan(180° - \theta) = -\tan\theta$; $\cot(180° - \theta) = -\cot\theta$

$\sec(180° - \theta) = -\sec\theta$; $\text{cosec}(180° - \theta) = \text{cosec}\,\theta$

(ii) **In form $(180° + \theta)$ / $(180° + \theta)$ रूप में**

$\sin(180° + \theta) = -\sin\theta$; $\cos(180° + \theta) = -\cos\theta$

$\tan(180° + \theta) = \tan\theta$; $\cot(180° + \theta) = \cot\theta$

$\sec(180° + \theta) = -\sec\theta$; $\text{cosec}(180° + \theta) = -\text{cosec}\,\theta$

(iii) **In form $(360° - \theta)$ or $(-\theta)$ / $(360° - \theta)$ या $(-\theta)$ रूप में,**

$\sin(-\theta) = -\sin\theta$, $\cos(-\theta) = \cos\theta$

$\tan(-\theta) = -\tan\theta$; $\cot(-\theta) = -\cot\theta$

$\sec(-\theta) = \sec\theta$; $\text{cosec}(-\theta) = -\text{cosec}\,\theta$

Remember/ याद रखें

- For $(90 \pm \theta)$ and $(270° \pm \theta)$, change $\sin\theta \leftrightarrow \cos\theta$, $\tan\theta \leftrightarrow \cot\theta$, $\text{cosec}\,\theta \leftrightarrow \sec\theta$ and sign changed according to quadrant.
  $(90 \pm \theta)$ तथा $(270° \pm \theta)$ के लिए परिवर्तन $\sin\theta \leftrightarrow \cos\theta$, $\tan\theta \leftrightarrow \cot\theta$, $\text{cosec}\,\theta \leftrightarrow \sec\theta$ तथा चिह्न चतुर्थांश के अनुसार परिवर्तित होंगे।
- For $(180 \pm \theta)$ and $(360° \pm \theta)$ no change in trigonometric ratio, only sign will be changed according to the quadrant.
  $(180 \pm \theta)$ तथा $(360° \pm \theta)$ के लिए त्रिकोणमितीय अनुपात में परिवर्तन नहीं होगा, केवल चिह्न चतुर्थांश के अनुसार परिवर्तित होंगे।

For Example / जैसे–

- $\sin(135^\circ) = \sin(90^\circ + 45^\circ) = \cos 45^\circ = \frac{1}{\sqrt{2}}$
- $\cos(-30^\circ) = \cos 30^\circ = \frac{\sqrt{3}}{2}$
- $\cos(300^\circ) = \cos(270^\circ + 30^\circ) = \sin 30^\circ = \frac{1}{2}$
- $\sin(180^\circ) = \sin(90^\circ + 90^\circ) = \cos 90^\circ = 0$
- $\cos(120^\circ) = \cot(90^\circ + 30^\circ) = -\tan 30^\circ = -\frac{1}{\sqrt{3}}$
- $\sec(150^\circ) = \sec(180^\circ - 30^\circ) = -\sec 30^\circ = -\frac{2}{\sqrt{3}}$

पूरक और सम्पूरक कोणों के त्रिकोणमितीय अनुपात पर आधारित उदाहरण निम्नलिखित हैं

Example What is the value of $\sin(-405^\circ)$? / $\sin(-405^\circ)$ का मान क्या है? *SSC CGL 2023*

Sol. $\sin(-405^\circ) = -\sin 405^\circ = -\sin(360^\circ + 45^\circ) = -\sin 45^\circ = -\frac{1}{\sqrt{2}}$

Relation Between Trigonometric Ratios
त्रिकोणमितीय अनुपातों के बीच सम्बन्ध

(i) $\sin\theta = \frac{1}{\text{cosec}\,\theta} \Rightarrow \text{cosec}\,\theta = \frac{1}{\sin\theta} \Rightarrow \sin\theta \cdot \text{cosec}\,\theta = 1$

(ii) $\cos\theta = \frac{1}{\sec\theta} \Rightarrow \sec\theta = \frac{1}{\cos\theta} \Rightarrow \cos\theta \cdot \sec\theta = 1$

(iii) $\tan\theta = \frac{1}{\cot\theta} \Rightarrow \cot\theta = \frac{1}{\tan\theta} \Rightarrow \tan\theta \cdot \cot\theta = 1$

(iv) $\tan\theta = \frac{\sin\theta}{\cos\theta} \Rightarrow \sin\theta = \tan\theta \cdot \cos\theta \Rightarrow \cos\theta = \frac{\sin\theta}{\tan\theta}$ *Imp.*

(v) $\cot\theta = \frac{\cos\theta}{\sin\theta} \Rightarrow \cos\theta = \cot\theta \cdot \sin\theta \Rightarrow \sin\theta = \frac{\cos\theta}{\cot\theta}$

Trigonometric Identities त्रिकोणमितीय सर्वसमिकाएँ

(i) $\sin^2\theta + \cos^2\theta = 1$

- $\sin\theta = \sqrt{1 - \cos^2\theta}$
- $\cos\theta = \sqrt{1 - \sin^2\theta}$
- $\sin^4\theta + \cos^4\theta = 1 - 2\sin^2\theta\cos^2\theta$
- $\sin^6\theta + \cos^6\theta = 1 - 3\sin^2\theta\cos^2\theta$

परीक्षा ब्रह्मास्त्र

त्रिकोणमितीय सर्वसमिकाएँ SSC परीक्षा की दृष्टि से M. Imp हैं। इस पर आधारित 1-2 प्रश्न अवश्य पूछे जाते हैं।

(ii) $\sec^2\theta - \tan^2\theta = 1$

- $\sec^2\theta = 1 + \tan^2\theta$ • $\tan^2\theta = \sec^2\theta - 1$
- $(\sec\theta + \tan\theta)(\sec\theta - \tan\theta) = 1$
- $\sec^4\theta + \tan^4\theta = 1 + 2\sec^2\theta \cdot \tan^2\theta$
- $(\sec\theta + \tan\theta) = \dfrac{1}{(\sec\theta - \tan\theta)}$
- $\sec^6\theta - \tan^6\theta = 1 + 3\sec^2\theta\tan^2\theta$

(iii) $\text{cosec}^2\theta - \cot^2\theta = 1$

- $\text{cosec}^2\theta = 1 + \cot^2\theta$ • $\cot^2\theta = \text{cosec}^2\theta - 1$
- $(\text{cosec}\,\theta + \cot\theta)(\text{cosec}\,\theta - \cot\theta) = 1$
- $\text{cosec}^4\theta + \cot^4\theta = 1 + 2\,\text{cosec}^2\theta\cot^2\theta$
- $(\text{cosec}\,\theta + \cot\theta) = \dfrac{1}{(\text{cosec}\,\theta - \cot\theta)}$
- $\text{cosec}^6\theta - \cot^6\theta = 1 + 3\,\text{cosec}^2\theta\cot^2\theta$

(iv) $\tan\theta + \cot\theta = \dfrac{1}{\sin\theta\cos\theta} = \sec\theta \cdot \text{cosec}\,\theta$

(v) $\sec^2\theta + \text{cosec}^2\theta = \sec^2\theta \cdot \text{cosec}^2\theta$

(vi) $\tan^2\theta - \sin^2\theta = \tan^2\theta \cdot \sin^2\theta$

(vii) $\cot^2\theta - \cos^2\theta = \cot^2\theta \cdot \cos^2\theta$

Interchange of Trigonometric Ratio Table
त्रिकोणमिति अनुपात परिवर्तन की तालिका

	$\sin\theta$	$\cos\theta$	$\tan\theta$	$\cot\theta$	$\sec\theta$	$\text{cosec}\,\theta$
$\sin\theta$	$\sin\theta$	$\sqrt{1-\cos^2\theta}$	$\dfrac{\tan\theta}{\sqrt{1+\tan^2\theta}}$	$\dfrac{1}{\sqrt{1+\cot^2\theta}}$	$\dfrac{\sqrt{\sec^2\theta-1}}{\sec\theta}$	$\dfrac{1}{\text{cosec}\theta}$
$\cos\theta$	$\sqrt{1-\sin^2\theta}$	$\cos\theta$	$\dfrac{1}{\sqrt{1+\tan^2\theta}}$	$\dfrac{\cot\theta}{\sqrt{1+\cot^2\theta}}$	$\dfrac{1}{\sec\theta}$	$\dfrac{\sqrt{\text{cosec}^2\theta-1}}{\text{cosec}\theta}$
$\tan\theta$	$\dfrac{\sin\theta}{\sqrt{1-\sin^2\theta}}$	$\dfrac{\sqrt{1-\cos^2\theta}}{\cos\theta}$	$\tan\theta$	$\dfrac{1}{\cot\theta}$	$\sqrt{\sec^2\theta-1}$	$\dfrac{1}{\sqrt{\text{cosec}^2\theta-1}}$
$\cot\theta$	$\dfrac{\sqrt{1-\sin^2\theta}}{\sin\theta}$	$\dfrac{\cos\theta}{\sqrt{1-\cos^2\theta}}$	$\dfrac{1}{\tan\theta}$	$\cot\theta$	$\dfrac{1}{\sec^2\theta-1}$	$\text{cosec}^2\theta-1$
$\sec\theta$	$\dfrac{1}{\sqrt{1-\sin^2\theta}}$	$\dfrac{1}{\cos\theta}$	$\sqrt{1+\tan^2\theta}$	$\dfrac{\sqrt{1+\cot^2\theta}}{\cot\theta}$	$\sec\theta$	$\dfrac{\text{cosec}^2\theta}{\text{cosec}^2\theta-1}$
$\text{cosec}\,\theta$	$\dfrac{1}{\sin\theta}$	$\dfrac{1}{\sqrt{1-\cos^2\theta}}$	$\dfrac{\sqrt{1+\tan^2\theta}}{\tan\theta}$	$\sqrt{1+\cot^2\theta}$	$\dfrac{\sec\theta}{\sqrt{\sec^2\theta-1}}$	$\text{cosec}\theta$

Trigonometric Ratios of Sum and Difference of Two Angles
दो कोणों के योग और अन्तर के त्रिकोणमितीय अनुपात

(i) $\sin(A \pm B) = \sin A \cos B \pm \cos A \sin B$

(ii) $\cos(A \pm B) = \cos A \cos B \mp \sin A \sin B$

(iii) $\tan(A + B) = \dfrac{\tan A + \tan B}{1 - \tan A \tan B}$

(iv) $\cot(A + B) = \dfrac{\cot A \cot B - 1}{\cot B + \cot A}$

(v) $\tan(A - B) = \dfrac{\tan A - \tan B}{1 + \tan A \tan B}$

(vi) $\cot(A - B) = \dfrac{\cot A \cot B + 1}{\cot B - \cot A}$

(vii) $\sin(A + B)\sin(A - B) = \sin^2 A - \sin^2 B = \cos^2 B - \cos^2 A$

(viii) $\cos(A + B)\cos(A - B) = \cos^2 A - \sin^2 B = \cos^2 B - \sin^2 A$

Conversion of Trigonometric Ratios from Product to Sum & Difference
त्रिकोणमितीय अनुपातों का गुणा से योग और अन्तर में परिवर्तन

- $2 \sin A \cos B = \sin(A + B) + \sin(A - B), (A > B)$
- $2 \cos A \sin B = \sin(A + B) - \sin(A - B), (A > B)$
- $2 \cos A \cos B = \cos(A + B) + \cos(A - B)$
- $2 \sin A \sin B = \cos(A - B) - \cos(A + B)$

Conversion of Sum & Difference Trigonometric Ratios, into Product
त्रिकोणमितीय अनुपातों के योग और अन्तर का गुणा में परिवर्तन

- $\sin C + \sin D = 2 \sin \dfrac{C + D}{2} \cos \dfrac{C - D}{2}$
- $\sin C - \sin D = 2 \cos \dfrac{C + D}{2} \sin \dfrac{C - D}{2}$
- $\cos C + \cos D = 2 \cos \dfrac{C + D}{2} \cos \dfrac{C - D}{2}$
- $\cos C - \cos D = 2 \sin \dfrac{C + D}{2} \sin \dfrac{D - C}{2}$

Trigonometric Ratios of Multiple of an Angle
कोण के गुणक के त्रिकोणमितीय अनुपात

(i) $\sin 2\theta = 2\sin\theta\cos\theta = \dfrac{2\tan\theta}{1+\tan^2\theta}$ *Imp.*

(ii) $\cos 2\theta = \cos^2\theta - \sin^2\theta = 2\cos^2\theta - 1 = 1 - 2\sin^2\theta = \dfrac{1-\tan^2\theta}{1+\tan^2\theta}$

(iii) $\tan 2\theta = \dfrac{2\tan\theta}{1-\tan^2\theta}$ (iv) $\cot 2\theta = \dfrac{\cot^2\theta - 1}{2\cot\theta}$

(v) $\sin 3\theta = 3\sin\theta - 4\sin^3\theta$ (vi) $\cos 3\theta = 4\cos^3\theta - 3\cos\theta$

(vii) $\tan 3\theta = \dfrac{3\tan\theta - \tan^3\theta}{1-3\tan^2\theta}$

(viii) $\cos\theta\cdot\cos(60^\circ-\theta)\cdot\cos(60^\circ+\theta) = \dfrac{1}{4}\cos 3\theta$

(ix) $\sin\theta\sin(60^\circ-\theta)\sin(60^\circ+\theta) = \dfrac{1}{4}\sin 3\theta$

(x) $\tan\theta\tan(60^\circ-\theta)\tan(60^\circ+\theta) = \tan 3\theta$

कोण के गुणक के त्रिकोणमितीय अनुपात पर आधारित उदाहरण को ध्यानपूर्वक समझें।

Example **What is the value of $\cos 20^\circ\cdot\cos 40^\circ\cdot\cos 80^\circ$?**

$\cos 20^\circ\cdot\cos 40^\circ\cdot\cos 80^\circ$ का मान क्या है?

Sol. $\because \cos\theta\cdot\cos(60^\circ-\theta)\cdot\cos(60^\circ+\theta) = \dfrac{1}{4}\cos 3\theta$

Here $\theta = 20^\circ$ $\therefore$ $\cos 20^\circ\cdot\cos 40^\circ\cdot\cos 80^\circ = \dfrac{1}{4}\cos 3\theta = \dfrac{1}{4}\cos 60^\circ = \dfrac{1}{4}\times\dfrac{1}{2} = \dfrac{1}{8}$

Trigonometric Ratios of Sub-multiple Angles
कोण के अपवर्त्य के त्रिकोणमितीय अनुपात

- $\sin\theta = 2\sin\dfrac{\theta}{2}\cos\dfrac{\theta}{2} = \dfrac{2\tan\theta/2}{1+\tan^2\theta/2}$
- $\cos\theta = \cos^2\theta/2 - \sin^2\theta/2 = 2\cos^2\theta/2 - 1 = 1 - 2\sin^2\theta/2 = \dfrac{1-\tan^2\theta/2}{1+\tan^2\theta/2}$
- $\tan\theta = \dfrac{2\tan\theta/2}{1-\tan^2\theta/2}$
- $\cot\theta = \dfrac{2\cot^2\theta/2 - 1}{2\cot\theta/2}$
- $1+\cos\theta = 2\cos^2\theta/2$
- $1-\cos\theta = 2\sin^2\theta/2$
- $\sin\dfrac{\theta}{2} = \sqrt{\dfrac{1-\cos\theta}{2}}$
- $\cos\dfrac{\theta}{2} = \sqrt{\dfrac{1+\cos\theta}{2}}$
- $\tan\dfrac{\theta}{2} = \sqrt{\dfrac{1-\cos\theta}{1+\cos\theta}}$

Some other Relation of Trigonometric Ratios

त्रिकोणमितीय अनुपातों के कुछ अन्य सम्बन्ध

- $\dfrac{\cot A + \tan B}{\cot B + \tan A} = \tan B \cot A$
- $\left(\dfrac{1 - \tan\theta}{1 - \cot\theta}\right)^2 = \tan^2\theta$
- $\sqrt{\dfrac{1 - \sin\theta}{1 + \sin\theta}} = \dfrac{1 - \sin\theta}{\cos\theta} = \sec\theta - \tan\theta$
- $\sqrt{\dfrac{1 + \cos\theta}{1 - \cos\theta}} = \dfrac{1 + \cos\theta}{\sin\theta} = \operatorname{cosec}\theta + \cot\theta$
- $\sqrt{\dfrac{\operatorname{cosec}\theta + 1}{\operatorname{cosec}\theta - 1}} = \sqrt{\dfrac{1 + \sin\theta}{1 - \sin\theta}} = \sec\theta + \tan\theta$
- $\sqrt{\dfrac{\sec\theta + \tan\theta}{\sec\theta - \tan\theta}} = \sec\theta + \tan\theta$
- $\sqrt{\dfrac{\cot\theta + \cos\theta}{\cot\theta - \cos\theta}} = \sec\theta + \tan\theta$
- $\sin 1^\circ \cdot \sin 2^\circ \cdot \sin 3^\circ .... \sin 180^\circ = 0$
- $\cos 1^\circ \cdot \cos 2^\circ \cdot \cos 3^\circ .... \cos 90^\circ = 0$
- $\tan 1^\circ \cdot \tan 2^\circ \cdot \tan 3^\circ .... \tan 89^\circ = 1$

त्रिकोणमितीय अनुपातों के मान पर आधारित उदाहरण को ध्यानपूर्वक समझें।

Example **Value of expression**
[cot 1° cot 2° cot 3° .... cot 178° cot 179°] is

व्यंजक [cot 1° cot 2° cot 3° .... cot 178° cot 179°] का मान है

SSC (10+2) 2023

Sol. $\cot 1^\circ \cot 2^\circ \cot 3^\circ \cot 4^\circ \ldots \cot 178^\circ \cot 179^\circ$

$= \cot 1^\circ \cot 2 \cot 3^\circ \ldots \cot 90^\circ \ldots \cot 178^\circ \cot 179^\circ$

$= \cot 1^\circ \cot 2^\circ \cot 3^\circ \ldots \times 0 .... \cot 178^\circ \cot 179^\circ = 0 \; (\because \cot 90^\circ = 0)$

Values of Trigonometric Ratios for Some Special Angles

कुछ विशेष कोणों के लिए त्रिकोणमितीय अनुपातों के मान

- $\sin 18^\circ = \dfrac{\sqrt{5} - 1}{4} = \cos 72^\circ$
- $\cos 18^\circ = \dfrac{\sqrt{10 + 2\sqrt{5}}}{4} = \sin 72^\circ$
- $\cos 36^\circ = \dfrac{\sqrt{5} + 1}{4} = \sin 54^\circ$
- $\sin 36^\circ = \dfrac{\sqrt{10 - 2\sqrt{5}}}{4} = \cos 54^\circ$
- $\sin 75^\circ = \cos 15^\circ = \dfrac{\sqrt{3} + 1}{2\sqrt{2}}$
- $\sin 15^\circ = \cos 75^\circ = \dfrac{\sqrt{3} - 1}{2\sqrt{2}}$
- $\tan 75^\circ = \cot 15^\circ = \dfrac{\sqrt{3} + 1}{\sqrt{3} - 1} = 2 + \sqrt{3}$
- $\tan 15^\circ = \cot 75^\circ = \dfrac{\sqrt{3} - 1}{\sqrt{3} + 1} = 2 - \sqrt{3}$
- $\sin 22\dfrac{1^\circ}{2} = \dfrac{\sqrt{2 - \sqrt{2}}}{2}$
- $\cos 22\dfrac{1^\circ}{2} = \dfrac{\sqrt{2 + \sqrt{2}}}{2}$
- $\tan 22\dfrac{1^\circ}{2} = \sqrt{2} - 1$
- $\cot 22\dfrac{1^\circ}{2} = \sqrt{2} + 1$

Maximum and Minimum Values of Trigonometric Ratios
त्रिकोणमितीय अनुपातों के महत्तम और न्यूनतम मान

1. $a\sin\theta + b\cos\theta \rightarrow$ Max./महत्तम $= \sqrt{a^2 + b^2}$
   $\rightarrow$ Min./न्यूनतम $= -\sqrt{a^2 + b^2}$

> त्रिकोणमितीय अनुपातों के महत्तम और न्यूनतम मान सभी परीक्षाओं की दृष्टि से M. Imp टॉपिक है।

2. If/यदि $a > b$
   $a\sin^2\theta + b\cos^2\theta$ → Max./महत्तम $= a$; → Min./न्यूनतम $= b$
   If/यदि $b > a$
   $a\sin^2\theta + b\cos^2\theta$ → Max./महत्तम $= b$; → Min./न्यूनतम $= a$
3. If a and b both are positive/यदि a और b दोनों धनात्मक हैं
   $a\sin^2\theta + b\,\text{cosec}^2\theta$
   $a\cos^2\theta + b\sec^2\theta$
   $a\tan^2\theta + b\cot^2\theta$
   ] Min./न्यूनतम $= 2\sqrt{ab}$
4. $a\sec^2\theta + b\,\text{cosec}^2\theta \rightarrow$ Min./न्यूनतम $= (\sqrt{a} + \sqrt{b})^2$
5. $\sin^{2m}\theta + \cos^{2n}\theta \rightarrow$ Max./महत्तम $= 1,\ m, n \in N$
6. $a\tan^2\theta + b\cos^2\theta \rightarrow$ Min./न्यूनतम $= 2\sqrt{ab}$
   Max./महत्तम $= \infty$
7. $\sin^n\theta \cdot \cos^n\theta$ → Max./महत्तम $= \frac{1}{2n}$, → Min./न्यूनतम $= \frac{-1}{2n}$

   When/जब, $n =$ odd/विषम
8. $\sin^n\theta \cdot \cos^n\theta$ → Max./महत्तम $= \frac{1}{2n}$, → Min./न्यूनतम $= 0$

   When/जब, $n =$ even/विषम
9. $\sin^{2n}\theta + \cos^{2n}\theta$ → Max./महत्तम $= 1$; → Min./न्यूनतम $= 0$

Trigonometric ratio त्रिकोणमितीय अनुपात	Min. न्यूनतम	Max. महत्तम
$\sin^n\theta$ ($n \rightarrow$ odd/विषम)	-1	$+1$
$\sin^n\theta$ ($n \rightarrow$ even/सम)	0	$+1$
$\cos^n\theta$ ($n \rightarrow$ odd/विषम)	-1	$+1$
$\cos^n\theta$ ($n \rightarrow$ even/सम)	0	1
$\tan^n\theta$ ($n \rightarrow$ odd/विषम)	$-\infty$	$+\infty$
$\tan^n\theta$ ($n \rightarrow$ even/सम)	0	∞

Trigonometric ratio त्रिकोणमितीय अनुपात	Min. न्यूनतम	Max. महत्तम
$\cot^n \theta$ ($n \to$ odd/विषम)	$-\infty$	$+\infty$
$\cot^n \theta$ ($n \to$ even/सम)	0	∞
$\sec^n \theta$ ($n \to$ odd/विषम)	$-\infty$	$+\infty$
$\sec^n \theta$ ($n \to$ even/सम)	1	∞
$\text{cosec}^n \theta$ ($n \to$ odd/विषम)	$-\infty$	$+\infty$
$\text{cosec}^n \theta$ ($n \to$ even/सम)	1	∞

त्रिकोणमितीय अनुपातों के महत्तम तथा न्यूनतम मान पर आधारित उदाहरण निम्नलिखित हैं

Example **Minimum value of $(4\cos\theta + 3)$ is**/$(4\cos\theta+3)$ का न्यूनतम मान है **UP TGT 2021**

Sol. Minimum value of $\cos\theta = -1$

Minimum value of $(4\cos\theta + 3) = 4 \times (-1) + 3 = -1$

Example **Maximum value of $(\sin x + \cos x)$ is**/$(\sin x + \cos x)$ का अधिकतम मान है **RRB NTPC 2021**

Sol. $(\sin x + \cos x)_{max} = \sqrt{1^2 + 1^2} = \sqrt{2}$

Example **What is the minimum value of $(9\sin^2\theta + 10\cos^2\theta)$?** $(9\sin^2\theta+10\cos^2\theta)$ का न्यूनतम मान क्या है? **RRB NTPC 2021**

Sol. $(9\sin^2\theta + 10\cos^2\theta)_{min} = (a\sin^2\theta + b\cos^2\theta)$ here, $a < b$

So, $(9\sin^2\theta + 10\cos^2\theta)_{min} = a = 9$

Example **Find the minimum and maximum value of $\sin^7\theta\cos^7\theta$.** / $\sin^7\theta\cos^7\theta$ का न्यूनतम और अधिकतम मान ज्ञात कीजिए। **RRB NTPC 2021**

Sol. Minimum value of $\sin^7\theta \cdot \cos^7\theta = -\frac{1}{2^7} = \frac{-1}{128}$

Maximum value of $\sin^7\theta \cdot \cos^7\theta = \frac{1}{2^7} = \frac{1}{128}$

Relation between Angles and Sides of a Triangle in Trigonometric Ratio Form

त्रिकोणमितीय अनुपात के रूप में त्रिभुज के कोणों और भुजाओं के बीच सम्बन्ध

1. Sine Rule/ज्या नियम

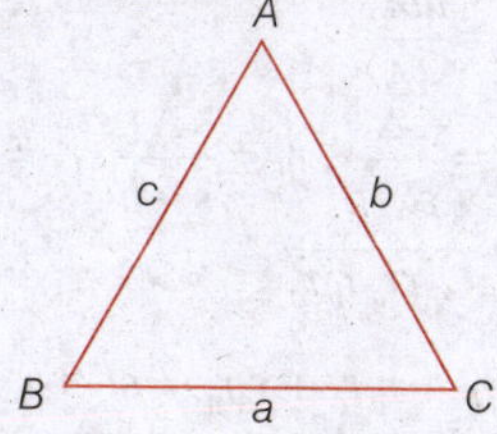

- $\frac{a}{\sin A} = \frac{b}{\sin B} = \frac{c}{\sin C} = k$ (constant/नियतांक)

  $a : b : c = k\sin A : k\sin B : k\sin C$

  $a : b : c = \sin A : \sin B : \sin C$
- $h = c\sin B = b\sin C$

2. Cosine Rule/कोज्या नियम

$$\cos A = \frac{b^2 + c^2 - a^2}{2bc}$$

$$a^2 = b^2 + c^2 - 2bc\cos A$$

$$\cos B = \frac{c^2 + a^2 - b^2}{2ca}$$

$$b^2 = a^2 + c^2 - 2ac\cos B$$

$$\cos C = \frac{a^2 + b^2 - c^2}{2ab} \qquad c^2 = a^2 + b^2 - 2ab\cos C$$

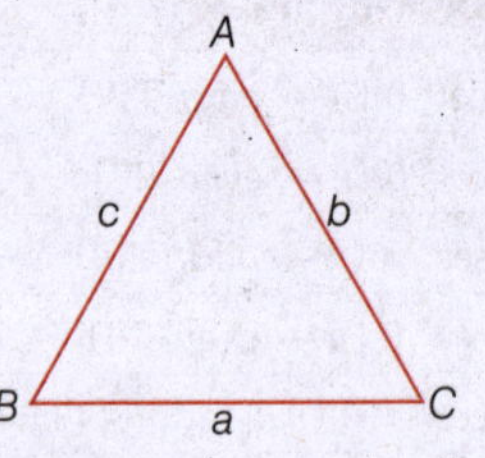

3. Semi Angle Formula for a Triangle/त्रिभुज के लिए अर्द्धकोण सूत्र

(i) $\sin\frac{A}{2} = \sqrt{\frac{(s-b)(s-c)}{bc}}; \sin\frac{B}{2} = \sqrt{\frac{(s-a)(s-c)}{ac}}; \sin\frac{C}{2} = \sqrt{\frac{(s-a)(s-b)}{ab}}$

(ii) $\cos\frac{A}{2} = \sqrt{\frac{s(s-a)}{bc}}; \cos\frac{B}{2} = \sqrt{\frac{s(s-b)}{ac}}; \cos\frac{C}{2} = \sqrt{\frac{s(s-c)}{ab}}$

(iii) $\tan\frac{A}{2} = \sqrt{\frac{(s-b)(s-c)}{s(s-a)}}; \tan\frac{B}{2} = \sqrt{\frac{(s-a)(s-c)}{s(s-b)}};$

$\tan\frac{C}{2} = \sqrt{\frac{(s-a)(s-b)}{s(s-c)}}$ Where/जहाँ, $s = \frac{a+b+c}{2}$

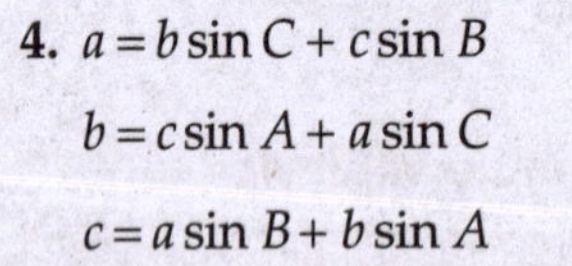

4. $a = b\sin C + c\sin B$

$b = c\sin A + a\sin C$

$c = a\sin B + b\sin A$

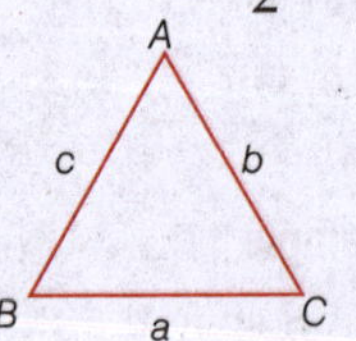

Area of a Triangle through Trigonometric Ratios

त्रिकोणमितीय अनुपातों से त्रिभुज का क्षेत्रफल

In ΔABC/ΔABC में,

(i) $\Delta = \frac{1}{2}ab\sin C = \frac{1}{2}bc\sin A = \frac{1}{2}ca\sin B$

(ii) $\Delta = \frac{a^2 \sin B \sin C}{\sin(B+C)} = \frac{b^2 \sin C \sin A}{\sin(C+A)} = \frac{c^2 \sin A \sin B}{\sin(A+B)}$

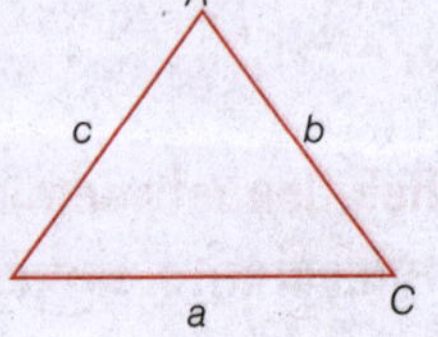

(iii) $R = \frac{abc}{4\Delta}$ (iv) $r = \frac{\Delta}{s}$

(v) $r_a = \frac{\Delta}{s-a}, r_b = \frac{\Delta}{s-b}, r_c = \frac{\Delta}{s-c}$

(vi) $\Delta = \sqrt{r r_a r_b r_c}$ (vii) $r_a + r_b + r_c = 4R + r$

(viii) $r_a r_b + r_b r_c + r_c r_a = s^2$ where/जहाँ, $s = \frac{a+b+c}{2}$

(ix) $r_a^2 + r_b^2 + r_c^2 = (4R + r)^2 - 2s^2$

Height and Distance ऊँचाई एवं दूरी

- **Line of Sight** The straight line joining the object seen by the observer's eye is called line of sight.

दृष्टि रेखा प्रेक्षक की आँख द्वारा देखी गई वस्तु को मिलाने वाली सरल रेखा दृष्टि रेखा कहलाती है।

- **Horizontal Line** The line which goes straight from the eye of the observer parallel to the ground is called horizontal line.

क्षैतिज रेखा ऐसी रेखा, जो प्रेक्षक की आँख से सीधी भूमि के समानान्तर जाती है, क्षैतिज रेखा कहलाती है।

Angle of Elevation उन्नयन कोण

Angle of elevation is the angle between the observer's line of sight and the horizontal line when the object being observed is above the horizontal line.

This angle can be formed while looking at some objects like flying balloon in sky, on top of building, an aeroplane etc.

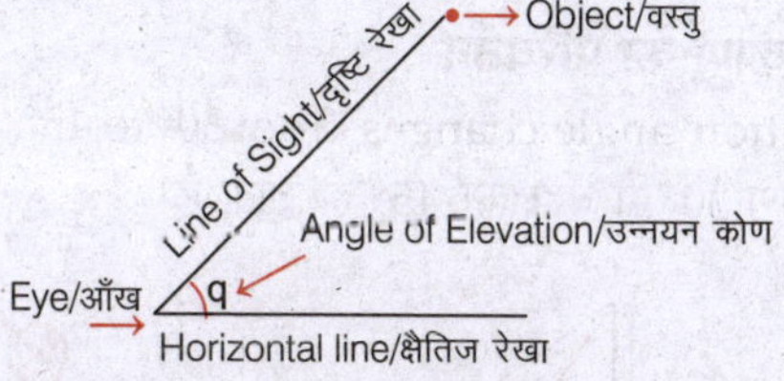

उन्नयन कोण प्रेक्षक की दृष्टि रेखा और क्षैतिज रेखा के बीच का कोण होता है, जब जिस वस्तु का अवलोकन किया जा रहा होता है, वह क्षैतिज रेखा से ऊपर होती है।

यह कोण कुछ वस्तुओं को देखने से बन सकता है जैसे–आकाश में उड़ते गुब्बारे, इमारत के शीर्ष, हवाई जहाज आदि।

Angle of Depression अवनमन कोण

When an object is situated below the horizontal line passing through the eye of the observer, then in this situation the angle made by the line joining the object and the line of sight with this horizontal line is called the angle of depression of the object.

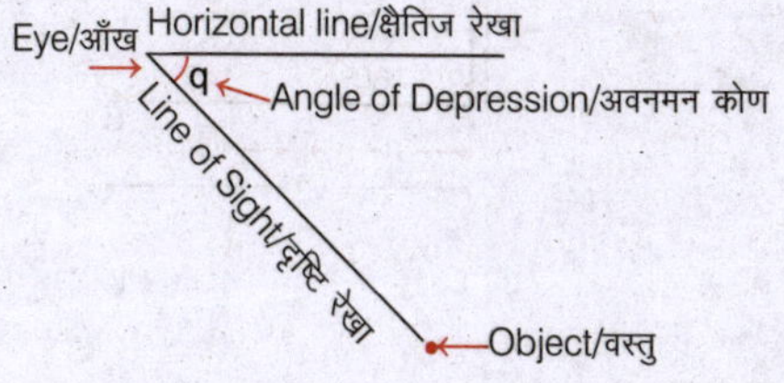

जब कोई वस्तु, देखने वाले की आँख से जाने वाली क्षैतिज रेखा के नीचे स्थित होती है, तो इस अवस्था में वस्तु व दृष्टि रेखा को मिलाने वाली रेखा इस क्षैतिज रेखा से जो कोण बनाती है, उसे वस्तु का अवनमन कोण कहते हैं।

e.g. An angle of depression is formed when a person looks at car parked on the road from the top of the tower or in other similar situations.

जैसे–जब एक व्यक्ति एक टॉवर के शीर्ष से सड़क पर खड़ी एक कार को देखता है अथवा इसी प्रकार की अन्य स्थितियों में अवनमन कोण बनता है।

Angle-Side Ratio कोण-पक्ष (भुजा) अनुपात

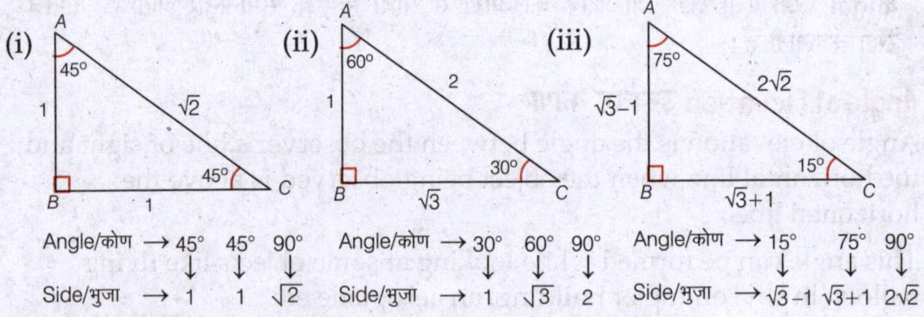

Change of Angle कोण का परिवर्तन

(i) When elevation angle changes from 30° to 45°.
जब उन्नयन कोण 30° से बदलकर 45° हो जाता है।

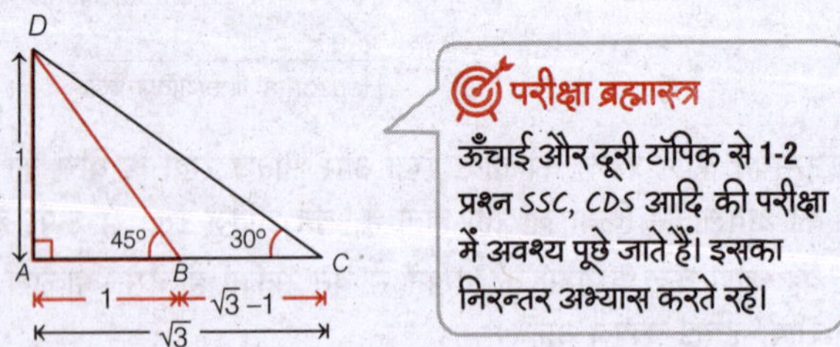

(ii) When elevation angle changes from 45° to 60°. / जब उन्नयन कोण 45° से बदलकर 60° हो जाता है।

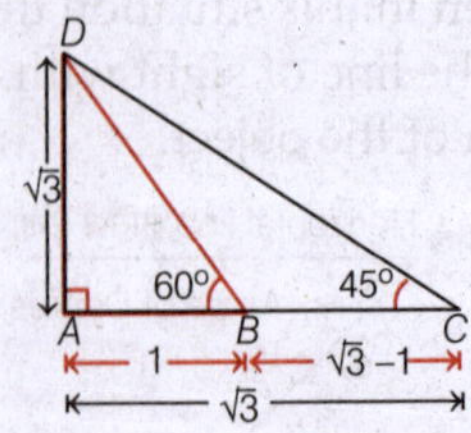

(iii) When elevation angle changes from 30° to 60°. /जब उन्नयन कोण 30° से बदलकर 60° हो जाता है।

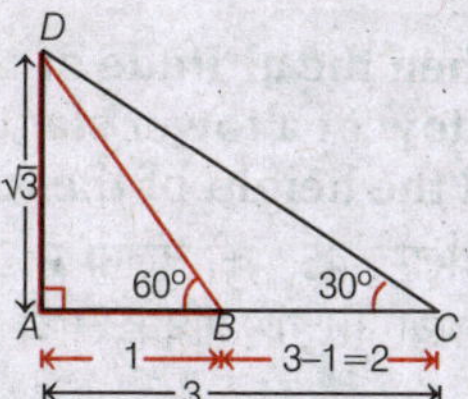

(iv) When elevation angle changes from 15° to 30°. /जब उन्नयन कोण 15° से बदलकर 30° हो जाता है।

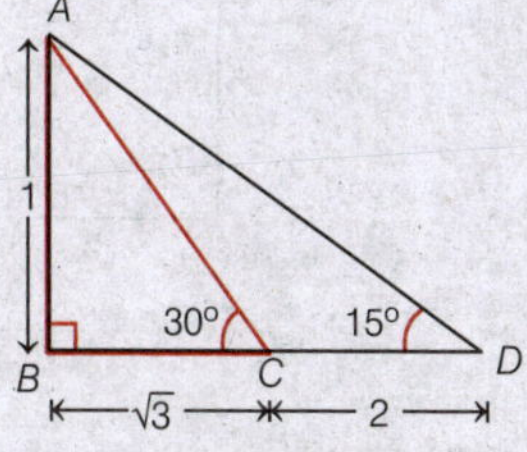

(v) When elevation angle changes from θ° to 2θ°. / जब उन्नयन कोण θ° से बदलकर 2θ° हो जाता है।

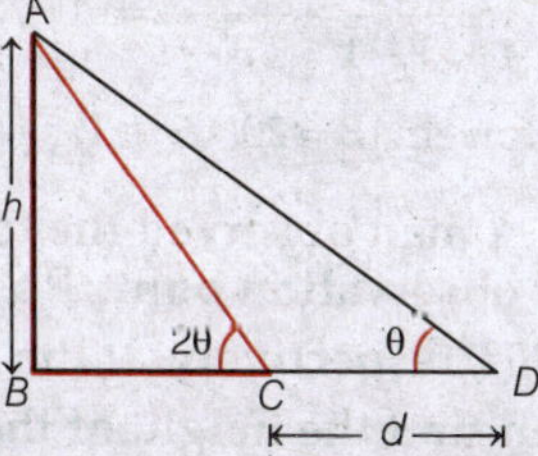

$$h^2 = y^2 - x^2$$

(vi) When elevation angle changes from θ_1 to θ_2. / जब उन्नयन कोण θ_1 से बदलकर θ_2 हो जाता है।

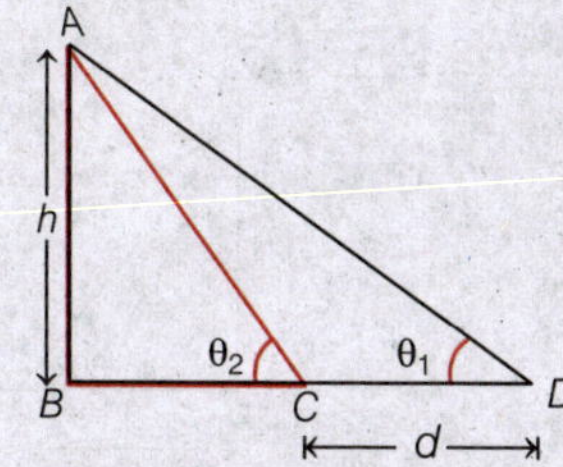

$$d = h(\cot\theta_1 - \cot\theta_2)$$

कोण के परिवर्तन पर आधारित महत्त्वपूर्ण उदाहरण निम्न हैं

Example **When the altitude of the Sun is 30° as compared to 45°, the shadow of a tower standing on ground level is 40 m longer. Find the height of the tower (in meter).**

जब सूर्य का शीर्षलम्ब 45° की तुलना में 30° होता है, तो समतल जमीन पर खड़े एक मीनार की छाया 40 मी अधिक लम्बी बनती है। मीनार की ऊँचाई (मीटर में) ज्ञात कीजिए। *SSC (10+2) 2024*

Sol. Given,

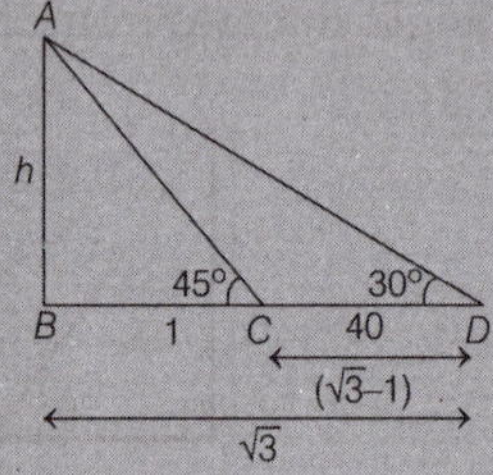

$\because \ (\sqrt{3}-1) = 40$

$$1 = \frac{40}{\sqrt{3}-1} \times \frac{\sqrt{3}+1}{\sqrt{3}-1} = 20(\sqrt{3}+1)\text{ m}$$

Height of tower $AB = 20(\sqrt{3}+1)$ m

Example **A man observed the top of an electric pole from two different observation point P and Q at angles of elevation of 60° and 30°, respectively. If the distance between P and Q is 10 m, then find the height of the pole (in meter).**

एक आदमी ने दो अलग-अलग अवलोकन बिन्दुओं P और Q से क्रमशः 60° और 30° के उन्नयन कोण पर बिजली के एक खम्भे के शीर्ष का अवलोकन किया। यदि P और Q के बीच की दूरी 10 मी है, तो खम्भे की ऊँचाई (मीटर में) ज्ञात कीजिए। *SSC (10+2) 2024*

Sol. Given,

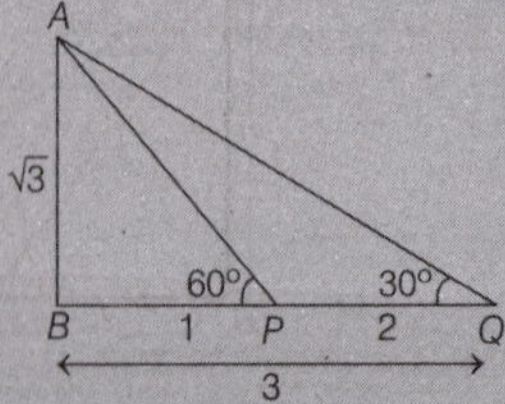

$PQ = 10$ m

$\because \ 2 \text{ units} = 10$ m

$\therefore \ 1 \text{ unit} = 5$ m

$\therefore \ \sqrt{3} \text{ units} = 5\sqrt{3}$ m

Height of pole $= 5\sqrt{3}$ m

Some Special Cases कुछ विशेष स्थितियाँ

(i) When angle of elevations are complementary to each other
यदि उन्नयन कोण एक-दूसरे के पूरक होते हैं।

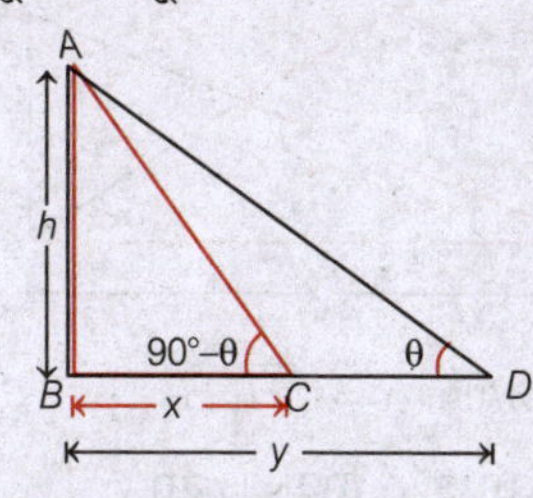

$$h = \sqrt{xy}$$

(ii) According to the following figure, /निम्न आकृति के अनुसार,

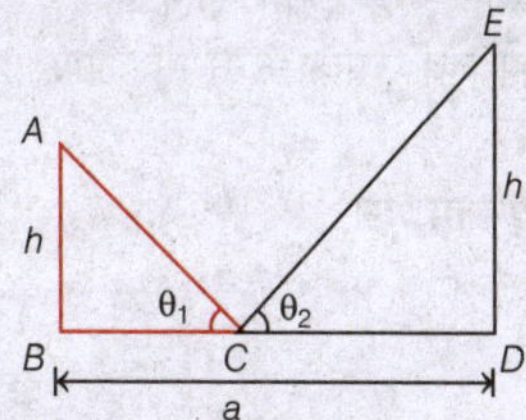

$$a = h(\cot\theta_1 + \cot\theta_2)$$

(iii) According to the following figure, / निम्न आकृति के अनुसार,

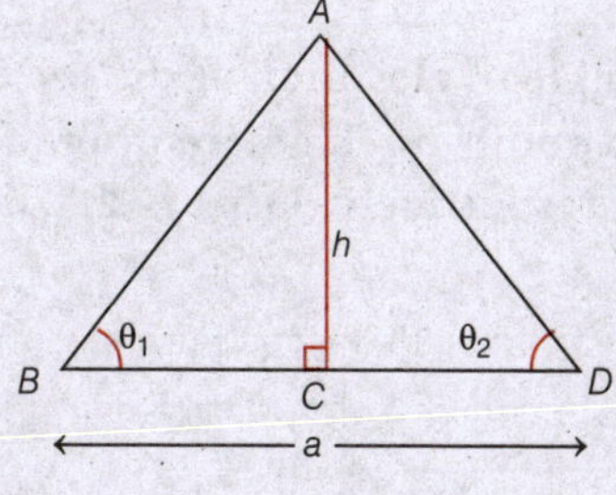

$$a = h(\cot\theta_1 + \cot\theta_2)$$

(iv) In the following figure/निम्न आकृति में,

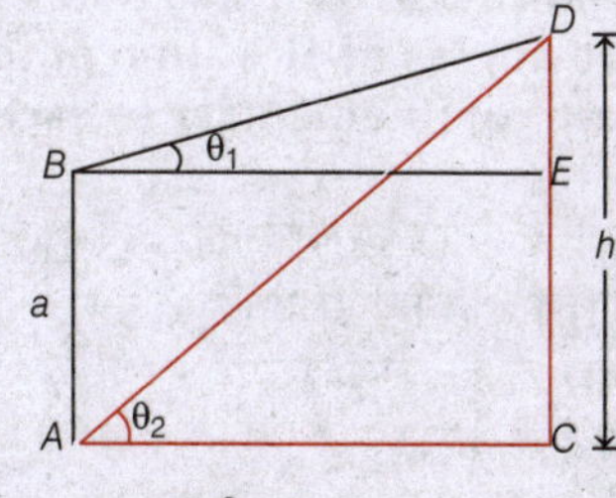

$$h = \frac{a\cot\theta_2}{\cot\theta_2 - \cot\theta_1}$$

(v) In the following figure/निम्न आकृति में,

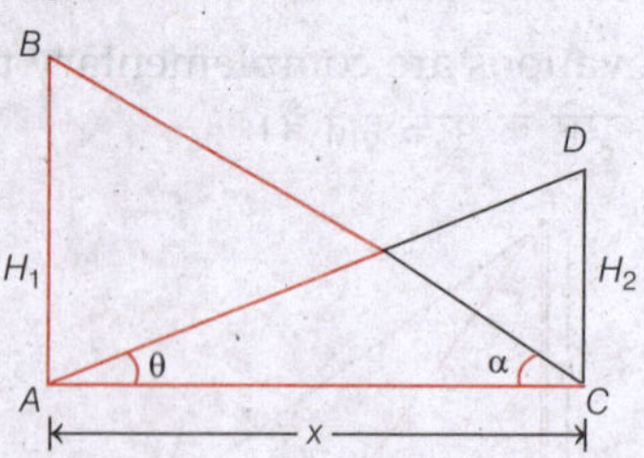

If/यदि $\theta + \alpha = 90°$, then/तब $x = \sqrt{H_1 H_2}$

(vi) When a balloon of radius 'r' makes an angle 'α' at eye of an observer and angle of elevation of it's center is 'β', then

जब एक गुब्बारा जिसकी त्रिज्या 'r' है देखने वाले के नेत्र पर कोण 'α' बना रहा हो एवं उसका उन्नयन कोण गुब्बारे के मध्य से 'β' हो, तब

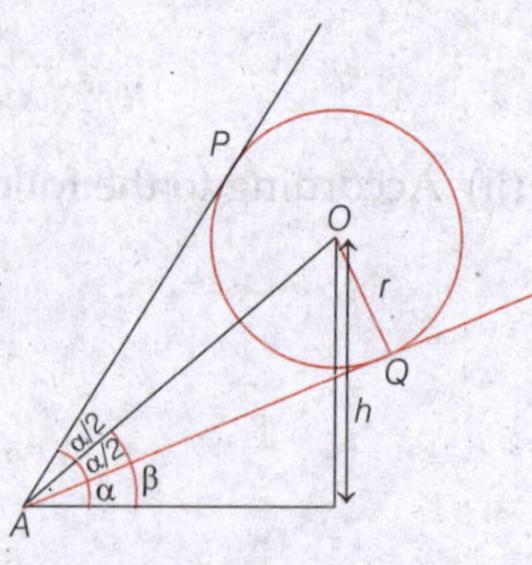

Minimum distance/न्यूनतम दूरी,

$AO = r \operatorname{cosec} \dfrac{\alpha}{2}$

Height/ऊँचाई, $(h) = r \operatorname{cosec}\left(\dfrac{\alpha}{2}\right) \cdot \sin\beta$

ऊँचाई और दूरी की विशेष स्थितियों पर आधारित उदाहरण को ध्यानपूर्वक समझें।

Example **The angle of elevation of the top of a $25\sqrt{3}$ m high tower from two points on the plain ground on opposite sides are 45° and 60°, respectively. What is the distance between two points?**

$25\sqrt{3}$ मी ऊँचें एक मीनार के शीर्ष का उसकी विपरीत दिशा में समतल जमीन पर दो बिन्दुओं से उन्नयन कोण क्रमशः 45° और 60° हैं। दो बिन्दुओं के बीच की दूरी क्या है?

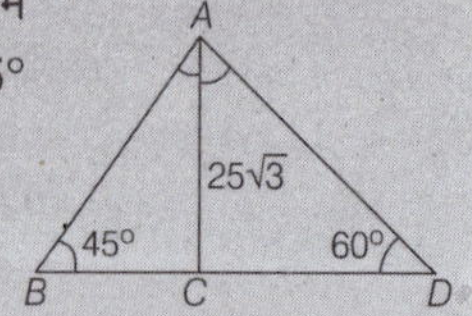

Sol. Here, $h = 25\sqrt{3}$

$$BD = h(\cot\theta_1 + \cot\theta_2) = 25\sqrt{3}$$

Example **The distance between two pole of length 25 m and 16 m is x m. If two angles of elevation of their top from the bottom of other are complementary to each other. Find the value of x.**

25 मी और 16 मी लम्बे दो स्तम्भ के बीच की दूरी x मी है। यदि एक-दूसरे के नीचे से उनके सम्बन्धित शीर्ष की ऊँचाई के दो कोण परस्पर पूरक हों, तो x का मान ज्ञात करें।

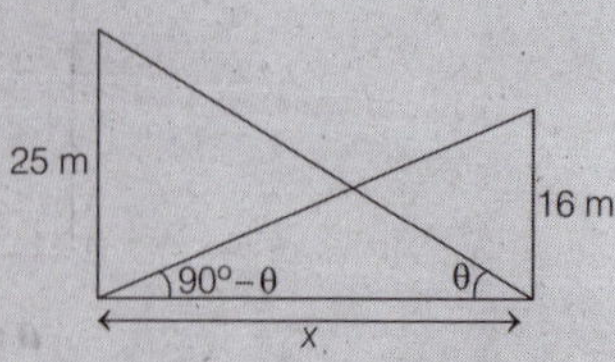

Sol. According to the question,

$$x = \sqrt{25 \times 16} \Rightarrow x = \sqrt{400} = 20 \text{ m}$$

Example **Balloon of radius '7' cm makes an angle 60° at eye of an observer and angle of elevation of centre of ballon from foot of the observer (*A*) is 45°. Find shortest distance from observer to centre of balloon and height of centre of balloon from ground.** ***CDS 2022***

एक गुब्बारा जिसकी त्रिज्या '7' सेमी है एक देखने वाले की आँख से 60° का कोण बना रहा है तथा गुब्बारे का उन्नयन कोण देखने वाले के पैर के साथ 45° है। गुब्बारे के मध्य से देखने वाले व्यक्ति की न्यूनतम दूरी व भूमि से उसकी दूरी ज्ञात कीजिए।

Sol. Minimum distance, $(AO) = r \operatorname{cosec} \frac{\alpha}{2} = 7 \times \operatorname{cosec} \frac{60}{2}$

$$= 7 \times 2 = 14 \text{ cm}$$

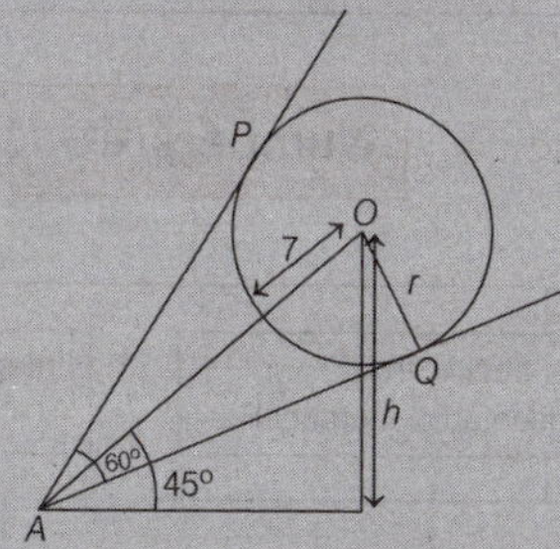

Height, $(h) = r \operatorname{cosec}\left(\frac{\alpha}{2}\right) \cdot \sin \beta = 14 \times \sin 45°$

$$= 14 \times \frac{1}{\sqrt{2}} = 7\sqrt{2} \text{ cm}$$

अध्याय 11

Number System

संख्या पद्धति

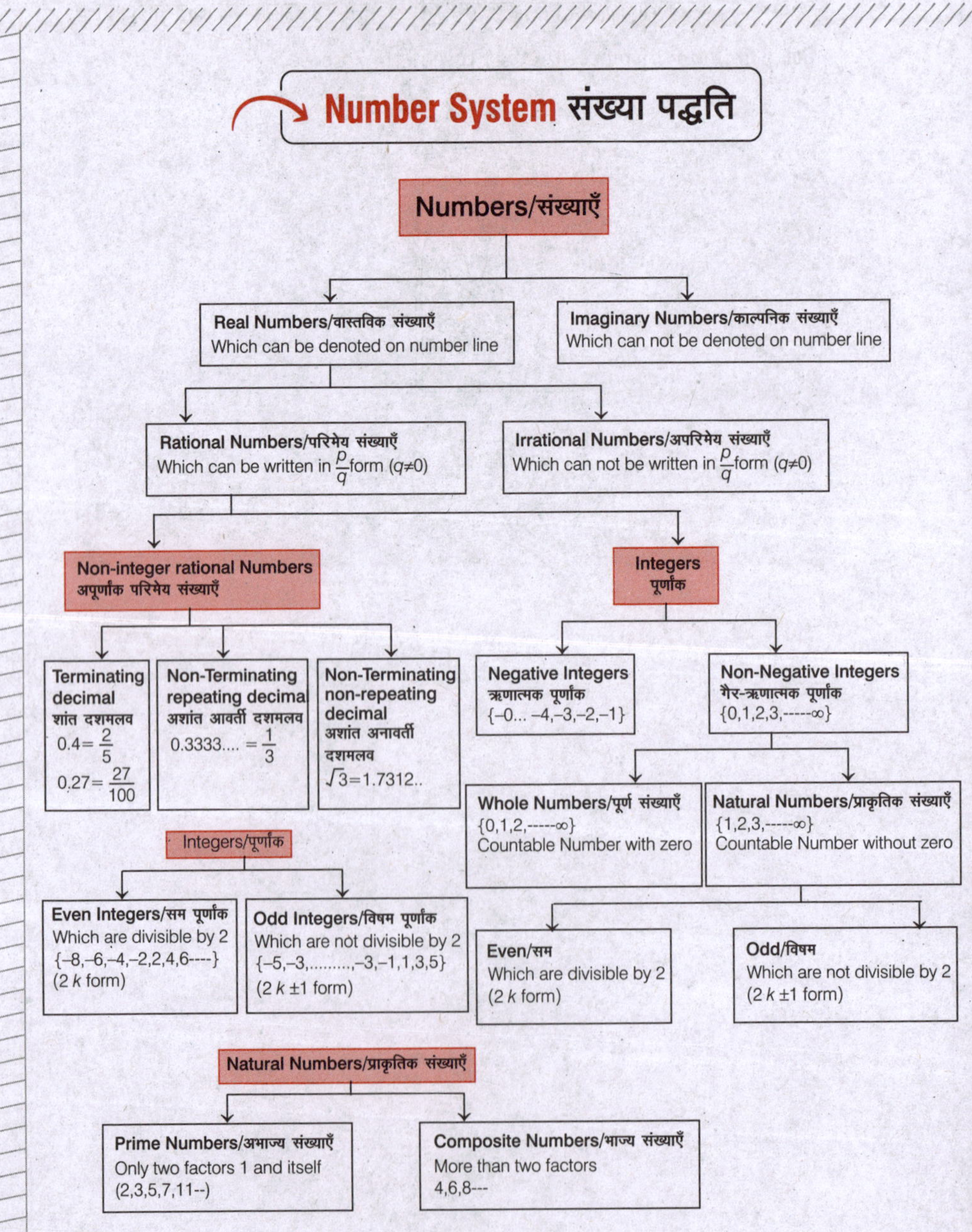

Different Systems to Write the Number संख्या को लिखने की विभिन्न प्रणाली

(i) Decimal System/दाशमिक (दशमलव) प्रणाली

We use 10 digits 0, 1, 2, 3, 4, 5, 6, 7, 8, 9 for writing a number. These digits are called numeral.

एक संख्या को लिखने के लिए हम दस अंकों 0, 1, 2, 3, 4, 5, 6, 7, 8, 9 का प्रयोग करते हैं। ये अंक संख्यांक कहलाते हैं।

(ii) Roman System/रोमन प्रणाली

We use some special digits for writing a number.

एक संख्या को लिखने के लिए हम कुछ विशेष अंकों का प्रयोग करते हैं।

$1 \rightarrow$ I, $2 \rightarrow$ II, $3 \rightarrow$ III, $5 \rightarrow$ V, $10 \rightarrow$ X; L $\rightarrow 50$, C $\rightarrow 100$, D $\rightarrow 500$, M $\rightarrow$ 1000.

(iii) Binary System/बाइनरी प्रणाली

We use two digits 0 and 1 for writing a number. These digits are called 'Bit'.

एक संख्या को लिखने के लिए हम दो अंकों 0 और 1 का प्रयोग करते हैं। ये अंक 'बिट' कहलाते हैं।

Prime Number अभाज्य संख्या

Prime number between 1 to 100/1 से 100 के बीच अभाज्य संख्याएँ = 2, 3, 5, 7, 11, 13, 17, 19, 23, 29, 31, 37, 41, 43, 47, 53, 59, 61, 67, 71, 73, 79, 83, 89, 97

Number of prime numbers/अभाज्य संख्याओं की संख्या

1-50 $\rightarrow$ 15   50-100 $\rightarrow$ 10

1-100 $\rightarrow$ 25   1-200 $\rightarrow$ 46

1-1000 $\rightarrow$ 168

Co-Prime Numbers सह-अभाज्य संख्याएँ

Two numbers in which no common factor *i.e*, their HCF = 1

दो संख्याएँ जिनमें कोई भी गुणनखण्ड उभयनिष्ठ न हों, अर्थात् उनका म.स. = 1

(25, 19) (11, 13) (13, 17) .... .

Twin-Prime Numbers/जुड़वाँ अभाज्य संख्याएँ

Two prime numbers with a gap of 2/2 के अन्तराल के साथ दो अभाज्य संख्याएँ

= (3, 5) (5, 7) (11, 13) (17, 19)...

Perfect Number/आदर्श संख्या

If the sum of all factors (excluding that number) is equal to that number, then that number is called perfect number.

यदि सभी गुणनखण्डों (उस संख्या को छोड़कर) का योग उस संख्या के बराबर हो, तो उसे आदर्श संख्या कहते हैं।

$6 \rightarrow 1, 2, 3, 6$ (factors/गुणनखण्ड)

Here/यहाँ $1 + 2 + 3 = 6$

Some Points Based on Numbers संख्याओं पर आधारित कुछ बिन्दु

- All intergers are rational numbers/सभी पूर्णांक परिमेय संख्याएँ हैं।
- 2 → Even prime number and smallest prime number
  सम अभाज्य संख्या और सबसे छोटी अभाज्य संख्या
- 3, 5, 7 → Only pair of consecutive odd prime numbers
  क्रमागत विषम अभाज्य संख्याओं का एकमात्र युग्म
- 0 → Neither positive nor negative, only integer
  तो धनात्मक और न ही ऋणात्मक है, केवल पूर्णांक है।
- 1 → Neither prime nor composite/न तो अभाज्य और न ही भाज्य
- 4 → Smallest composite number/सबसे छोटी भाज्य संख्या
- 9 → Smallest odd composite number/सबसे छोटी विषम भाज्य संख्या
- 101 → Smallest 3 digit prime number/सबसे छोटी 3 अंकों की अभाज्य संख्या
- 997 → Largest 3 digit prime number/सबसे बड़ी 3 अंकों की अभाज्य संख्या
- odd × odd → odd/विषम × विषम → विषम
- odd × even → even/विषम × सम → सम
- odd ± odd → even/विषम ± विषम → सम
- odd ± even → odd/विषम ± सम → विषम

Divisibility Rules विभाज्यता के नियम

1. **Divisibility Rule of 2, 4, 8, 16/2, 4, 8, 16 से विभाज्यता के नियम**
   - 2 → Last digit should be divisible by 2
     अन्तिम अंक 2 से विभाज्य होना चाहिए।
   - 4 → The number formed by last 2 digits should be divisible by 4
     अन्तिम दो अंकों से बनी संख्या 4 से विभाज्य होनी चाहिए।
   - 8 → The number formed by last 3 digits should be divisible by 8
     अन्तिम तीन अंकों से बनी संख्या 8 से विभाज्य होनी चाहिए।
   - 16 → The number formed by last 4 digits should be divisible by 16
     अन्तिम चार अंकों से बनी सख्या 16 से विभाज्य होनी चाहिए।

> **अतिसम्भावित!**
> विभाज्यता के नियम SSC, RRBs, IAS CSAT परीक्षाओं की दृष्टि से अतिमहत्वपूर्ण है। इस टॉपिक से 1-2 प्रश्न इन परीक्षाओं में अवश्य पूछे जाते हैं।

2. **Divisibility Rule of 3 and 9/3 और 9 से विभाज्यता के नियम**
   - 3 → Sum of digits should be divisible by 3
     अंकों का योग 3 से विभाज्य होना चाहिए।
   - 9 → Sum of digits should be divisible by 9
     अंकों का योग 9 से विभाज्य होना चाहिए।

3. **Divisibility Rule of 5, 25, 125/5, 25, 125 से विभाज्यता के नियम**
   - 5 → Last digit should be 0 and 5/ अन्तिम अंक 0 और 5 होने चाहिए।
   - 25 → The number formed by last two digits should be divisible by 25
     अन्तिम दो अंकों से बनी संख्या 25 से विभाज्य होनी चाहिए।
   - 125 → The number formed by last three digits should be divisible by 125
     अन्तिम तीन अंकों से बनी संख्या 125 से विभाज्य होनी चाहिए।
4. **Divisibility Rule of 6/6 से विभाज्यता का नियम**
   - 6 → If a number is divisible by 2 and 3 both, then that number will also be divisible by 6.
     यदि कोई संख्या 2 और 3 दोनों से विभाजित हो, तब वह 6 से भी विभाजित होगी।
5. **Divisibility Rule of 7/7 से विभाज्यता का नियम**
   - Multiply last digit by 2 and subtract from remaining number. Repeat this process until number become less, then check divisibility of 7.
     संख्या के अन्तिम अंक को 2 से गुणा करें और शेष संख्या से घटा दें। यह प्रक्रिया तब तक दोहराए जब तक संख्या छोटी ना हो जाए और फिर सात से विभाज्यता जाँचे।

     Ex. $17339 \rightarrow 1733 - 2 \times 9 = 1733 - 18 = 1715$
     $1715 \rightarrow 171 - 2 \times 5 = 171 - 10 = 161$
     $161 \rightarrow 16 - 2 \times 1 = 16 - 2 = 14$ (divisible by 7/7 से विभाज्य)
6. **Divisibility Rule of 11/11 से विभाज्यता का नियम**
   - If the difference between the sum of the digits at odd places and sum of the digits at even places is zero or multiple of 11.
     यदि विषम स्थानों के अंकों के योग और सम स्थानों के अंकों के योग के बीच का अन्तर शून्य या 11 का गुणक है।
7. **Divisibility Rule of 12/12 से विभाज्यता का नियम**
   - If a number is divisible by 3 and 4 both, then that number will also be divisible by 12.
     यदि कोई संख्या 3 और 4 दोनों से विभाजित होती है, तब वह संख्या 12 से भी विभाजित होगी।
8. **Divisibility Rule of 13/13 से विभाज्यता का नियम**
   - Multiply last digit by 4 and add to the remaining number. Repeat this process until number becomes less, then check divisibility of 13.
     संख्या के अन्तिम अंक को 4 से गुणा करें और शेष संख्या में जोड़ दें। यह प्रक्रिया तब तक दोहराएँ जब तक संख्या छोटी ना हो जाए, फिर 13 से विभाज्यता जाँचे।

     Ex. $2314 \rightarrow 231 + 4 \times 4 = 231 + 16 = 247$
     $247 \rightarrow 24 + 7 \times 4 = 24 + 28 = 52$ (divisible by 13/13 से विभाज्य)

9. **Divisibility Rule of 17/17 से विभाज्यता का नियम**

- Multiply last digit by 5 and subtract from the remaining number. Repeat this process until number becomes less, then check divisibility of 17.
  संख्या के अन्तिम अंक को 5 से गुणा करें और शेष संख्या से घटा दें। यह प्रक्रिया तब तक दोहराएँ जब तक संख्या छोटी ना हो जाए, फिर 17 से विभाज्यता जाँचे।
  Ex. $5389 \to 538-5\times9=538-45=493$
  $493 \to 49-5\times3=49-15=34$ (divisible by 17/17 से विभाज्य)

Remember/ याद रखें

- Any number of the form ABAB is always divisible by 101
  ABAB के रूप में कोई भी संख्या हमेशा 101 से विभाज्य होती है।
  Ex. $4747=47\times101$
- Any number of the form ABCABC is always divisible by 1001, 7, 11, 13 ABCABC के रूप में कोई भी संख्या हमेशा 1001, 7, 11, 13 से विभाज्य होती है।
  Ex. $39039=39\times1001=7\times11\times13=1001$

विभाज्यता से पूछे जाने वाले प्रश्नों पर आधारित उदाहरण नीचे दिए गए हैं

Example **What is the minimum value of x, such that 5- digit number $725x3$ is divisible by 9?**
x का सबसे छोटा मान क्या है, ताकि 5-अंकों की संख्या $725x3$, 9 से विभाज्य हो जाए? *SSC CONSTABLE (GD) 2023*

Sol. Number = $725x3$
Sum of digit $=7+2+5+x+3=(17+x)$
Required minimum value of $x=1$
So, $17+1=18$ is divisible by 9,
$\therefore$ Minimum value of $x=1$

Example **In number 9592450, 9592330, 9592885 and 9592741, which number is divisible by 11?**
संख्या 9592450, 9592330, 9592885 और 9592741 में से कौन-सी संख्या 11 से विभाज्य है? *SSC CGL (TIER I) 2023*

Sol. If number = 9592330
Sum of odd digit's place $=9+9+3+0=21$
Sum of even digit's place $=5+2+3=10$
Required difference $=21-10=11$
Which is multiple of 11.
So, number 9592330 is divisible by 11.

Remainder Rules शेषफल के नियम

1. Remainder Theorem/शेषफल प्रमेय

Divisor) Dividend (Quotient
Remainder

Dividend = Divisor × Quotient + Remainder
भाज्य = भाजक × भागफल + शेषफल

> **अवश्य याद रखें!**
> शेषफल के नियम पर आधारित 1 प्रश्न SSC, RRBs IAS CSAT etc. परीक्षाओं में निरन्तर पूछे जा रहे हैं। अत: विद्यार्थी इन्हें अवश्य याद रखें।

2. (i) $(a^n - b^n)$ is divisible by $(a-b)$ for all values of n.
n के सभी मानों के लिए $(a^n - b^n)$, $(a-b)$ से विभाजित है।

(ii) $(a^n - b^n)$ is divisible by $(a+b)$ for even values of n.
n के केवल सम मानों के लिए $(a^n - b^n)$, $(a+b)$ से विभाज्य है।

(iii) $(a^n + b^n)$ is divisible by $(a+b)$ for odd values of n.
n के केवल विषम मानों के लिए $(a^n + b^n)$, $(a+b)$ से विभाज्य है।

(iv) $a^n + b^n + c^n + d^n$ is divisible by $a+b+c+d$, if n is odd.
$a^n + b^n + c^n + d^n$, $a+b+c+d$ से विभाज्य है, यदि n विषम है।

3. Fermat's Theorem/फरमट की प्रमेय

$$\frac{a^{P-1}}{P} \Rightarrow 1 \text{ (Remainder/शेषफल)}$$

Where/जहाँ, a, P

$a \to$ Co-prime/सह अभाज्य

$P \to$ Prime number/अभाज्य संख्या

Ex. $$\frac{64^{30}}{31} = \frac{64^{(31-1)}}{31} = \frac{a^{P-1}}{P}$$

Hence, remainder = 1 (शेषफल)

4. Wilson's Theorem/विल्सन की प्रमेय

(i) If P is any prime number/यदि P कोई अभाज्य संख्या है

$\frac{(P-1)!}{P} \to$ Remainder/शेषफल $= -1$ or/या $(P-1)$

(ii) $\frac{(ax+k)^n}{a} \to$ Remainder/शेषफल $= k^n$

(iii) $\frac{(ax+1)^n}{a} \to$ Remainder/शेषफल $= 1$

(iv) $\frac{(ax-k)^n}{a} \to$ Remainder/शेषफल $= (-k)^n$

(v) $\frac{(ax-1)^n}{a} \to$ Remainder $= (-1)^n$

5. Euler's Theorem/यूलर की प्रमेय

$$\frac{a^{\phi(N)}}{N} \rightarrow R = 1$$

$a \rightarrow$ Co-prime/सह-अभाज्य

$N \rightarrow$ Natural number/प्राकृतिक संख्या

$$\phi(N) = N\left(1-\frac{1}{p}\right)\left(1-\frac{1}{q}\right)\left(1-\frac{1}{r}\right).....$$

p, q, r .... are prime factors of N.

p, q, r .... N के अभाज्य गुणनखण्ड हैं।

6. If two numbers are divided by same divisor, the remainders are r_1 and r_2, respectively. If sum of these two numbers are divided by the same divisor, the remainder is r_3.

Then, divisor $= r_1 + r_2 - r_3$

यदि दो संख्याओं को एक ही भाजक द्वारा भाग देने पर शेषफल क्रमशः r_1 व r_2 आता है तथा यदि दोनों संख्याओं के योग को उसी भाजक द्वारा भाग देने पर शेषफल r_3 है, तो भाजक $= r_1 + r_2 - r_3$

Note $\frac{4^n}{6} \rightarrow$ Remainder/शेषफल $= 4$ and/तथा $\frac{10^n}{6} \rightarrow$ Remainder/शेषफल $= 4$

शेषफल पर आधारित उदाहरण निम्नलिखित है

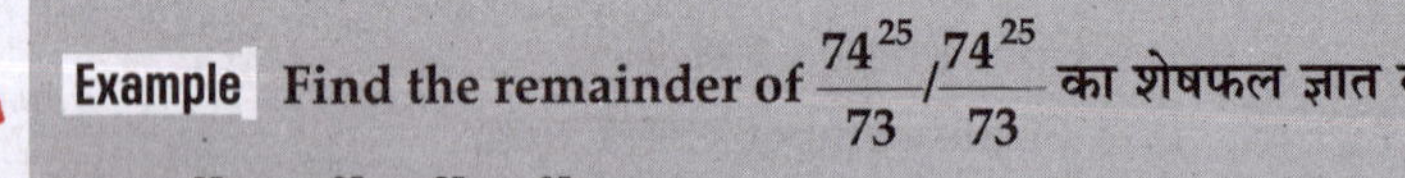

Example **Find the remainder of $\frac{74^{25}}{73}$/$\frac{74^{25}}{73}$ का शेषफल ज्ञात करें।**

Sol. $\frac{74^{25}}{73} = \frac{74^{25} - 1^{25} + 1^{25}}{74 - 1} = 1$

$\because$ $(a^n - b^n)$ is divisible by $(a - b)$ for all values of n.

$\therefore$ Required remainder $= 1$

Number of Factors
गुणनखण्डों की संख्या

1. Number of prime factors of
$$a^p \times b^q \times c^r \times d^s = p+q+r+s,$$
where a, b, c, d are prime number and p, q, r, s are positive integers.
$a^p \times b^q \times c^r \times d^s$ के अभाज्य गुणनखण्डों की संख्या $= p+q+r+s$
जहाँ a, b, c, d अभाज्य संख्याएँ तथा p, q, r, s धनात्मक पूर्णांक हैं।

2. If $N = a^p b^q c^r d^s$, then number of divisors of N
$$=(p+1)(q+1)(r+1)(s+1)$$
Where, a, b, c and d are prime numbers.
यदि $N = a^p b^q c^r d^s$, तब N के भाजकों की संख्या $=(p+1)(q+1)(r+1)(s+1)$
जहाँ, a, b, c और d अभाज्य संख्याएँ हैं।

3. Number of positive factors of $a^p + a^q + a^r + a^s$
$= \{(p+q+r+s) \times 2\} = \{$(Number of frequency of terms of a) $\times 2\}$.
Where, a is any integer.
$a^p + a^q + a^r + a^s$ के धनात्मक गुणनखण्डों की संख्या
$= \{(p+q+r+s) \times 2\} = \{(a$ के पदों की बारम्बारता$) \times 2\}$
जहाँ, a पूर्णांक है।

4. If/यदि $N = a^p \times b^q \times c^r \times .....$ then/तो
Sum of factors/सभी गुणनखण्डों का योग $=(a^0 + a^1 + ... a^p)$
$(b^0 + b^1 + b^q)(c^0 + c^1 + ... c^r)....$
$$= \left(\frac{a^{p+1} - 1}{a-1}\right) \times \left(\frac{b^{q+1} - 1}{b-1}\right)\left(\frac{c^{r+1} - 1}{c-1}\right).....$$

5. If/यदि $N = a^p \times b^q \times c^r \times ....$then/तो
Product of factors/सभी गुणनखण्डों का गुणनफल
$$= N^{\frac{1}{2}(p+1)(q+1)(r+1).....}$$

परीक्षा ब्रह्मास्त्र
गुणनखण्डें की संख्या ज्ञात करना SSC, RRBs एवं सभी परीक्षाओं की दृष्टि से M. Imp. टॉपिक हैं, इन परीक्षाओं में 1 प्रश्न अवश्य पूछा जाता है।

निम्न उदाहरण की सहायता से गुणनखण्डों के योग पर आधारित प्रश्नों को आसानी से हल कर सकते हैं

Example **Sum of factors of 120 is/120 के गुणनखण्डों का योग है।**

Sol. $\because 120 = 2^3 \times 3^1 \times 5^1$

$$\therefore \quad N = \left(\frac{2^{3+1} - 1}{2-1}\right)\left(\frac{3^{1+1} - 1}{3-1}\right)\left(\frac{5^{1+1} - 1}{5-1}\right)$$

$$= 15 \times \frac{8}{2} \times 6 = 360$$

Unit Digit इकाई अंक

To find the unit digit of an expression we have to operate on the unit digits of all the number rather than entire numbers.

किसी व्यंजक के इकाई अंक ज्ञात करने के लिए, हमें केवल सभी संख्याओं के इकाई अंक पर संक्रिया करनी है न कि पूरी संख्याओं पर।

Example **Find the unit digit of $123 \times 255 \times 311$**
$123 \times 255 \times 311$ का इकाई अंक है

Sol. Unit digit of $(123 \times 255 \times 311)$
$= \text{Unit digit of } (3 \times 5 \times 1) = \text{Unit digit of } 15 = 5$

Concept of Cyclicity चक्रीयता की अवधारणा

- After a certain cycle of power the unit digit repeats itself.
  घातों के एक निश्चित चक्र के बाद इकाई अंक स्वयं को दोहराना शुरू करता है।

Number/संख्या	Cyclicity/ चक्रीयता	Power cycle/ घात का चक्र
1	1	1
2	4	2, 4, 8, 6
3	4	3, 9, 7, 1
4	2	4, 6
5	1	5
6	1	6
7	4	7, 9, 3, 1
8	4	8, 4, 2, 6
9	2	9, 1
0	1	0

- The units digit of an expression can be calculated by dividing the power by four and find the unit's digit by remainder.
  किसी व्यंजक के इकाई अंक की गणना घात में 4 से भाग देने पर प्राप्त शेषफल से करते हैं।

इकाई अंक ज्ञात करने पर आधारित उदाहरण निम्नलिखित दिया गया है

Example **Let $x = (433)^{24} - (377)^{38} + (166)^{54}$, then unit digit of x is**

मान लीजिए $x = (433)^{24} - (377)^{38} + (166)^{54}$, तो x का इकाई अंक है।

Sol. $x = (433)^{24} - (377)^{38} + (166)^{54} = (3)^{24} - (7)^{38} + (6)^{54}$
$= (3)^4 - 7^2 + 6^2 = 1 - 9 + 6 = -2 = 10 - 2 = 8$

Conditions to Find Unit Digit इकाई अंक ज्ञात करने की शर्तें

1. (i) $5\times$ odd/विषम $\rightarrow$ Unit digit/इकाई अंक $=5$
   (ii) $5\times$ even/सम $\rightarrow$ Unit digit/इकाई अंक $=0$
2. Any power of $0, 1, 5, 6 \rightarrow$ Unit digit will be same
   0, 1, 5, 6 की कोई भी घात $\rightarrow$ इकाई अंक समान रहेगा।
3. $(4)^{\text{odd/विषम}} \rightarrow$ Unit digit/इकाई अंक $=4$
   $(4)^{\text{even/सम}} \rightarrow$ Unit digit/इकाई अंक $=6$
   $(9)^{\text{odd/विषम}} \rightarrow$ Unit digit/इकाई अंक $=9$
   $(9)^{\text{even/सम}} \rightarrow$ Unit digit/इकाई अंक $=1$
4. Rule of 2, 3, 7, 8/2, 3, 7, 8 का नियम
   Unit digit repeat after four power.
   चार घात के बाद इकाई अंक की पुनरावृत्ति होती है।

Number of Zeros शून्यों की संख्या

If we have expression as $11\times14\times15\times32$
So, expression can be written as, $2^6\times3\times5\times7\times11$
We know that zero can be performed by (2×5).
Hence, number of zero's in expression = (one 5, one 2) = One zero

यदि हमारे पास व्यंजक है $11\times14\times15\times32$
अतः व्यंजक को इस प्रकार लिखा जा सकता है, $2^6\times3\times5\times7\times11$
हम जानते हैं, शून्य 2 और 5 अर्थात् (2×5) के संयोजन से बनाया जा सकता है।
अतः व्यंजक में शून्य की संख्या = (एक 5, एक 2) = एक शून्य

HCF and LCM म.स. और ल.स.

HCF (Highest Common Factor) महत्तम समापवर्तक

HCF is the largest positive integer that divides each of the given integers/म.स. सबसे बड़ा धनात्मक पूर्णांक है, जो दिए गए पूर्णांकों में से प्रत्येक को विभाजित करता है।

Methods to Find HCF म.स. ज्ञात करने की विधियाँ

(i) Factorisation Method/गुणनखण्ड विधि

Take common prime factors of minimum power in all numbers and product of these will be the HCF.

सभी संख्याओं में न्यूनतम घातकों वाले उभयनिष्ठ अभाज्य गुणनखण्डों को लीजिए तथा इनका गुणनफल ही म.स. होगा।

(ii) Difference Method/अन्तर विधि

HCF of two numbers = H/ दो संख्याओं का म.स. = H

Numbers = Hx, Hy/संख्याएँ = Hx, Hy

$x, y \rightarrow$ Co-prime/$x, y \rightarrow$ सह-अभाज्य

$\therefore$ HCF = Factor of difference $H(x-y)$

म. स. = अन्तर $H(x-y)$ का गुणनखण्ड

LCM (Least Common Multiple) लघुत्तम समापवर्तक

LCM of two or more numbers is the least value that is completely divisible by each of the given numbers.

किन्हीं दो या दो से अधिक संख्याओं का लघुत्तम समापवर्त्य वह न्यूनतम मान है, जो दी गई संख्याओं में से प्रत्येक पूर्णतः विभाज्य है।

Method to Find LCM ल.स. ज्ञात करने की विधि

Factorisation Method/गुणनखण्ड विधि

Take prime factor of maximum power in all numbers and their multiplication will be the LCM.

सभी संख्याओं में से अधिकतम घात वाले अभाज्य गुणनखण्डों को लें तथा इनका गुणनफल ही ल.स. होगा।

Some Special Cases Based on HCF and LCM

म.स. और ल.स. पर आधारित कुछ विशेष स्थितियाँ

(i) The smallest number which when divided by a, b, c leaves remainder 'r' in each case

वह छोटी से छोटी संख्या जिसे a, b, c से भाग देने पर प्रत्येक स्थिति में 'r' शेष बचता है

Number/संख्या = LCM/ल.स. $(a, b, c) + r$ *M. Imp.*

(ii) The smallest number which when divided by $a, b, c,$ respectively gives remainder x, y, z such that

वह छोटी से छोटी संख्या जिसे क्रमशः a, b, c से भाग देने पर शेषफल x, y, z प्राप्त होता है

Number/संख्या = LCM/ल.स. $(a, b, c) - d$ *M. Imp.*

Where/जहाँ, $d = a - x = b - y = c - z$

(iii) The greatest number which when divided by x, y and z gives remainder a, b and c, respectively.

वह बड़ी-से-बड़ी संख्या जिसे x, y और z से भाग देने पर शेषफल क्रमशः a, b और c बचे

Number = HCF of $(x-a), (y-b), (z-c)$ *M. Imp.*

संख्या = $(x-a), (y-b), (z-c)$ का म.स.

(iv) The greatest number which when divided by x, y and z gives the same remainder in each case.

वह बड़ी-से-बड़ी संख्या जिसे x, y और z से भाग देने पर प्रत्येक दशा में समान शेष बचें।

Number/संख्या = HCF/म.स. $(|x-y|, |y-z|, |z-x|)$ *M. Imp.*

(v) **Product of two numbers = HCF × LCM**

दो संख्याओं का गुणनफल = म.स. × ल.स.

म.स. व ल.स. पर आधारित उदाहरण निम्नलिखित दिए गए हैं

Example 1 **Find the least number which, when divided by 15, 20, 36 and 48 seperately gives 3 as remainder in each case.**

वह न्यूनतम संख्या ज्ञात कीजिए, जिसे 15, 20, 36 और 48 से अलग-अलग विभाजित करने पर प्रत्येक स्थिति में शेषफल के रूप में 3 प्राप्त होता है। *SSC CGL 2019*

Sol. Required number

$= \text{LCM of } (15, 20, 36, 48) + 3 = 720 + 3 = 723$

Example 2 **Find the least number, which when divided by 15, 25 and 30 gives the remainder 11, 21 and 26, respectively.**

वह छोटी-से-छोटी संख्या क्या होगी, जिसे जब 15, 25 व 30 से भाग दिया जाता है, तो शेषफल क्रमशः 11, 21 व 26 बचता है?

Sol. $d = 15 - 11 = 25 - 21 = 30 - 26 = 4$

Required least number = LCM of (15, 25, 30) $- d = 150 - 4 = 146$

Example 3 **What is the LCM of two number, whose product is 432 and HCF is 6?**

दो संख्याओं का ल. स. क्या है, जिसका गुणनफल 432 और म.स. 6 है? *SSC MTS 2022*

Sol. LCM × HCF = Product of two numbers

$\therefore \quad \text{LCM} = \frac{432}{6} = 72$

HCF and LCM of Fractions भिन्नों का म.स. तथा ल.स.

(i) HCF of fractions/भिन्नों का म.स.

$$= \frac{\text{HCF of Numerators / अंशों का म.स.}}{\text{LCM of Denominators / हरों का ल.स.}}$$

(ii) LCM of fractions/भिन्नों का ल.स.

$$= \frac{\text{LCM of Numerators / अंशों का ल.स.}}{\text{HCF of Denominators / हरों का म.स.}}$$

भिन्नों के ल.स. व म.स. पर आधारित उदाहरण निम्न है

Example 1 **Find the LCM of $\frac{3}{4}, \frac{5}{7}, \frac{10}{9}$.**

$\frac{3}{4}, \frac{5}{7}, \frac{10}{9}$ का लघुत्तम समापवर्त्य ज्ञात कीजिए।

Sol. LCM of fraction $= \frac{\text{LCM of }(3, 5, 10)}{\text{HCF of }(4, 7, 9)} = \frac{30}{1} = 30$

Fraction भिन्न

(i) **When Fraction is Pure Recurring Decimal**
जब भिन्न शुद्ध आवर्त दशमलव हो

Firstly write down the repeated digits only once in numerator and then place as many 9 in the denominator as the number of digits repeating.

सबसे पहले, पुनरावृत्त अंकों को केवल एक बार अंश में लिखते हैं और फिर हर में पुनरावृत्त अंकों की संख्या के समान संख्या में 9 लिखते हैं।

Ex. 0.838383............$0.\overline{83} = \frac{83}{99}$

(ii) **When Fraction is Mixed Recurring Decimal Fraction**
जब भिन्न मिश्रित आवर्त दशमलव हो

In the numerator, take the difference between the number formed by all the digits after decimal point and the number formed by non-repeating digits. In the denominator, place as many 9 as there are repeating digits and after 9 put as many zeros as the number of non-repeating digits

अंश में, दशमलव के बाद आने वाले सभी अंकों से बनी संख्या और उन अंकों से बनी संख्या जिनकी पुनरावृत्ति न हो, का अन्तर लेते हैं। हर में, पुनरावृत्त अंकों की संख्या के समान संख्या में 9 लिखकर उसके बाद बिना पुनरावृत्त अंकों की संख्या के समान संख्या में शून्य लिखते हैं।

Ex. 0.531531531............$0.5\overline{31} = \frac{531-5}{990} = \frac{526}{990}$

आवर्त दशमलव पर आधारित उदाहरण निम्न है

Example **Express $0.4\overline{32}$ in the form of $\frac{p}{q}$, where p and q are integers and $q \neq 0$.**

$0.4\overline{32}$ को $\frac{p}{q}$ के रूप में व्यक्त करें, जहाँ p तथा q पूर्णांक हैं तथा $q \neq 0$.

RRC GROUP D 2022

Sol. $0.4\overline{32} = \frac{432-4}{990} = \frac{428}{990} = \frac{214}{495}$

Simplification सरलीकरण

VBODMAS Rule/नियम

Order to solve/हल करने का क्रम

Do first सबसे पहले
- V → Vinculum/रेखा कोष्ठक (–)
- B → Brackets/कोष्ठक
- O → of/का
- D → Divide/भाग (÷)
- M → Multiply/गुणा (×)
- A → Addition/योग (+)
- S → Subtraction/घटाव (–)

Do last सबसे बाद में

Types of bracket and solving order
कोष्ठक के प्रकार एवं हल करने का क्रम

(i) ( ) → Small bracket (छोटा कोष्ठक)
(ii) { } → Curly bracket (मंझला कोष्ठक)
(iii) [ ] → Big bracket (बड़ा कोष्ठक)

The operation have to be carried out in this order.
संक्रियाओं को इसी क्रम में हल करते हैं

BODMAS नियम पर आधारित उदाहरण निम्नलिखित है

Example 1 The value of expression/व्यंजक का मान है

$$1\frac{2}{5}-\left[3\frac{3}{4}\div\left\{1\frac{1}{4}\div\frac{1}{2}\times\left(1\frac{1}{2}\times3\frac{1}{3}\div1\frac{1}{3}\right)\right\}\right] \text{ is}$$

IBPS CLERK (PRE) 2023

Sol. $\frac{7}{5}-\left[\frac{15}{4}\div\left\{\frac{5}{4}\div\frac{1}{2}\times\left(\frac{3}{2}\times\frac{10}{3}\div\frac{4}{3}\right)\right\}\right]$

$$=\frac{7}{5}-\left[\frac{15}{4}\div\left\{\frac{5}{4}\div\frac{1}{2}\times\left(\frac{3}{2}\times\frac{10}{3}\times\frac{3}{4}\right)\right\}\right]=\frac{7}{5}-\left[\frac{15}{4}\div\left\{\frac{5}{4}\div\frac{1}{2}\times\frac{15}{4}\right\}\right]$$

$$=\frac{7}{5}-\left[\frac{15}{4}\div\left\{\frac{5}{4}\times2\times\frac{15}{4}\right\}\right]=\frac{7}{5}-\left[\frac{15}{4}\div\left\{\frac{75}{8}\right\}\right]=\frac{7}{5}-\left[\frac{15}{4}\times\frac{8}{75}\right]$$

$$=\frac{7}{5}-\frac{2}{5}=\frac{7-2}{5}=\frac{5}{5}=1$$

अध्याय 12

Ratio and Proportion
अनुपात तथा समानुपात

Ratio अनुपात

A relation between two quantities of similar nature is called **ratio**. It written as $a:b$ and read as a ratio b.

समान प्रकृति की दो राशियों के बीच का सम्बन्ध, **अनुपात** कहलाता है। इसे $a:b$ के रूप में लिखा जाता है तथा a अनुपात b पढ़ा जाता है।

Types of Ratio अनुपात के प्रकार

(i) Duplicate ratio of $a:b = a^2:b^2$
$a:b$ का वर्गानुपात $= a^2:b^2$

(ii) The subduplicate ratio of $a:b = \sqrt{a}:\sqrt{b}$
$a:b$ का वर्गमूलानुपात $= \sqrt{a}:\sqrt{b}$

(iii) The triplicate ratio of $a:b = a^3:b^3$
$a:b$ का घनानुपात $= a^3:b^3$

(iv) The subtriplicate ratio of $a:b = \sqrt[3]{a}:\sqrt[3]{b}$
$a:b$ का घनमूलानुपात $= \sqrt[3]{a}:\sqrt[3]{b}$

Comparison of Ratios अनुपातों की तुलना

Let $a:b$ and $c:d$ be two ratios, then
यदि $a:b$ तथा $c:d$ दो अनुपात हैं, तब

(i) If/यदि $ad > bc$, then/तब $a:b > c:d$

(ii) If/यदि $ad < bc$, then/तब $a:b < c:d$

(iii) If/यदि $ad = bc$, then/तब $a:b = c:d$

अनुपातों की तुलना पर आधारित उदाहरण निम्न है

Example **Which is greater between 1 : 3 and 3 : 5?**
1 : 3 तथा 3 : 5 के बीच कौन-सा बड़ा है?

Sol. Here, $a:b = 1:3$, $c:d = 3:5$

$\therefore \quad ad = 5, bc = 9$

$\because \quad ad < bc$, then $1:3 < 3:5$

Based on Equivalent Ratio समतुल्य अनुपात पर आधारित

- If/यदि $A : B = m : n$, $B : C = p : q$

Then/तब, $A : B : C = pm : pn : nq$

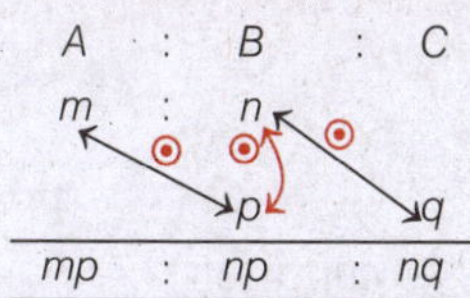

इस नियम पर आधारित अधिकांश परीक्षाओं में प्रश्न पूछे जाते हैं। इसके प्रयोग से प्रश्न को हल करने में समय की बचत होती है।

Note Here, ⊙ implies product/यहाँ, ⊙ गुणनफल को निरूपित करता है।

- If/यदि $A : B = m : n$, $B : C = p : q$, $C : D = r : s$

Then/तब, $A : B : C : D = mpr : npr : nqr : nqs$ M. Imp

$$A : D = \frac{m}{n} \times \frac{p}{q} \times \frac{r}{s}$$

- If/यदि $xA = yB = yB = zC$

Then/तब, $A : B : C = \frac{1}{x} : \frac{1}{y} : \frac{1}{z}$

समतुल्य अनुपात पर आधारित उदाहरण निम्न हैं

Example **If $15x = 24y = 32z$, then $x : y : z$ is equal to**

यदि $15x = 24y = 32z$, तब $x : y : z$ बराबर है *SSC MTS 2023*

Sol. If $15x = 24y = 32z$

Then, $x : y : z = \frac{1}{15} : \frac{1}{24} : \frac{1}{32} = 32 : 20 : 15$

Example **If $A : B = 1.2 : 2$, $B : C = 1.5 : 2$ and $C : D = 2 : 3$, then $B : D$ is equal to**

यदि $A : B = 1.2 : 2$, $B : C = 1.5 : 2$ तथा $C : D = 2 : 3$ है, तब $B : D$ बराबर है

RRB NTPC 2022

Sol.

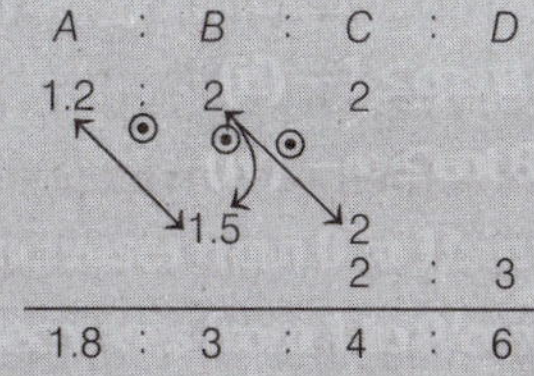

$= 9 : 15 : 20 : 30$

$\therefore \quad B : D = 15 : 30 = 1 : 2$

Division of a Quantity राशि का विभाजन

- If an amount R is to be divided between A and B in the ratio $m:n$, then
यदि एक राशि R को A और B के बीच $m:n$ के अनुपात में विभाजित किया जाना है, तब

(i) Part of A/A का भाग $= \frac{m}{m+n} \times R$

(ii) Part of B/B का भाग $= \frac{n}{m+n} \times R$

(iii) Difference of part A and B/A और B के भागों में अन्तर $= \frac{m-n}{m+n} \times R$

राशि के विभाजन पर आधारित उदाहरण निम्नलिखित है

Example **If 750 is divided into two parts such that the ratio of one to the other is in the ratio 8 : 7, then find the smallest number.**

यदि 750 को दो भागों में इस प्रकार विभाजित किया जाता है कि एक का अनुपात दूसरे से 8 : 7 है, तब सबसे छोटी संख्या ज्ञात कीजिए। ***SSC MTS 2023***

Sol. The smallest number $= \frac{7}{8+7} \times 750 = \frac{7}{15} \times 750 = 350$

Proportion समानुपात

Proportion is the equality of two ratios.
As $a:b = c:d$, we write $a:b::c:d$
Here, a, b, c and d are in proportion.

समानुपात दो अनुपातों की समानता है।
जैसे $a:b = c:d$ हम इसे लिख सकते हैं $a:b::c:d$
यहाँ a, b, c तथा d समानुपात में हैं।

Important Rules Related to Proportion
समानुपात से सम्बन्धित महत्त्वपूर्ण नियम

(i) **Componendo Rule** : If $a:b::c:d$, then $(a+b):b::(c+d):d$

योगानुपात नियम : यदि $a:b::c:d$, तब $(a+b):b::(c+d):d$

(ii) **Dividendo Rule** : If $a:b::c:d$, then $(a-b):b::(c-d):d$

अन्तरानुपात नियम : यदि $a:b::c:d$, तब $(a-b):b::(c-d):d$

(iii) **Componendo and Dividendo Rule** : If $a:b::c:d$, then
$(a+b):(a-b)::(c+d):(c-d)$

योगान्तर अनुपात का नियम : यदि $a:b::c:d$,
तब $(a+b):(a-b)::(c+d):(c-d)$

3. Mean Proportion मध्यानुपाती

- Mean proportional of a and $b = \sqrt{ab}$
  a और b का मध्यानुपाती $= \sqrt{ab}$

Third Proportional and Fourth Proportional तृतीयानुपाती और चतुर्थानुपाती

- Third proportional of a and $b = \frac{b^2}{a}/a$ और b का तृतीयानुपाती $= \frac{b^2}{a}$
- Fourth proportional of a, b and $c = \frac{bc}{a}$. a, b और c का चतुर्थानुपाती $= \frac{bc}{a}$

मध्यानुपाती, तृतीयानुपाती और चतुर्थानुपाती पर आधारित उदाहरण निम्नलिखित हैं

Example **The mean proportion between 7 and 112 is**
7 और 112 का मध्यानुपाती है **SSC CPO 2023**

Sol. Mean proportion between 7 and 112

$$= \sqrt{7 \times 112} = \sqrt{7 \times 7 \times 16} = 28$$

Example **Find the third proportional of 25 and 45.**
25 और 45 का तृतीयानुपाती ज्ञात कीजिए।

Sol. Assume third proportional be x.

$$\therefore \quad 25 : 45 :: 45 : x \Rightarrow x = \frac{45 \times 45}{25} = 81$$

Example **Find the fourth proportional to 0.48, 0.84 and 32.**
0.48, 0.84 तथा 32 का चतुर्थानुपाती ज्ञात कीजिए। **SSC CPO 2023**

Sol. Fourth proportional $(x) = \frac{0.84 \times 32}{0.48} = 56$

Word Problems शब्द समस्या

- Let 'x' be a number which is subtracted from a, b, c and d to make them proportional, then
  यदि 'x' एक संख्या है, जिसे a, b, c और d से घटाकर इन्हें समानुपातिक बनाया जाता है, तब
  $$\frac{a-x}{b-x} = \frac{c-x}{d-x} \Rightarrow x = \frac{ad-bc}{(a+d)-(b+c)}$$
- Let 'x' be a number which is added to a, b, c and d to make them proportional, then
  यदि 'x' एक संख्या है, जिसे a, b, c और d में जोड़कर इन्हें समानुपातिक बनाया जाता है, तब
  $$\frac{a+x}{b+x} = \frac{c+x}{d+x} \Rightarrow x = \frac{bc-ad}{(a+d)-(b+c)}$$

शब्द समस्या पर आधारित उदाहरण निम्नलिखित है

Example **If x is subtracted from each 24, 40, 33 and 57, the numbers, so obtained are in proportion. The ratio of $(5x + 12)$ to $(4x + 15)$ is**

24, 40, 33 तथा 57 प्रत्येक से x घटाने के पश्चात् संख्याएँ समानुपातिक हो जाती हैं। $(5x + 12)$ तथा $(4x + 15)$ का अनुपात है

SSC CGL (TIER-I) 2021

Sol. Here, $a = 24, b = 40, c = 33, d = 57$

$$\therefore \quad x = \frac{(24)(57) - (40)(33)}{(24 + 57) - (40 + 33)} = \frac{1368 - 1320}{81 - 73} = \frac{48}{8} = 6$$

$$\therefore \text{ Required ratio} = \frac{5 \times 6 + 12}{4 \times 6 + 15} = \frac{42}{39} = 14 : 13$$

Coin Problem सिक्के की समस्या

Example **Atul purchased bread costing ₹ 20 and gave a ₹ 100 note to the shopkeeper. The shopkeeper gave the balance money in coins of denomination 2, 5 and 10. If these coins are in the ratio 5 : 4 : 1, then how many ₹ 5 coins did the shopkeeper give?**

अतुल ने ₹ 20 की एक ब्रैड खरीदी और दुकानदार को ₹ 100 का नोट दिया। दुकानदार ने बची हुई धनराशि 2, 5 एवं 10 के सिक्कों के रूप में दी। यदि ये सिक्के 5 : 4 : 1 के अनुपात में हैं, तो ₹ 5 के सिक्कों की संख्या ज्ञात कीजिए।

SSC CGL (TIER-I) 2021

Sol. Denomination ⟶ ₹ 2 : ₹ 5 : ₹ 10

Number of coins ⟶ 5 : 4 : 1

Amount ratio ⟶ 10 : 20 : 10

Total amount = 40 units = ₹ 80 [$\because$ 1 unit = ₹ 2]

Hence, number of ₹ 5 coins = $4 \times 2 = 8$

अध्याय 13

Average औसत

- Average is defined as the sum of the given quantities divided by total number of quantities. Average is also called arithmetic mean.
औसत को दी गई राशियों के योग को राशियों की कुल संख्या से विभाजित करके परिभाषित किया जाता है। औसत को अंकगणितीय माध्य भी कहा जाता है।

$$\text{Average/औसत} = \frac{\text{Sum of all quantities/सभी राशियों का योग}}{\text{Total number of quantities/राशियों की कुल संख्या}}$$

- If the average of n_1 numbers is a_1 and the average of n_2 numbers is a_2, then
यदि n_1 संख्याओं का औसत a_1 तथा n_2 संख्याओं का औसत a_2 है, तब
average of total numbers n_1 and n_2/कुल संख्याओं n_1 और n_2 का औसत

$$= \frac{n_1 a_1 + n_2 a_2}{n_1 + n_2}$$

1. To Find Average by Deviation Method
विचलन विधि से औसत ज्ञात करना

Step I Assume a term which is nearest or among the given data.
एक ऐसा पद मान लें, जो दिए गए आँकड़ों के सबसे निकट या बीच में हो।

Step II Calculate the deviation between the observation and assumed average. (this deviation is the difference between the given terms and assumed average).
अवलोकन और अनुमानित औसत के बीच विचलन की गणना करें (यह विचलन दिए गए पदों और अनुमानित औसत के बीच का अन्तर है।)

Step III To calculate actual average, add or subtract the net deviation in the assumed average.
वास्तविक औसत की गणना करने के लिए, कल्पित औसत में शुद्ध विचलन जोड़े या घटाएँ।

> ***Note*** Net deviation/शुद्ध विचलन
>
> $$= \frac{\text{Total of all the deviations (Step II) सभी विचलनों का कुल योग}}{\text{Total number of observations/सभी विचलनों की कुल संख्या}}$$

विचलन विधि से औसत ज्ञात करने पर आधारित उदाहरण निम्न हैं

Example **The weights (in kg) of five girls of a class are 49, 42, 61, 55 and 58. What is the average weight (in kg) of these five girls?**

एक कक्षा की पाँच लड़कियों का वजन (किग्रा में) 49, 42, 61, 55 और 58 हैं। तो पाँचों लड़कियों का औसत वजन (किग्रा में) क्या है? *SSC CGL 2023*

Sol. Assume average = 55

$\therefore$ Total deviation is

$$\begin{array}{ccccc} 42 & 49 & 55 & 58 & 61 \\ \downarrow & \downarrow & \downarrow & \downarrow & \downarrow \\ -13 & -6 & 0 & 3 & 6 = -10 \end{array}$$

$\therefore$ Required average weight $= 55 + \left(\frac{-10}{5}\right)$

$= 55 - 2 = 53$ kg

Example **Ramesh scored 60, 110, 80 and 140 runs in four matches. What is his average score in the four matches?**

रमेश ने चार मैचों में 60, 110, 80 और 140 रन बनाए। चार मैचों में उसका औसत स्कोर क्या है?

Sol. Assume average = 80

$$\begin{array}{cccc} 60 & 80 & 110 & 140 \\ \downarrow & \downarrow & \downarrow & \downarrow \\ -20 & 0 & +30 & +60 = +70 \end{array}$$

$\therefore$ Actual average $= 80 + \left(\frac{70}{4}\right) = 80 + 17.5 = 97.5$

2. Average of Consecutive Numbers क्रमिक संख्याओं का औसत

- The average of 'n' consecutive natural numbers starting from 1

  1 से शुरू होने वाली 'n' क्रमागत प्राकृतिक संख्याओं का औसत $= \frac{n+1}{2}$

- The average of squares of 'n' consecutive natural numbers starting from 1

  1 से शुरू होने वाली 'n' क्रमागत प्राकृतिक संख्याओं के वर्गों का औसत $= \frac{(n+1)(2n+1)}{6}$

- The average of cubes of first 'n' consecutive natural numbers

  प्रथम n क्रमागत प्राकृत संख्याओं के घनों का औसत $= \frac{n(n+1)^2}{4}$

- The average of first 'n' consecutive even natural numbers *M. Imp.*

  प्रथम 'n' क्रमागत सम प्राकृत संख्याओं का औसत $= (n+1)$

- The average of first 'n' consecutive odd natural numbers

  प्रथम 'n' क्रमागत विषम प्राकृत संख्याओं का औसत $= n$

- The average of consecutive numbers or average of numbers in AP $= \frac{\text{First number + Last number}}{2}$

  क्रमागत संख्याओं का औसत या AP में संख्याओं का औसत $= \frac{\text{पहली संख्या + अन्तिम संख्या}}{2}$

  M. Imp

- The average of first 'n' multiples of certain number 'x'

  किसी संख्या x के पहले n गुणकों का औसत $= \frac{x(n+1)}{2}$

- Average of square of first n even numbers

  पहली n सम संख्याओं के वर्गों का औसत $= \frac{2(n+1)(2n+1)}{3}$

> क्रमागत संख्याओं के औसत के सूत्र सभी परीक्षाओं SSC, Railway, ... के लिए M. Imp हैं। इन सूत्रों का गहनता से अध्ययन करें।

- Average of cube of first n even numbers

  पहली n सम संख्याओं के घन का औसत $= 2n(n+1)^2$

- Average of square of first n odd numbers

  पहली n विषम संख्याओं के वर्गों का औसत $= \frac{(2n+1)(2n-1)}{3}$

- Average of cube of first n odd numbers

  पहली n विषम संख्याओं के घनों का औसत $= n(2n^2-1)$

- Average of 1 to n odd numbers $= \frac{\text{Last odd number} + 1}{2}$

  1 से n तक की विषम संख्याओं का औसत $= \frac{\text{अन्तिम विषम संख्या} + 1}{2}$

- Average of 1 to n even numbers $= \frac{\text{Last even number}}{2} + 1$

  1 से n तक की सम संख्याओं का औसत $= \frac{\text{अन्तिम सम संख्या}}{2} + 1$

क्रमिक संख्याओं के औसत पर आधारित उदाहरण निम्न है

Example **Find the average of first 20 multiples of 9.**

9 के प्रथम 20 गुणजों का औसत ज्ञात कीजिए। ***DP HEAD CONSTABLE 2023***

Sol. Required average $= \frac{9(20+1)}{2} = \frac{189}{2} = 94.5$

3. Average Age, Height, Income (When exclude/include any member)
उम्र, ऊँचाई, आय का औसत (जब कोई सदस्य शामिल/निकाला जाए)

- 't' yr before, the average age of N members of a family was 'x' yr. If during this period 'n' children have increased in the family but average age (present) remains same, then
't' वर्ष पहले, एक परिवार के N सदस्यों की औसत आयु 'x' वर्ष थी। यदि इस अवधि में परिवार में 'n' बच्चे बढ़े परन्तु औसत आयु (वर्तमान) वही रहे, तब
Present age of n children/n बच्चों की वर्तमान आयु $= nx - Nt$

- If in a group of N persons, a new person comes at the place of a person of 'x' yr, so that average age increase by 't' yr. Then,
यदि N व्यक्तियों के समूह में, 'x' वर्ष के व्यक्ति के स्थान पर एक नया व्यक्ति आता है, तो औसत आयु 't' वर्ष बढ़ जाती है, तब
(i) **The age of new person/नए व्यक्ति की आयु** $= x + Nt$
(ii) If the average age decrease by 't' yr after entry of new person, then
यदि एक नए व्यक्ति के प्रवेश के बाद औसत आयु 't' वर्ष कम हो जाती है, तब
The age of the new person /नए व्यक्ति की आयु $= x - Nt$

- The average age of a group of N students is x years. If n students join, the average age of the group increases by 't' yr, then
N छात्रों के एक समूह की औसत आयु 'x' वर्ष है। यदि 'n' छात्र शामिल होते हैं, तो समूह की औसत आयु 't' वर्ष बढ़ जाती है, तब
(i) **Average age of n new students/n नए छात्रों की औसत आयु**
$$= x + \left(\frac{N}{n} + 1\right)t$$
(ii) If the average age of the group decreases by 't' yr, then
यदि समूह की औसत आयु 't' वर्ष कम हो जाती है, तब
Average age of new students/नए छात्रों की औसत आयु
$$= x - \left(\frac{N}{n} + 1\right)t$$

- If the average age (height) of n persons is x yr (cm) and from them m persons went out whose average age (height) is 'y' yr (cm) and same number of persons joined whose average age (height) is 'z' yr (cm), then
यदि 'n' व्यक्तियों की औसत आयु (ऊँचाई) x वर्ष (सेमी) है और उनमें से m व्यक्ति निकले, जिनकी औसत आयु (ऊँचाई) 'y' वर्ष (सेमी) है और उतने ही व्यक्ति शामिल हुए जिनकी औसत आयु (ऊँचाई) 'z' वर्ष (सेमी) है, तब
Average age (height) of n person/n व्यक्तियों की औसत आयु (ऊँचाई)
$$= \frac{nx - my + mz}{n} = \left\{ x - \frac{m(y - z)}{n} \right\} \text{ year/वर्ष (cm/सेमी)}$$

- If a new member is added in a group, then
  यदि किसी समूह में कोई नया सदस्य जोड़ा जाता है, तब
  Age (or income) of added member = Average age (or income) $\pm x(n+1)$
  जोड़े गए सदस्य की आयु (या आय) = औसत आयु (या आय) $\pm x(n+1)$
- If a member leaves the group, then
  Income (or age) of left member = Average income (or age) $\pm x(n-1)$
  Where, x = Increase (+) or decrease (–) in average income (or age)
  n = Number of members.
  यदि कोई सदस्य समूह छोड़ देता है, तब
  छोड़कर गए सदस्य की आय (या आयु) = औसत आय (या आयु) $\pm x(n-1)$
  जहाँ, x = औसत आय (या आयु) में वृद्धि (+) या कमी (–)
  n = सदस्यों की संख्या
- If in a group, one member is replaced by a new member, then
  यदि किसी समूह में एक सदस्य को नए सदस्य से प्रतिस्थापित कर दिया जाए, तब
  Age of new member/ नये सदस्य की आयु
  = (Age of replaced member) $\pm xn$

जब समूह में कोई नया व्यक्ति शामिल किया जाए पर आधारित उदाहरण निम्न हैं

Example **The average age of four players is 18.5 yr. If the age of the coach is also included, the average age increases by 20%. The age of the coach is**
चार खिलाड़ियों की औसत आयु 18.5 वर्ष है। यदि कोच की आयु भी शामिल कर ली जाए, तो औसत आयु 20% बढ़ जाती है। कोच की आयु है

UPSSSC (PET) 2021

Sol. Required average $= 18.5 + (4+1)\left(\frac{20}{100} \times 18.5\right)$

$= 18.5 + 18.5 = 37$ yr

Example **5 yr ago, the average age of P, Q, R and S was 45 yr. After T joined them, the average age of all the five is 48 yr. How old is T?**
5 वर्ष पहले, P, Q, R और S की औसत आयु 45 वर्ष थी। T भी उनके साथ जुड़ने के बाद, सभी पाँचों की औसत आयु 48 वर्ष है। T की आयु कितनी है?

Sol. Actual average $= 45 + 5 = 50$ yr
Age of added member $(T) = 50 - 2(4+1) = 50 - 2(5) = 40$ yr

4. Average of Numbers, when any Number Replacement Increasing/Decreasing संख्याओं का औसत, जब कोई संख्या प्रतिस्थापित, गलत पढ़ना, हटाना/शामिल हो

- If in any series having common difference 'd' and average 'k', 'x' numbers are added in forward or backward, then

  यदि किसी श्रेणी में, जिसका सार्व अन्तर 'd' तथा औसत 'k' हो और इसमें आगे से अथवा पीछे से 'x' संख्याएँ जोड़ें, तब

  New average/नया औसत $= k \pm \frac{xd}{2}$

> **परीक्षा ब्रह्मास्त्र**
> ये सूत्र परीक्षा की दृष्टि से महत्त्वपूर्ण हैं। इन सूत्रों के प्रयोग से समय की बचत होती है।

- In a series of even or odd numbers having average 'k' when we add 'x' numbers in forward or backward, then

  सम अथवा विषम संख्याओं की श्रेणी जिसका औसत 'k' है, उसमें यदि हम आगे अथवा पीछे से 'x' संख्याएँ जोड़ें, तब

  New average/नया औसत $= k \pm x$

- In a series of natural numbers having average 'k', when we add 'x' numbers in forward or backward, then

  'k' औसत वाली प्राकृतिक संख्याओं की किसी श्रेणी में यदि आगे अथवा पीछे से 'x' संख्याएँ जोड़ी जाएँ, तब

  New average/नया औसत $= k \pm \frac{x}{2}$

- If average of 'n' numbers is 'a' but the average becomes 'b' when one number is eliminated, then

  यदि 'n' संख्याओं का औसत 'a' हो परन्तु यदि एक संख्या को विस्थापित कर दिया जाए, तब औसत 'b' हो जाता है।

  Value of eliminated number/विस्थापित संख्या का मान $= n(a - b) + b$

- If average of 'n' numbers is 'a' but the average becomes 'b' when a new number is added, then

  यदि n संख्याओं का औसत 'a' हो, परन्तु यदि एक संख्या को और शामिल कर दिया जाए, तब औसत 'b' हो जाता है, तो

  Value of added number/शामिल की गई संख्या का मान $= n(b - a) + b$

- If we have n numbers, out of which some numbers ($a_1, a_2, a_3 \ldots$) are replaced by some other new numbers and in this way, if the average increase or decrease by 'b', then value of new numbers.

  यदि हमारे पास n संख्याएँ हैं, जिसमें से कुछ संख्याओं ($a_1, a_2, a_3 \ldots$) को कुछ अन्य नई संख्याओं द्वारा प्रतिस्थापित कर दिया जाए तथा इस प्रक्रिया में, यदि औसत में 'b' की वृद्धि अथवा कमी हो जाए, तब

  नई संख्याओं का मान $= a \pm nb$ ('+' increase/वृद्धि, '–' decrease/कमी)

- If the average of 'm' numbers is 'x' and out of these 'm' numbers the average of 'n' numbers is 'y', then the average of remaining numbers will be
यदि 'm' संख्याओं का औसत 'x' है और इन 'm' संख्याओं में से 'n' संख्याओं का औसत 'y' है, तब शेष संख्याओं का औसत

$$\text{Average/औसत} = \frac{mx - ny}{m - n}$$

- If the average of 'n' numbers is 'x' and from these average of first 'm' numbers is 'y' and the average of last 'm' numbers is 'z', then
यदि 'n' संख्याओं का औसत 'x' है और इनमें से पहली 'm' संख्याओं का औसत 'y' है और अन्तिम 'm' संख्याओं का औसत 'z' है, तो

$$m\text{th number/}m\text{वीं संख्या} = m(y + z) - nx$$

$$(m + 1)\text{th number/}(m + 1)\text{वीं संख्या} = nx - m(y + z)$$

- If average of 'n' numbers is 'm' but later on it was found that a number 'a' was misread as 'b'.
यदि 'n' संख्याओं का औसत 'm' है, परन्तु बाद में यह पाया गया कि एक संख्या 'a' को गलती से 'b' पढ़ लिया गया था।

$$\text{Actual average/वास्तविक औसत} = m + \frac{(a - b)}{n}$$

- If the average of 'n' numbers is 'm' but later on it was found that two numbers 'a' and 'b' misread as 'p' and 'q'. Then,
यदि 'n' संख्याओं का औसत 'm' है, परन्तु बाद में यह पाया गया कि दो संख्याएँ 'a' और 'b' को गलती से 'p' और 'q' पढ़ लिया गया था। तब,

$$\text{The correct average/सही औसत} = m + \frac{(a + b - p - q)}{n}$$

संख्याओं का औसत गलत पढ़ने, हटाने पर आधारित उदाहरण निम्न हैं

Example **Average of 15 numbers is 53. A number is removed and the average of 14 numbers now becomes 55. Which number was removed?**
15 संख्याओं का औसत 53 है। यदि एक संख्या हटा दी जाए, तो 14 संख्याओं का औसत 55 हो जाता है। कौन-सी संख्या हटाई गई? *SSC (10+2) 2023*

Sol. Required number $= 53 - 14(55 - 53) = 53 - 28 = 25$

Example **The average of 36 numbers was found to be 45. Later, it was detected that 84 was misread as 48. Find the correct average of the given number.**
36 संख्याओं का औसत 45 पाया गया। बाद में पता चला कि 84 को गलती से 48 पढ़ लिया गया था। दी गई संख्या का सही औसत ज्ञात कीजिए। *SSC CGL 2023*

Sol. $\therefore$ Actual average $= 45 + \left(\frac{84 - 48}{36}\right) = 45 + 1 = 46$

अध्याय 14

Percentage
प्रतिशतता

- The meaning of the term per cent is per hundred or hundredth part.
  प्रतिशत का अर्थ है प्रति सौ या सौवाँ भाग।
  It is denoted by symbol % *e.g.*, $68\% = \frac{68}{100}$
  इसे चिह्न % द्वारा दर्शाया जाता है। जैसे– $68\% = \frac{68}{100}$

Conversion Table परिवर्तन तालिका

भिन्न (Fraction)	प्रतिशत (Per cent)	भिन्न (Fraction)	प्रतिशत (Per cent)
1	100%	$\frac{1}{13}$	7.69% or $7\frac{9}{13}\%$
$\frac{1}{2}$	50%	$\frac{1}{14}$	7.14% or $7\frac{1}{7}\%$
$\frac{1}{3}$	33.33% or $33\frac{1}{3}\%$	$\frac{1}{15}$	6.66% or $6\frac{2}{3}\%$
$\frac{1}{4}$	25%	$\frac{1}{16}$	6.25% or $6\frac{1}{4}\%$
$\frac{1}{5}$	20%	$\frac{1}{17}$	5.88% or $5\frac{15}{17}\%$
$\frac{1}{6}$	16.66% or $16\frac{2}{3}\%$	$\frac{1}{18}$	5.55% or $5\frac{5}{9}\%$
$\frac{1}{7}$	14.28% or $14\frac{2}{7}\%$	$\frac{1}{19}$	5.26% or $5\frac{5}{19}\%$
$\frac{1}{8}$	12.5% or $12\frac{1}{2}\%$	$\frac{1}{20}$	5%
$\frac{1}{9}$	11.11% or $11\frac{1}{9}\%$	$\frac{1}{25}$	4%
$\frac{1}{10}$	10%	$\frac{1}{40}$	2.5% or $2\frac{1}{2}\%$
$\frac{1}{11}$	9.09% or $9\frac{1}{11}\%$	$\frac{1}{80}$	1.25% or $1\frac{1}{4}\%$
$\frac{1}{12}$	8.33% or $8\frac{1}{3}\%$	$\frac{1}{100}$	1%

Conversion of Compound Percentage Fraction to Simplest Fraction

मिश्रित प्रतिशत भिन्न का साधारण भिन्न में परिवर्तन

- $108\frac{1}{3}\% \Rightarrow 100\% + 8\frac{1}{3}\% = 1 + \frac{1}{12} = \frac{13}{12}$
- $116\frac{2}{3}\% \Rightarrow 100\% + 16\frac{2}{3}\% = 1 + \frac{1}{6} = \frac{7}{6}$
- $305\frac{5}{9}\% \Rightarrow 300\% + 5\frac{5}{9}\% = 3 + \frac{1}{18} = \frac{55}{18}$
- $504\frac{16}{21}\% \Rightarrow 500\% + 4\frac{16}{21}\% = 5 + \frac{1}{21} = \frac{106}{21}$

1. Basic Concept मूल अवधारणा

प्रतिशत की मूल अवधारणा M. Imp है। इस पर आधारित Basic प्रश्न सभी परीक्षाओं में पूछे जाते हैं।

- $x\%$ of $y = \frac{xy}{100}$ / y का $x\% = \frac{xy}{100}$
- To find x is what per cent less or more than y.

  x, y से कितना प्रतिशत (%) कम या ज्यादा है ज्ञात करना।

  Required per cent /अभीष्ट प्रतिशत $= \frac{|(y-x)|}{y} \times 100\%$

मूल अवधारणा पर आधारित उदाहरण निम्न हैं

Example **The value of 75% is equal to**/75% का मान बराबर है

Sol. $75\% = \frac{75}{100} = \frac{3}{4}$

Example **Find 80 is what per cent less than 120?**

ज्ञात कीजिए 80, 120 से कितना प्रतिशत कम है?

Sol. Required % $= \left(\frac{120-80}{120}\right) \times 100 = \frac{40}{120} \times 100\% = 33.33\%$

2. Percentage Increase/Decrease of Quantity मात्रा में प्रतिशत वृद्धि/कमी

- If an amount is increased by $x\%$ and then it is reduced by $x\%$ again, then percentage change will be decrease of $\frac{x^2}{100}\%$.

  यदि किसी राशि में $x\%$ की वृद्धि की जाती है और फिर इसे $x\%$ घटाया जाता है, तब इसमें $\frac{x^2}{100}\%$ की कमी होगी।

- If a number is increased/decreased by $a\%$ and then it is increased/decreased by $b\%$, then resultant change in percentage will be

  यदि एक संख्या में $a\%$ की वृद्धि/कमी की जाती है और फिर इसमें $b\%$ की वृद्धि/कमी की जाती है, तब इसमें परिणामी परिवर्तन $= \left[\pm a \pm b + \frac{(\pm a)(\pm b)}{100}\right]\%$

  [Where/जहाँ, '+' → Increase/वृद्धि and/और '–' → Decrease/कमी]

मात्रा में वृद्धि/कमी पर आधारित उदाहरण निम्नलिखित हैं

Example **A number when decreased by $22\frac{1}{2}$% becomes 217. The number is**

एक संख्या $22\frac{1}{2}$% की कमी के बाद 217 हो जाती है। संख्या है

Sol. As, $22\frac{1}{2}\% = \frac{45}{200} = \frac{9}{40}$ and $(40-9)$ units $\rightarrow 217$

$\therefore \quad 40 \rightarrow \frac{217}{31} \times 40 = 280$

Example **The income of a worker is first increased by 5% and then it is decreased by 5%. What is the per cent change in his income?**

एक कार्यकर्त्ता की आय में पहले 5% की वृद्धि होती है और फिर 5% की कमी होती है। उसकी आय में कितने प्रतिशत परिवर्तन हुआ?

Sol. Decreased per cent $= \left(\frac{5}{10}\right)^2\% = \frac{25}{100}\% = 0.25\%$

Example **The price of an article is first increased by 20% and later on the price were decreased by 25% due to reduction in sales. Find the net per cent change in final price of the article.**

एक वस्तु की कीमत में पहले 20% की वृद्धि होती है और बाद में बिक्री में कमी से कीमत में 25% की कमी होती है। वस्तु के अन्तिम मूल्य में शुद्ध प्रतिशत परिवर्तन ज्ञात कीजिए।

Sol. Required change $= \left[20 - 25 + \frac{(20)(-25)}{100}\right]\%$

$= [-5 - 5]\% = -10\%$

$=$ 10% decrease

3. If Two Quantities are More or Less Than Third Quantity

यदि दो राशि तीसरी राशि से ज्यादा या कम हो

- If two numbers are x% and y% less than the third number, respectively. Then, first number as a percentage of second

  यदि दो संख्याएँ तीसरी संख्या से क्रमशः x% और y% कम हैं, तब दूसरी संख्या के प्रतिशत के रूप में पहली संख्या $= \left(\frac{100-x}{100-y}\right) \times 100\%$

- If two numbers are x% and y% more than the third number, respectively. Then, the first number as percentage of second

  यदि दो संख्याएँ तीसरी संख्या से क्रमशः x% और y% अधिक हैं, तो दूसरी संख्या के प्रतिशत के रूप में पहली संख्या $= \left(\frac{100+x}{100+y}\right) \times 100\%$

यदि दो राशि तीसरी राशि से ज्यादा या कम हो पर आधारित उदाहरण निम्नलिखित है

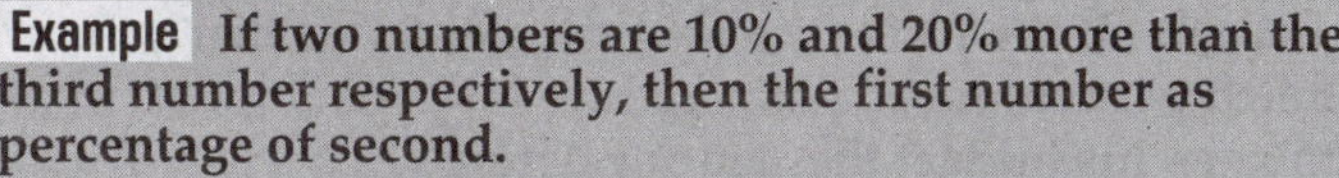

Example **If two numbers are 10% and 20% more than the third number respectively, then the first number as percentage of second.**

यदि दो संख्याएँ तीसरी संख्या से क्रमशः 10% तथा 20% अधिक है, तब पहली संख्या का दूसरी संख्या से प्रतिशत ज्ञात कीजिए।

Sol. Required per cent $= \left(\dfrac{100+10}{100+20}\right) \times 100\% = \dfrac{110}{120} \times 100\% = 91.66\%$

4. Based on Comparison of Two or More Quantities
दो या दो से अधिक राशियों की तुलना पर आधारित

- If the cost of an article is increased by $x\%$, then how much consumption of article, to decrease so that expenditure remains same is given by
यदि किसी वस्तु की कीमत में $x\%$ की वृद्धि की जाती है, तो वस्तु की खपत को कितना कम किया जाए, ताकि व्यय समान रहे

परीक्षा ब्रह्मास्त्र

दो या दो से अधिक राशियों की तुलना पर आधारित प्रश्न *SSC, Railway, UPSC (CSAT)...* परीक्षाओं में अवश्य पूछे जाते हैं।

If the income of a man is $x\%$ more than the income of an another man, then income of another man is less in comparison to the first man.
यदि एक आदमी की आय दूसरे आदमी की तुलना में $x\%$ अधिक है, तब दूसरे आदमी की आय पहले आदमी की तुलना में कम है $= \left(\dfrac{x}{100+x} \times 100\right)\%$

- If the cost of an article is decreased by $x\%$, then the increase in consumption of article to maintain the expenditure will be
यदि किसी वस्तु की लागत में $x\%$ की कमी की जाती है, तो व्यय को समान रखने के लिए वस्तु की खपत में वृद्धि होगी

or/या

If A is $x\%$ less than B, then B is more than A
यदि A, B से $x\%$ कम है, तब B, A से अधिक है $= \left(\dfrac{x}{100-x} \times 100\right)\%$

- On increasing/decreasing the cost of a certain article by $x\%$, a person can buy a kg article less/more in ₹y, then
किसी वस्तु के मूल्य में $x\%$ की वृद्धि/कमी करने पर, एक व्यक्ति ₹y में a किग्रा वस्तु कम/अधिक खरीद सकता है, तब
Increased/decreased cost of the article

$$\text{वस्तु की लागत में वृद्धि/कमी} = \frac{xy}{100 \times a}$$

and initial cost/और प्रारम्भिक लागत $= \dfrac{xy}{(100 \pm x)a}$

दो या दो से अधिक राशियों की तुलना पर आधारित उदाहरण निम्नलिखित हैं

Example **If the price of wheat rises by 25%, then by how much per cent must a man reduce his consumption in order to keep his budget the same as before ?**

यदि आटे का मूल्य 25% बढ़ जाता है, तब एक व्यक्ति को अपनी खपत में कितने प्रतिशत की कमी करनी चाहिए, जिससे उसका बजट पहले के समान हो? *CDS 2018*

Sol. Required percentage $= \frac{25}{125} \times 100\% = 20\%$

Example **If the income of A is 10% more than that of B and the income of B is 20% less than that of C, then ratio of income of A, B and C, respectively.**

यदि A की आय, B की आय से 10% अधिक है तथा B की आय C की आय से 20% कम है, तब A, B और C की आय का अनुपात क्रमशः होगा

HSSC CET 2022

Sol.

A	B	C (Assumed)
88	80	100

(B → A: +10%; C → B: –20%)

$\therefore$ Required ratio $= 88 : 80 : 100$

$= 22 : 20 : 25$

5. Population or Cost जनसंख्या या लागत

- If the population/cost of a certain town/article is P and annual increment rate is $r\%$, then
  यदि किसी कस्बे/वस्तु की जनसंख्या/लागत P है और वार्षिक वृद्धि दर $r\%$ है, तो
  (i) After t yr population/cost
  t वर्षों के बाद जनसंख्या/लागत $= P\left(1 + \frac{r}{100}\right)^t$
  (ii) Before t yr population/cost
  t वर्ष पहले जनसंख्या/लागत $= \frac{P}{\left(1 + \frac{r}{100}\right)^t}$

अतिसम्भावित!
जनसंख्या/लागत टॉपिक (SSC, Railway...) परीक्षाओं की दृष्टि से M. Imp है। इस पर 1 प्रश्न अवश्य पूछा जाता है।

- If the population of a town/cost of article is P and it decreases/reduces at the rate of $r\%$ annually, then
  यदि किसी कस्बे की जनसंख्या/वस्तु की लागत P है और यह $r\%$ वार्षिक दर से घटती है, तब
  (i) After t yr population/cost/ t वर्षों के बाद जनसंख्या/लागत $= P\left(1 - \frac{r}{100}\right)^t$
  (ii) Before t yr population/cost/t वर्ष पहले जनसंख्या/लागत $= \frac{P}{\left(1 - \frac{r}{100}\right)^t}$

- If the present population of a town is P and the population increases or decreases at the rate of R_1 %, R_2 % and R_3 % in first, second and third year, respectively. Then, the population of town after 3 yr
यदि किसी कस्बे की वर्तमान जनसंख्या P है और जनसंख्या क्रमशः पहले, दूसरे और तीसरे वर्ष में R_1 %, R_2 % और R_3 % की दर से बढ़ती या घटती है। तब, 3 वर्ष बाद शहर की जनसंख्या

$$= P\left(1 \pm \frac{R_1}{100}\right)\left(1 \pm \frac{R_2}{100}\right)\left(1 \pm \frac{R_3}{100}\right)$$

Note '+' → Increase/वृद्धि, '–' → Decrease/कमी

जनसंख्या/लागत पर आधारित उदाहरण निम्न हैं

Example **The population of a town is 45000. If it is inreasing at the rate of 10% every year. Then, population of city after 2 yr is**
एक शहर की जनसंख्या 45000 है। यदि यह प्रतिवर्ष 10% की दर से बढ़ रही है, तो शहर की जनसंख्या दो वर्ष बाद कितनी होगी?

Sol. $\therefore$ Population after 2 yr $= 45000 \times \left(1 + \frac{10}{100}\right)^2$

$$= 45000 \times \frac{11}{10} \times \frac{11}{10} = \boxed{54450}$$

Example **The population of a city is 352314. It increases annually at the rate of 15% p.a. What was its population 2 yr ago?** **SSC CPO 2023**
एक शहर की जनसंख्या 352314 है। यह प्रतिवर्ष 15% की दर से बढ़ती है। दो वर्ष पूर्व जनसंख्या क्या थी?

Sol. Required population $= \frac{352314}{\left(1 + \frac{15}{100}\right)^2} = \frac{352314 \times 20 \times 20}{23 \times 23} = 266400$

Alternate Method
Assume population 2 yr ago = 100

A	I yr	II yr
100	115	132.25
	+15%	+15%

$\because$ 132.25 → 352314

$\therefore$ 100 → $\boxed{266400}$

6. Election चुनाव

- If a candidate got a% votes in a poll and he won or defeated by x votes, then total number of votes which was casted in poll
यदि एक उम्मीदवार को एक मतदान में a% वोट मिले और वह x वोटों से जीता या पराजित हुआ, तब मतदान में डाले गए वोटों की कुल संख्या $= \frac{50x}{100 - a}$ *M. Imp.*

Example **If a candidate got 70% votes in a poll and he won by 1200 votes, then total number of votes which was casted in poll?**
यदि एक उम्मीदवार को 70% मत मिले और वह 1200 मतों से जीतता है, तब डाले गए कुल मतों की संख्या क्या थी?

Sol. Total number of votes $= \dfrac{50 \times 1200}{(100-70)} = \dfrac{50 \times 1200}{30} = 2000$

Example **In the election, 8% of the votes did not cast their votes. Two candidates contested the election in which the winning candidate got 48% of the total votes and won election by 1100 votes. Find the total number of voters in the election. (Given that all the votes casted were valid)**
एक चुनाव में 8% मतदाताओं ने मतदान नहीं किया। दो उम्मीदवारों के मध्य चुनाव लड़ा गया, जिनमें विजेता उम्मीदवार को कुल मतों का 48% मत मिला और वह चुनाव 1100 मतों से जीत जाता है। चुनाव में मतों की संख्या ज्ञात कीजिए। (दिया है सभी डाले गए मत वैध हैं) *SSC CGL (TIER-I) 2023*

Sol. Total votes (100)
↓
92
Winning candidate ↙ ↘ Losing candidate
48 44

$\because \quad (48-44)\% \rightarrow 1100$

$\Rightarrow \quad 100\% \rightarrow \dfrac{1100}{4} \times 100 = \boxed{27500}$

7. Examination परीक्षा

- In an examination, a% candidates failed in Maths and b% candidates failed in English. If c% candidates failed in both the subjects, then
एक परीक्षा में, a% अभ्यर्थी गणित में तथा b% अभ्यर्थी अंग्रेजी में अनुत्तीर्ण हुए। यदि c% अभ्यर्थी दोनों विषय में अनुत्तीर्ण हैं, तो

  (i) Passed candidates in both the subjects M. Imp.
  दोनों विषय में उत्तीर्ण अभ्यर्थी $= [100-(a+b-c)]\%$

  (ii) Percentage of candidates who failed in either subject
  किसी भी विषय में असफल होने वाले अभ्यर्थियों का प्रतिशत $= (a+b-c)\%$

- In a certain examination, passing marks are a%. If any candidate obtains 'b' marks and fails by 'c' marks, then/ एक निश्चित परीक्षा में उत्तीर्ण अंक a% हैं। यदि कोई अभ्यर्थी 'b' अंक प्राप्त करता है और 'c' अंक से अनुत्तीर्ण हो जाता है, तब

  Total marks/कुल अंक $= \dfrac{100\,(b+c)}{a}$

Example **A student required 20% marks to pass in Psychology. He secured 20 marks and failed by 5 marks. What is the maximum marks? / एक छात्र को मनोविज्ञान में उत्तीर्ण होने के लिए 20% अंकों की आवश्यकता है। वह 20 अंक लाता है और 5 अंकों से फेल हो जाता है। अधिकतम अंक क्या हैं?** *RRB NTPC 2021*

Sol. Maximum marks $= \dfrac{(20+5) \times 100}{20} = 125$

अध्याय 15

Mixture मिश्रण

When two or more components are mixed in a certain ratio, resultant component is called **mixture**.

जब दो या दो से अधिक घटकों को एक निश्चित अनुपात में मिलाया जाता है, तब परिणामी घटक को **मिश्रण** कहा जाता है।

1. Based on Increasing the Quantity of One Substance

एक पदार्थ की मात्रा में वृद्धि करने पर आधारित

- There is $x\%$ milk in 'a' unit mixture of milk and water. The amount of milk that should be added to increase the percentage of milk from $x\%$ to $y\%$ is given by

  दूध और पानी के 'a' इकाई मिश्रण में $x\%$ दूध है। दूध की मात्रा, जो दूध के प्रतिशत को $x\%$ से $y\%$ तक बढ़ाने के लिए मिलाई जानी चाहिए।

  Quantity of milk/दूध की मात्रा $= \dfrac{a(y-x)}{100-y}$ *M. Imp.*

- The amount of acid/milk is $x\%$ in a units mixture. How much water should be mixed in it so that percentage amount of acid/milk would be $y\%$.

  'a' इकाई मिश्रण में अम्ल/दूध की मात्रा $x\%$ है। इसमें कितना पानी मिलाया जाए, कि अम्ल/दूध की प्रतिशत मात्रा $y\%$ हो जाए।

  Amount of water/पानी की मात्रा $= \dfrac{a(x-y)}{y}$ *M. Imp.*

Example **If 60 L of a milk solution has 40% milk in it, then how much milk should be added to make the milk 60% in the solution?**

यदि 60 लीटर दूध के एक मिश्रण में 40% दूध है, तो मिश्रण में दूध की मात्रा 60% करने के लिए इसमें और कितना दूध मिलाना चाहिए? *SSC CPO 2023*

Sol. Here, $a = 60$ L; $x = 40\%$; $y = 60\%$

$$\text{Quantity of milk} = \frac{a(y-x)}{100-y} = \frac{60 \times (60-40)}{(100-60)}$$

$$= \frac{60 \times 20}{40} = 30 \text{ L}$$

Example **20 L of mixture contains 60% alcohol. How many liters of water should be added, so that the mixture obtained contains 30% alcohol?**

20 लीटर मिश्रण में 60% एल्कोहॉल है। कितना लीटर पानी और मिलाया जाए कि प्राप्त मिश्रण में 30% एल्कोहॉल हो?

Sol. Here, $x = 60\%$, $y = 30\%$, $a = 20$ L

$$\text{Amount of water} = \frac{a(x-y)}{y} = \frac{20 \times (60-30)}{30} = \frac{20 \times 30}{30} = 20\text{ L}$$

2. Based on Replacing Some Quantity of Mixture

मिश्रण की कुछ मात्रा का प्रतिस्थापन करने पर आधारित

- If from x L of liquid A, p L is withdrawn and same quantity of liquid B is added. Again from mixture, q L mixture is withdrawn and same quantity of liquid B is added. Again from mixture, r L mixture is withdrawn and same quantity of liquid B is added, then in final mixture, quantity of liquid A

  यदि x लीटर द्रव A में से p लीटर निकाल लिया जाता है और उतनी ही मात्रा में द्रव B मिला दिया जाता है। पुनः मिश्रण से, q लीटर मिश्रण निकालकर समान मात्रा में तरल B मिला दिया जाता है। पुनः मिश्रण से, r लीटर मिश्रण निकालकर समान मात्रा में तरल B मिलाया जाता है। तब, अन्तिम मिश्रण में द्रव A की मात्रा $= x\left(\frac{x-p}{x}\right)\left(\frac{x-q}{x}\right)\left(\frac{x-r}{x}\right)\ldots$

- If only one process is repeated n times, then
  यदि केवल एक प्रक्रिया को n बार दोहराया जाता है, तब

A in final mixture/अन्तिम मिश्रण में $A = x\left(1 - \frac{p}{x}\right)^n$

यह सूत्र UPSC, (CSAT), SSC, Railway परीक्षाओं के लिए M. Imp. है। इस पर 1 प्रश्न पूछा जाता है

Example **A bottle contains 20 L of liquid A. 4 L of liquid A is taken out of it and replaced by same quantity of liquid B. Again, 4 L of the mixture is taken out and replaced by same quantity of liquid B. What is the ratio of quantity of liquid A to that of liquid B in the final mixture?**

एक बोतल में द्रव A की 20 लीटर मात्रा है। इसमें से द्रव A की 4 लीटर मात्रा बाहर निकाल दी जाती है और द्रव B की उतनी ही मात्रा वापस डाल दी जाती है। फिर से मिश्रण की 4 लीटर मात्रा बाहर निकाल दी जाती है और द्रव B की उतनी ही मात्रा वापस डाल दी जाती है। अन्तिम मिश्रण में द्रव A की मात्रा का द्रव B की मात्रा से अनुपात क्या है?

UPSC CSAT 2020

Sol. Given, $x = 20$ L; $p = 4$ L; $n = 2$

$$\therefore A \text{ in final mixture} = x\left(1 - \frac{p}{x}\right)^n = 20\left(1 - \frac{4}{20}\right)^2 = 20 \times \frac{4}{5} \times \frac{4}{5} = \frac{64}{5}$$

$$B \text{ in final mixture} = 20 - \frac{64}{5} = \frac{36}{5}$$

$$\therefore \text{The ratio of } A \text{ and } B \text{ in final mixture} = \frac{64}{5} : \frac{36}{5} = 16 : 9$$

3. Based on Vapourisation वाष्पीकरण पर आधारित

- There is $x\%$ water in 'a' units mixture of sugar and water. The quantity of water vapourised, such that decrease in the percentage of water is from $x\%$ to $y\%$.

  चीनी और पानी के 'a' इकाई मिश्रण में $x\%$ चीनी है। वाष्पीकृत पानी की मात्रा, जिससे चीनी का प्रतिशत $x\%$ से $y\%$ हो जाता है।

$$\text{Quantity/मात्रा} = \frac{a(x \sim y)}{y}$$

Example **What quantity of water should be taken out (vapourised) from 29 L of liquid sugar, so that percentage of sugar becomes 17% to 29%?**

29 लीटर तरल चीनी से कितनी मात्रा में पानी निकाला जाना चाहिए (वाष्पीकृत हो जाए), ताकि चीनी का प्रतिशत 17% से 29% हो जाए?

Sol. According to the question, $a = 29\text{ L};\ x = 17\%;\ y = 29\%$

$$\therefore \quad \text{Quantity} = \frac{a(x \sim y)}{y} = \frac{29 \times (17 \sim 29)}{29} = \frac{29 \times 12}{29} = 12\text{ L}$$

4. Based on Obtaining a New Mixture by Mixing Two or More Mixtures
दो या अधिक मिश्रणों को मिलाकर नया मिश्रण प्राप्त करने पर आधारित

Example **Milk and water in two vessels A and B are in the ratio 5 : 3 and 2 : 3, respectively. In what ratio should the liquids in both the vessels be mixed to obtain a new mixture in vessel C containing half milk and half water?**

A और B दो बर्तनों में दूध और पानी क्रमशः 5 : 3 और 2 : 3 के अनुपात में हैं। बर्तन C में आधा दूध और आधा पानी वाला एक नया मिश्रण प्राप्त करने के लिए दोनों बर्तनों के द्रवों को किस अनुपात में मिलाया जाना चाहिए? *SSC (10+2) 2023 (SHIFT-I)*

Sol. According to the question,

Vessel (A) — Milk — $\frac{5}{8}$ Vessel (B) — Milk — $\frac{2}{5}$

Vessel → (C) — Milk — $\frac{1}{2}$

$$\left(\frac{1}{2} - \frac{2}{5}\right) : \left(\frac{5}{8} - \frac{1}{2}\right)$$

$$\left(\frac{5-4}{10}\right) : \left(\frac{5-4}{8}\right)$$

$$\frac{1}{10} : \frac{1}{8}$$

The ratio of vessels A and $B = \frac{1}{10} : \frac{1}{8} = 4 : 5$

5. Mixing Different Mixtures to Form a Single Mixture

विभिन्न मिश्रणों को मिश्रित करके एकल मिश्रण तैयार करना

Example The capacities of three containers P, Q and R are in the proportion of 5 : 4 : 8. They are completely filled with a mixture of milk and water in the proportions of 1 : 4, 2 : 3 and 3 : 5, respectively. If the mixtures of all the three containers are mixed together, then what will be the proportion of milk to water in the final mixture?

तीन कण्टेनरों P, Q और R की क्षमता 5 : 4 : 8 के अनुपात में हैं। वे क्रमशः 1 : 4, 2 : 3 और 3 : 5 के अनुपात में दूध और पानी के मिश्रण से पूरी तरह से भरे हुए हैं। यदि तीनों कण्टेनरों के मिश्रण को एकसाथ मिला दिया जाता है, तो अन्तिम मिश्रण में दूध और पानी का अनुपात कितना होगा? *SSC MTS 2021*

Sol. According to the question,

After mixing the all three containers,

$$\text{Milk} = \frac{1}{5} \times 5 + \frac{2}{5} \times 4 + \frac{3}{8} \times 8 = 4 + \frac{8}{5} = \frac{28}{5}$$

$$\text{Water} = \frac{4}{5} \times 5 + \frac{3}{5} \times 4 + \frac{5}{8} \times 8 = 9 + \frac{12}{5} = \frac{57}{5}$$

Now, $\text{Milk} : \text{Water} = \frac{28}{5} : \frac{57}{5} = 28 : 57$

6. Based on Mixing Two Substances of Different Price

दो भिन्न मूल्यों वाले पदार्थों को मिलाने पर आधारित

- If a new mixture is prepared by mixing a cheaper object and dearer object, then

  यदि सस्ती वस्तु और महँगी वस्तु को मिलाकर एक नया मिश्रण तैयार किया जाता है, तब

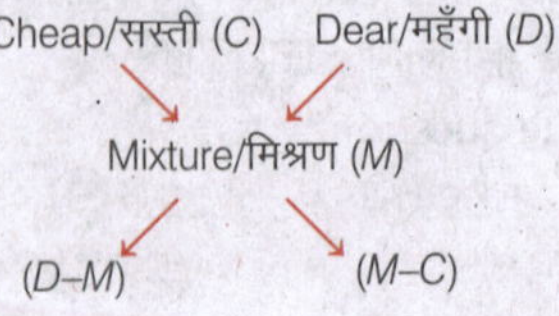

$$\frac{\text{Quantity of cheap object/ सस्ती वस्तु की मात्रा}}{\text{Quantity of dear object/ महँगी वस्तु की मात्रा}} = \frac{D - M}{M - C}$$ M. Imp.

Here, $D \rightarrow$ Cost price of dear object/महँगी वस्तु का क्रय मूल्य

$C \rightarrow$ Cost price of cheap object/सस्ती वस्तु का क्रय मूल्य

$M \rightarrow$ Cost of mixture/मिश्रण का मूल्य

Example **How much chicory, costing ₹ 100/kg should be added to 40 kg of coffee costing ₹ 45/kg, so that the mixture will be worth ₹ 60/kg?**

₹ 100/किग्रा मूल्य वाली कितनी चिकोरी को ₹ 45/किग्रा मूल्य वाली 40 किग्रा कॉफी में मिलाई जानी चाहिए, ताकि मिश्रण का मूल्य ₹ 60/किग्रा हो जाए? *SSC (10+2) 2023*

Sol. According to the question,

Price ① ₹100/kg   Price ② ₹45/kg

₹60/kg

15 : 40

3 : 8

$\because \quad 8 \text{ units} = 40\,\text{kg} \Rightarrow 3 \text{ units} = \frac{40 \times 3}{8} = \boxed{15\,\text{kg}}$

Example **A grocery seller mixes two varieties of tea costing ₹ 300 per kg and ₹ 160 per kg in such a ratio that by selling this mixture at the rate of ₹ 187 per kg the shopkeeper earns 10% profit. In what respective ratio does the grocery seller mixes the two varieties of tea?**

एक किराना विक्रेता ₹ 300 प्रति किग्रा और ₹ 160 प्रति किग्रा की दो किस्मों की चाय को इस अनुपात में मिलाता है कि मिश्रण को ₹ 187 प्रति किग्रा की दर से बेचने पर दुकानदार 10% का लाभ अर्जित करता है। किराना विक्रेता दोनों किस्म की चाय को क्रमशः किस अनुपात में मिलाता है? *SSC MTS 2023*

Sol. According to the question,

$\text{CP of mixture} = \frac{187 \times 100}{110} = ₹\ 170/\text{kg}$

Apply Alligation Method,

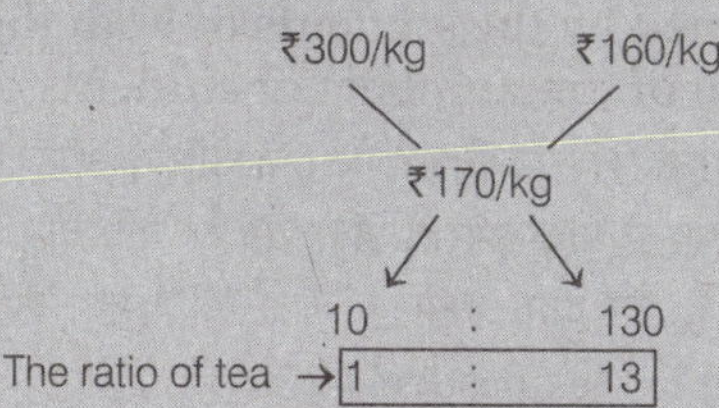

7. Based on Percentage Increase/Decrease प्रतिशत वृद्धि/कमी पर आधारित

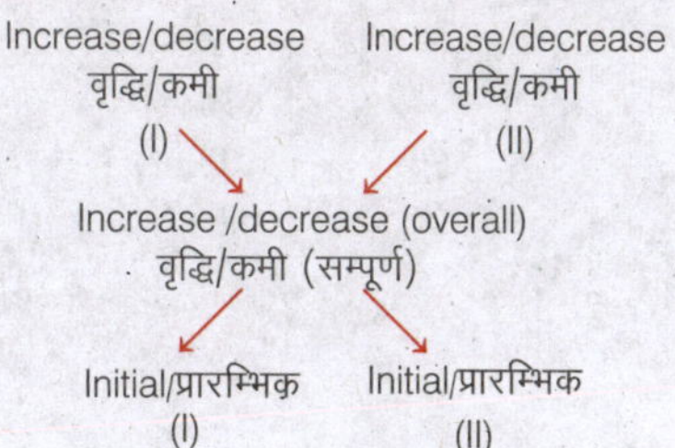

Example **If the income of a family is increased by 20% and expenditure is increased by 10% and its saving is increased by 50%. Find the ratio of income and expenditure of that family.**

यदि एक परिवार की आय में 20% की वृद्धि की जाती है और व्यय में 10% की वृद्धि की जाती है और उसकी बचत में 50% की वृद्धि की जाती है, तब परिवार की आय और व्यय का अनुपात ज्ञात कीजिए।

Sol. According to the question,

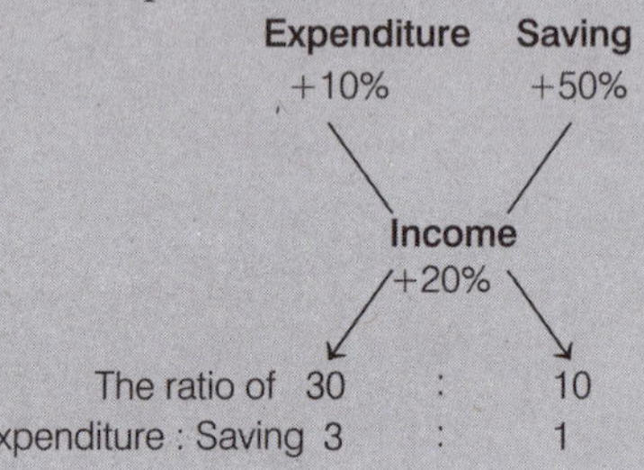

∴ The ratio of Income : Expenditure = (3 + 1) : 3 = 4 : 3

8. Based on Average औसत पर आधारित

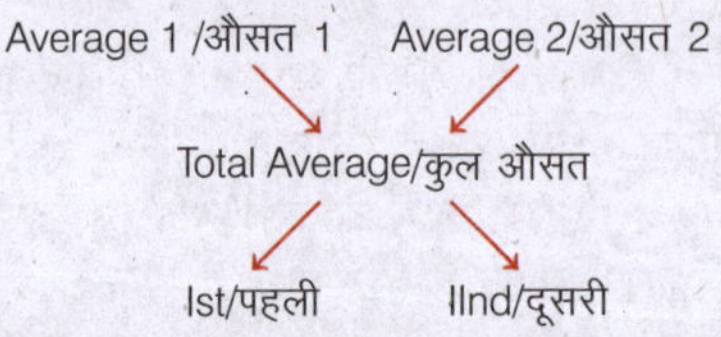

Example **In the first class, the average marks obtained in a certain subject is 25 and in another second class, the average marks obtained in the same subject is 70. If the average marks obtained by the students of both the classes is 40, then find the ratio of the numbers of students of both the classes.**

पहली कक्षा में, एक निश्चित विषय में प्राप्त औसत अंक 25 है और एक अन्य दूसरी कक्षा में, उसी विषय में प्राप्त औसत अंक 70 है। यदि दोनों कक्षाओं के छात्रों द्वारा प्राप्त औसत अंक 40 है, तो दोनों कक्षा के छात्रों की संख्या का अनुपात ज्ञात कीजिए।

Sol. According to the question,

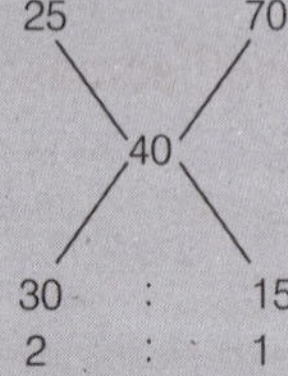

Required ratio = 2 : 1

अध्याय 16

Profit and Loss

लाभ तथा हानि

Profit and Loss are important terms related to monetary transaction in business and trade.

लाभ और हानि व्यापार और वाणिज्य में मौद्रिक लेन-देन से सम्बन्धित महत्त्वपूर्ण शब्द हैं।

- **Cost Price (CP)/क्रय मूल्य** The price at which an article is purchased or manufactured.
  जिस मूल्य पर किसी वस्तु को खरीदा/निर्माण किया जाता है।
- **Selling Price (SP)/विक्रय मूल्य** The price at which an article is sold.
  जिस मूल्य पर किसी वस्तु को बेचा जाता है।
- **Profit (*P*)/लाभ** When a selling price of an article or item is more than its cost price.
  जब वस्तु या सामान का विक्रय मूल्य उसके क्रय मूल्य से अधिक हो।
- **Loss (*L*)/हानि** When SP of an article is less than CP.
  जब वस्तु का विक्रय मूल्य उसके क्रय मूल्य से कम हो।
- **Overhead Expense/अतिरिक्त शुल्क** The extra expenditure on purchased goods apart from the actual cost price.
  वस्तु को खरीदने पर वस्तु के वास्तविक मूल्य के अतिरिक्त लगने वाला खर्च, अतिरिक्त शुल्क कहलाता है।

1. Basic Concept मूल अवधारणा

- If SP > CP, then **Profit = SP − CP**
  यदि SP > CP, तब **लाभ = विक्रय मूल्य − क्रय मूल्य**
- If SP < CP, then **Loss = CP − SP**
  यदि SP < CP, तब **हानि = क्रय मूल्य − विक्रय मूल्य**

Note If/यदि SP = CP, then/तब No profit no loss/न लाभ न हानि

- If an article is sold on r% profit, then
  यदि कोई वस्तु r% लाभ पर बेची जाती है, तब

$$SP = CP\left[\frac{100 + \text{Profit\%}/\text{लाभ\%}}{100}\right]$$

or/या $$CP = SP\left[\frac{100}{100 + \text{Profit\%}/\text{लाभ\%}}\right]$$

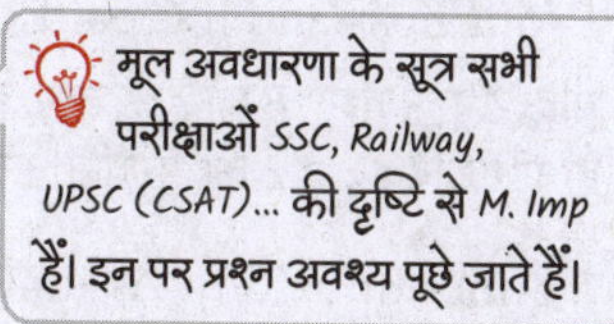

- If an object is sold on $r\%$ loss, then
  यदि एक वस्तु $r\%$ हानि पर बेची जाती है, तब

$$SP = CP\left[\frac{100 - \text{Loss\%/ हानि\%}}{100}\right]$$

or/या $$CP = SP\left[\frac{100}{100 - \text{Loss\%/ हानि\%}}\right]$$

Example **Find the CP, when SP is ₹ 200 and loss is 35%.**

क्रय मूल्य ज्ञात कीजिए, यदि विक्रय मूल्य ₹ 200 तथा हानि 35% हैं।

Sol. $$CP = SP \times \left(\frac{100}{100 - L\%}\right)$$

$$CP = 200 \times \frac{100}{65} = ₹\ 307.6$$

Example **When a plot is sold for ₹ 18700, the owner loses 15%. At what price must the plot be sold in order to gain 15%?**

एक भूखण्ड को ₹ 18700 में बेचने पर मालिक को 15% की हानि होती है। 15% का लाभ पाने के लिए भूखण्ड को कितने में बेचा जाना चाहिए? ***HSSC CET 2022***

Sol. Let

CP = 100 —(15% loss)→ 85 = SP

CP = 100 —(15% profit)→ 115 = SP

$$\therefore \quad 85 \rightarrow 18700$$

$$\Rightarrow 115 \rightarrow \frac{18700}{85} \times 115 = ₹\ 25300$$

2. When an Article is Sold and Bought Multiple Times

जब एक वस्तु एक से अधिक बार बेची और खरीदी जाती है

- If A sells an article to B at $a\%$ profit and B sells it to C at $b\%$ profit, then
  यदि A एक वस्तु B को $a\%$ लाभ/हानि पर बेचता है और B इसे C को $b\%$ लाभ/हानि पर बेचता है, तब

$$\text{Total profit/कुल लाभ} = \left[(\pm a) + (\pm b) + \frac{(\pm a)(\pm b)}{100}\right]\%$$ *M. Imp*

- If A sells an article to B at $a\%$ Profit/Loss and B sells it to C at $b\%$ Profit/Loss and if C paid ₹ x, then
  यदि A एक वस्तु B को $a\%$ लाभ/हानि पर बेचता है और B इसे C को $b\%$ लाभ/हानि पर बेचता है और यदि C, ₹ x का भुगतान करता है, तो

$$\therefore \text{Amount paid by } A / A \text{ द्वारा भुगतान की गई राशि} = x \times \left(\frac{100}{100 \pm a}\right)\left(\frac{100}{100 \pm b}\right)$$

Example **If P sells an article to Q at 10% profit and Q sells it to R at 20% profit, then find the total profit per cent.**

यदि P एक वस्तु Q को 10% के लाभ पर बेचता है और Q, R को 20% के लाभ पर बेचता है, तब कुल लाभ प्रतिशत ज्ञात कीजिए।

Sol. Total profit $= \left[+10 + 20 + \dfrac{(10)(20)}{100}\right]\%$

$$= \left[30 + \frac{200}{100}\right]\% = 32\%$$

Example **A bought an article and spent ₹ 550 on its repairs. He then sold it to B at a profit of 10%. B sold it to C at a loss of 20%. C finally sold it for ₹ 6864 at a profit of 30%. How much did A pay for the article?**

A एक वस्तु खरीदता है और ₹ 550 मरम्मत पर खर्च करता है। तब, वह 10% के लाभ पर B को बेचता है। B यही वस्तु C को 20% की हानि पर बेचता है। अन्ततः C, 30% के लाभ पर वस्तु को ₹ 6864 में बेच देता है। A द्वारा वस्तु के लिए चुकाई गई कीमत क्या थी? *UP POLICE SI 2021*

Sol. Let amount paid by $A = x$

Then, $x + 550 = 6864 \times \left(\dfrac{100}{100+10}\right) \times \left(\dfrac{100}{100-20}\right) \times \left(\dfrac{100}{100+30}\right)$

$$x + 550 = 6864 \times \frac{100}{110} \times \frac{100}{80} \times \frac{100}{130}$$

$$x + 550 = 6000$$

$$\therefore \quad x = 6000 - 550 = ₹\ 5450$$

3. Based on CP and SP relation CP और SP के सम्बन्ध पर आधारित

- If cost price of x articles is equal to selling price of y articles, then
  यदि x वस्तुओं का क्रय मूल्य y वस्तुओं के विक्रय मूल्य के बराबर हो, तब

  (i) Profit%/लाभ% $= \dfrac{x-y}{y} \times 100\%$,
  (when/जब $x > y$)

  (ii) Loss%/हानि% $= \dfrac{y-x}{y} \times 100\%$,
  (when/जब $x < y$)

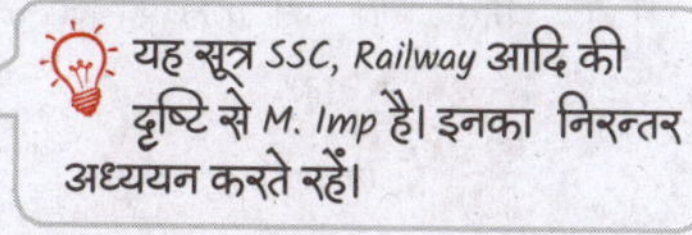

- On selling 'x' articles, the profit or loss is equal to selling price of 'y' articles, then
  x वस्तुओं को बेचने पर लाभ या हानि y वस्तु के विक्रय मूल्य के बराबर है, तब

$$\text{Profit\%/लाभ\%} = \frac{y \times 100}{x-y}\ \%$$

$$\text{Loss\%/हानि\%} = \frac{y}{x+y} \times 100\%$$

Example **If the cost price of 25 articles is equal to the selling price of 35 articles, find profit/loss per cent.**

यदि 25 वस्तुओं का क्रय मूल्य, 35 वस्तुओं के विक्रय मूल्य के बराबर है, तब लाभ/हानि प्रतिशत की गणना कीजिए। *SSC (10+2) 2020*

Sol. Here, $x = 25, y = 35$

$$\therefore \quad \text{Loss \%} = \frac{35 - 25}{35} \times 100\% \qquad [\because x < y]$$

$$= \frac{2 \times 100}{7}\% = 28.57\%$$

Example **If on selling 10 articles the profit is equal to selling of 2 articles, then find profit per cent.**

10 वस्तुओं को बेचने पर 2 वस्तुओं के विक्रय मूल्य के बराबर लाभ होता है, तब लाभ प्रतिशत ज्ञात कीजिए।

Sol. $\text{Profit\%} = \left[\frac{y \times 100}{x - y}\right]\% = \left[\frac{2 \times 100}{10 - 2}\right]\% = 25\%$

4. When Two Items are Sold at Same or Different Price

जब दो वस्तुओं को समान या अलग-अलग मूल्य पर बेचा जाता है

- If a man sells two similar objects, one at a loss of $x\%$ and another at gain of $x\%$, then he always incures loss in this transaction, where

  यदि एक व्यक्ति दो समान वस्तुओं को, पहली को $x\%$ हानि पर और दूसरी को $x\%$ लाभ पर बेचता है तब उसे इस लेन-देन में सदैव हानि होती है, जहाँ

  $$\text{Loss\%} = \frac{x^2}{100}\%$$

  M. Imp

- A man sells his items at a profit/loss of $x\%$. If he had sold it for ₹ R more, he would have gained/loss $y\%$, then

  एक आदमी अपनी वस्तुओं को $x\%$ के लाभ/हानि पर बेचता है। यदि उसने इसे ₹ R अधिक में बेचा होता, तब उसे $y\%$ की लाभ/हानि होती, तब

  $$CP = \frac{R}{(y \pm x)} \times 100$$

  ['+' = When one is profit and other is loss
  जब एक लाभ तथा दूसरा हानि हो
  '–' = When both are either profit or loss
  जब दोनों लाभ या हानि हो]

Example **A trader sells two computers at the same price, making a profit of 30% on one and a loss of 30% on the other. What is the net loss or profit percentage on the transaction?**

एक व्यापारी दो कम्प्यूटरों को समान मूल्य पर बेचता है, एक पर वह 30% का लाभ व दूसरे पर 30% की हानि कमाता है। पूरी लेन-देन में उसका शुद्ध लाभ/हानि प्रतिशत क्या है? *CDS 2019*

Sol. Required loss percentage $= \frac{(30)^2}{100}\% = 9\%$

5. When m Items are Bought for ₹ x and n Items are Sold for ₹ y

जब m वस्तुओं का क्रय मूल्य ₹ x तथा n वस्तुओं का विक्रय मूल्य ₹ y हो

- If a man purchase 'a' items for ₹ x and sells 'b' items for ₹ y, then
  यदि एक आदमी ₹ x में a वस्तुएँ खरीदता है और ₹ y में b वस्तुएँ बेचता है, तब

$$\text{Profit\%/लाभ \%} = \frac{ay - bx}{bx} \times 100, \text{ when/जब } ay > bx$$

$$\text{Loss\%/हानि \%} = \frac{bx - ay}{bx} \times 100, \text{ when/जब } ay < bx$$

Example **A vendor buys bananas at 4 for ₹ 3 and sells at 3 for ₹ 4. What will be the net results?**

एक विक्रेता ₹ 3 में 4 केले खरीदता है और ₹ 4 में 3 केले बेचता है। शुद्ध परिणाम क्या होगा?

SSC CGL (TIER-II) 2018

Sol. Profit per cent $= \left(\frac{4 \times 4 - 3 \times 3}{3 \times 3}\right) \times 100 = 77.77\%$ $[\because ay > bx]$

6. Dishonest Dealer बेईमान दुकानदार

- A dishonest shopkeeper sells his goods at CP, but uses false weight, then his profit

$$= \frac{\text{True weight} - \text{False weight}}{\text{False weight}} \times 100\%$$

$$= \frac{\text{Error}}{\text{True value} - \text{Error}} \times 100\%$$

एक बेईमान दुकानदार अपने माल को क्रय मूल्य पर बेचता है, परन्तु गलत बाट (वजन) का उपयोग करता है, तब उसका लाभ

$$= \frac{\text{सही वजन} - \text{गलत वजन}}{\text{गलत वजन}} \times 100\%$$

$$= \frac{\text{त्रुटि}}{\text{सही मान} - \text{त्रुटि}} \times 100\%$$

Example **A dishonest shopkeeper sells rice at the cost price but he uses a faulty weight of 925 gm for 1000 gm. Find his gain percentage.**

एक बेईमान दुकानदार चावल को क्रय मूल्य पर बेचता है परन्तु वह गलत तराजू जिसमें 1000 ग्राम के स्थान पर 925 ग्राम का उपयोग करता है। उसका लाभ प्रतिशत ज्ञात कीजिए।

SSC CONSTABLE (GD) 2023

Sol. Profit % $= \frac{75}{(1000 - 75)} \times 100\% = \frac{75}{925} \times 100\%$

$= 8.1088\% \approx 8.11\%$

अध्याय 17

Discount छूट

- **Marked Price** (MP) The labelled/list price is called marked price.
  अंकित मूल्य वस्तु पर अंकित अथवा छपा हुआ मूल्य अंकित मूल्य कहलाता है।
- **Discount** Reduction in the marked price of an article is called discount.
  छूट अंकित मूल्य में कमी ही छूट कहलाती है।

1. Basic Concept मूल अवधारणा

- Discount = MP − SP/छूट = अंकित मूल्य − विक्रय मूल्य
- Discount % = $\frac{\text{Discount}}{\text{MP}} \times 100\%$ /छूट% = $\frac{\text{छूट}}{\text{अंकित मूल्य}} \times 100\%$
- $\frac{\text{CP/क्रय मूल्य}}{\text{MP/अंकित मूल्य}} = \frac{100 - d\%}{100 + P\%}$

  Where/जहाँ, d = Discount/छूट, P = Profit/लाभ
- SP = $\frac{\text{MP}(100 - d\%)}{100}$/विक्रय मूल्य = $\frac{\text{अंकित मूल्य } (100 - d\%)}{100}$

  or/या MP = $\frac{\text{SP} \times 100}{(100 - d\%)}$/अंकित मूल्य = $\frac{\text{विक्रय मूल्य} \times 100}{(100 - d\%)}$

  Where/जहाँ, $d \rightarrow$ Discount/छूट
- **Ratio Method** Problems based on discount can be easily solved by ratio method.
  अनुपात विधि छूट पर आधारित समस्याओं को अनुपात विधि से आसानी से हल किया जा सकता है।

Example **After allowing a discount of 10% an article is sold for ₹ 900. Find marked price.**

10% की छूट देने के बाद एक वस्तु को ₹ 900 में बेचा जाता है। अंकित मूल्य ज्ञात कीजिए।

Sol. As, 10% = $\frac{-1}{10}$ $\begin{matrix} -1 \rightarrow \text{Discount} \\ 10 \rightarrow \text{MP} \end{matrix}$ (−) SP

∴ SP = 10 − 1 = 9 → ₹ 900

⇒ 10 → ₹ 1000

Example **An item is sold for ₹ 680 by allowing a discount of 15% on its marked price. Find the marked price of the item.**
एक वस्तु को अंकित मूल्य पर 15% की छूट देने के बाद ₹ 680 में बेचा जाता है। वस्तु का अंकित मूल्य ज्ञात कीजिए।

Sol. $15\% = \frac{-15}{100} = \frac{-3}{20}$ ↗ $-3 \to$ Discount, ↘ $20 \to$ MP

$\therefore$ SP $= 20 - 3 = 17 \to$ ₹ 680

$20 \to \frac{680}{17} \times 20 =$ ₹ 800

Example **A bike is sold for ₹ 87500 by allowing a discount of 44% on its marked price. The marked price of the bike is**
अंकित मूल्य पर 44% की छूट देने के बाद एक बाइक को ₹ 87500 में बेचा जाता है। बाइक का अंकित मूल्य है *SSC CGL (TIER II) 2023*

Sol. $44\% = \frac{44}{100} = \frac{-11}{25}$ ↗ $-11 \to$ Discount, ↘ $25 \to$ MP

$\therefore$ SP = MP − Discount = 25 − 11 = 14

$\therefore$ $14 \to$ ₹ 87500

$\Rightarrow$ $25 \to$ ₹ 156250 (Marked Price)

2. Successive Discount क्रमिक छूट

- If successive discount, $x\%$ and $y\%$, then
  यदि क्रमागत छूट $x\%$ तथा $y\%$, तब
  Net discount/एकल छूट या शुद्ध छूट
  $$= \left(x + y - \frac{xy}{100}\right)\%$$

परीक्षा ब्रह्मास्त्र
क्रमिक छूट टॉपिक सभी परीक्षाओं की दृष्टि से *M. Imp* है। इससे 1 प्रश्न परीक्षा में अवश्य पूछा जाता है।

- If successive discount/यदि क्रमागत छूट, $x\%, y\%, z\%$, then/तब
  Net discount/शुद्ध (एकल) छूट $= (x + y + z) - \frac{xy + yz + zx}{100} + \frac{xyz}{10000}$
- If y articles are given free on purchasing x articles, then
  x वस्तुएँ खरीदने पर y वस्तुएँ मुफ्त दी जाती हैं, तब
  $$\text{Discount/छूट} = \frac{y}{x + y} \times 100\%$$
- If marked price is given and also discount ($D_1\%, D_2\%, D_3\%$......) are given, then seling price is
  यदि अंकित मूल्य दिया हो और छूट ($D_1\%, D_2\%, D_3\%$......) भी दिया हो, तब विक्रय मूल्य है
  SP/विक्रय मूल्य = MP (अंकित मूल्य)
  $$\times \left(\frac{100 - D_1\%}{100}\right)\left(\frac{100 - D_2\%}{100}\right)\left(\frac{100 - D_3\%}{100}\right)...$$

The above method is known as series method.

Example **What will be a single equivalent discount for successive discounts of 10% and 5% on marked price of an article?**

यदि किसी वस्तु के अंकित मूल्य पर 10% तथा 5% की क्रमिक छूट दी जाए, तो एकल समतुल्य छूट क्या होगी?

Sol. Single equivalent discount

$$= \left(10 + 5 - \frac{(10)(5)}{100}\right)\% = (15 - 0.5)\% = 14.5\%$$

Example **The marked price of an article is ₹ 29500. It is sold at two successive discounts of 22% and 10%. Find the selling price of that item.**

एक वस्तु का अंकित मूल्य ₹ 29500 है। यदि वस्तु को दो क्रमिक छूटों 22% तथा 10% पर बेचा जाता है, तब वस्तु का विक्रय मूल्य ज्ञात कीजिए।

DSSSB LAB ASSISTANT 2024

Sol. Here, $x = 22\%$, $y = 10\%$

$$\therefore \quad \text{Net discount} = \left[22 + 10 - \frac{(22)(10)}{100}\right]\%$$

$$= (32 - 2.2)\% = 29.8\%$$

$$\therefore \quad SP = 29500 \times \frac{70.2}{100} = ₹\ 20709$$

3. Discounting the Marked up Price अंकित मूल्य में घटोत्तरी

- A tradesman marks his goods $r\%$ above his cost price. If he allows his customers a discount of r_1 % on the marked price, then the profit or loss per cent

  एक व्यापारी अपने सामान पर क्रय मूल्य से $r\%$ अधिक मूल्य अंकित करता है। यदि वह अपने ग्राहकों को अंकित मूल्य पर r_1 % की छूट देता है, तब लाभ या हानि प्रतिशत

$$= \frac{r \times (100 - r_1)}{100} - r_1$$

- The marked price of an article is fixed in such a way that after allowing a discount of $r\%$ a profit of $R\%$ is obtained. Then, the marked price of the article is $\left(\frac{r + R}{100 - r} \times 100\right)\%$ more than its cost price.

  एक वस्तु का अंकित मूल्य इस प्रकार निर्धारित किया जाता है, कि $r\%$ की छूट देने के बाद $R\%$ का लाभ प्राप्त होता है। तब, वस्तु का अंकित मूल्य उसके क्रय मूल्य से $\left(\frac{r + R}{100 - r} \times 100\right)\%$ अधिक होगा।

Example **After allowing 10% discount on the marked price of an article, a dealer makes a profit of 5%. What is the marked price, if the cost price of the article is ₹ 300?**

वस्तु के अंकित मूल्य पर 10% की छूट देने के बाद एक विक्रेता 5% का लाभ अर्जित करता है। यदि वस्तु का क्रय मूल्य ₹ 300 है, तब अंकित मूल्य क्या है?

SSC CPO 2019

Sol. Method-1

Here, $r = 10\%$, $R = 5\%$

$$\therefore \quad MP = CP + CP\left(\frac{r+R}{100-r} \times 100\right)\% = 300 + 300\left(\frac{10+5}{90} \times 100\right)\%$$

$$MP = ₹(300 + 50) = ₹\,350$$

Method-2

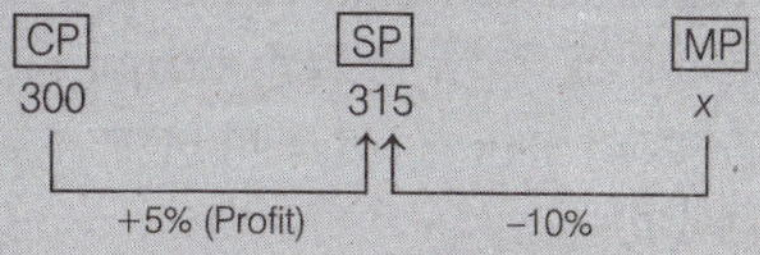

CP	SP	MP
300	315	x

CP → SP: +5% (Profit); MP → SP: −10%

$$\therefore \quad MP = \left(315 \times \frac{100}{90}\right) = ₹\,350$$

Example **A dealer marks his goods 20% above the cost price. He then allows a discount on it and makes a profit of 8%. Find the rate of discount offered by the dealer.**

एक विक्रेता अपनी वस्तुओं का अंकित मूल्य क्रय मूल्य से 20% अधिक अंकित करता है। इसके बाद वह इस पर छूट देता है तथा 8% का लाभ अर्जित करता है। विक्रेता द्वारा दी जाने वाली छूट/बट्टा की गणना कीजिए।

RRB NTPC 2021

Sol. Assume

CP	MP	SP
100	120	108

CP → MP: +20%; CP → SP: +8% (Profit)

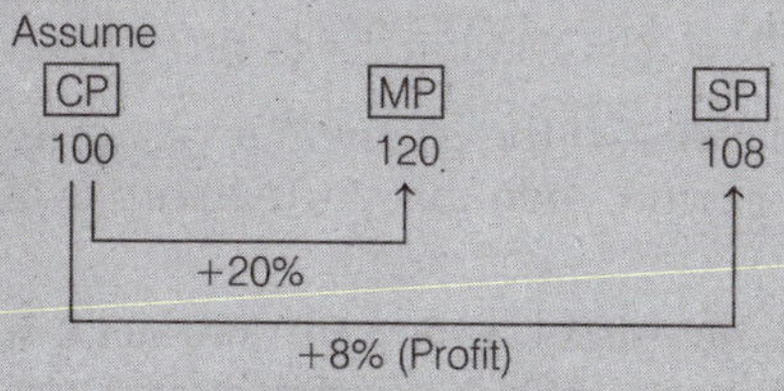

$$\therefore \text{ Discount } \% = \left(\frac{120-108}{120}\right) \times 100\% = 10\%$$

अध्याय 18

Partnership

साझेदारी

An association of two or more persons who invest their money together in order to run a business and after certain time, they receive profit or loss in the ratio of their investments is called partnership.

दो या दो से अधिक व्यक्तियों का संगठन जो किसी व्यवसाय को चलाने के लिए मिलकर अपना धन निवेश करते हैं तथा निश्चित समय के बाद अपने निवेश के अनुपात में लाभ या हानि प्राप्त करते हैं, साझेदारी कहलाती है।

Profit = Capital × Time/लाभ = पूँजी × समय

Types of Partnership साझेदारी के प्रकार

- **Simple Partnership** If the partners invest different money for same time period or same money for different time period, then this partnership is called simple partnership.

  साधारण साझेदारी यदि साझेदार एक ही समय अवधि के लिए अलग-अलग धनराशि का निवेश करते हैं, अथवा अलग-अलग समय अवधि के लिए एक ही धनराशि का निवेश करते हैं, तो इस साझेदारी को साधारण साझेदारी कहा जाता है।

- **Compound Partnership** If the partners invest different money for different time period, then this partnership is called compound partnership.

  मिश्रित साझेदारी यदि साझेदार अलग-अलग समय अवधि के लिए, अलग-अलग धनराशि का निवेश करते हैं, तो इस साझेदारी को मिश्रित साझेदारी कहा जाता है।

Types of Partners साझेदारों के प्रकार

- **Active Partner** A partner who invests money as well as takes part in business operations.

  सक्रिय साझेदार वह साझेदार जो धन का निवेश करने के साथ-साथ व्यवसाय संचालन में भी भाग लेता है।

- **Sleeping Partner** A partner who invests money and does not take part in business operations.

  निष्क्रिय साझेदार वह साझेदार जो धन का निवेश तो करता है, परन्तु व्यवसाय संचालन में भाग नहीं लेता है।

1. Simple Partnership साधारण साझेदारी

If the money invested by partner is different $I_1, I_2, I_3, \ldots$ for the same time period, then profit or loss $P_1, P_2, P_3, \ldots$ is distributed among the partners in the ratio of their investments

यदि साझेदार द्वारा एक निश्चित समय के लिए निवेश की गई राशि $I_1, I_2, I_3, \ldots$ हों, तब उनके लाभ या हानि $P_1, P_2, P_3, \ldots$ को उनके निवेश के अनुपात के आधार पर वितरित किया जाता है।

$$P_1 : P_2 : P_3 : \ldots = I_1 : I_2 : I_3 : \ldots$$ *M. Imp.*

साधारण साझेदारी पर आधारित उदाहरण निम्न हैं

Example **A and B enter into partnership such that 5 times of A's investment is equal to 7 times of B's investment. If the total profit is ₹ 12000, then what is the share of A?**

A और B इस प्रकार साझेदारी करते हैं कि A के निवेश का 5 गुना B के निवेश के 7 गुना के बराबर है। यदि कुल लाभ ₹ 12000 है, तब A का हिस्सा क्या है?

SSC MTS 2023

Sol. $5A = 7B \Rightarrow A : B = 7 : 5$ (+)

$\Rightarrow \quad 12 \rightarrow 12000$

$1 \rightarrow 1000$

$\therefore \quad 7 \rightarrow \boxed{₹\,7000}$

Example **A and B invest ₹ 42000 and ₹ 56000 respectively, in a business. At the end of the year, they make a profit of ₹ 87220. Find B's share in the profit.**

एक व्यवसाय में A और B क्रमशः ₹ 42000 और ₹ 56000 का निवेश करते हैं। वर्ष के अन्त में वे ₹ 87220 का लाभ कमाते हैं। लाभ में B के हिस्से की गणना कीजिए।

Sol. Ratio of investments by A and $B = I_1 : I_2 = ₹\,42000 : ₹\,56000$

$= 3 : 4 = P_1 : P_2$

$\therefore \quad B$'s share $= ₹\left(\frac{4}{7} \times 87220\right) = ₹\,49840$

2. Compound Partnership मिश्रित साझेदारी

If the money invested by the partners is different as $I_1, I_2, I_3, \ldots$ for different time periods, $t_1, t_2, t_3, \ldots$ in a business, then
Ratio of their profits = Ratio of product of money invested and respective time period

यदि साझेदारों द्वारा अलग-अलग धनराशि $I_1, I_2, I_3, \ldots$ अलग-अलग समय अवधि $t_1, t_2, t_3, \ldots$ के लिए एक व्यवसाय में निवेश किया जाता है, तब
उनके लाभों का अनुपात = निवेशित धनराशि और सम्बन्धित समयावधि के गुणनफल का अनुपात

i.e./अर्थात् $P_1 : P_2 : P_3, \ldots = I_1 \times t_1 : I_2 \times t_2 : I_3 \times t_3 \ldots$

मिश्रित साझेदारी पर आधारित 1 प्रश्न सभी परीक्षाओं SSC, Railway, Banking .... आदि में पूछा जाता है।

मिश्रित साझेदारी पर आधारित उदाहरण निम्न हैं

Example **A invest ₹ X in a business. After 4 months, B joined him with ₹ $2X$ and A double his investment. If at the end of the year, total profit was ₹ 13950, then find the profit share of A.**

A ने एक व्यवसाय में ₹ X का निवेश किया। 4 महीने बाद B, ₹ $2X$ के साथ व्यवसाय में शामिल हुआ और A ने अपने निवेश को दोगुना कर दिया। यदि वर्ष के अन्त में कुल लाभ ₹ 13950 था, तब लाभ में A का हिस्सा ज्ञात करें।

IBPS CLERK (PRE) 2020

Sol. Ratio of profits of A and B = Ratio of share of A and B

$$= (X \times 4 + 2X \times 8) : (2X \times 8)$$

$$= 20X : 16X = 5 : 4$$

$\therefore$ Profit share of $A = \frac{5}{9} \times 13950 = \boxed{₹ 7750}$

Example **Anuj started a business by investing ₹ 30000. After 6 months, Ajay joined him with a capital of ₹ 45000. After another 3 months, Ankit joined him with a capital of ₹ 60000. At the end of the year, profit was ₹ 18000. Find the share of Anuj in the profit.**

अनुज ने एक व्यवसाय ₹ 30000 के साथ शुरू किया। 6 महीने के बाद अजय ₹ 45000 निवेश के साथ व्यवसाय में जुड़ गया। अगले 3 महीने बाद, अंकित ₹ 60000 की धनराशि के साथ व्यवसाय में शामिल हुआ। वर्ष के अन्त में कुल लाभ ₹ 18000 था, लाभ में अनुज का हिस्सा ज्ञात कीजिए। *SSC (10+2) 2023*

Sol. According to the question,

$$P_{(\text{Anuj})} : P_{(\text{Ajay})} : P_{(\text{Ankit})} = I_1 \times t_1 : I_2 \times t_2 : I_3 \times t_3$$

$$= (30000 \times 12) : (45000 \times 6) : (60000 \times 3) = 4 : 3 : 2$$

$\therefore$ Anuj's share $= \frac{4}{9} \times 18000 = \boxed{₹ 8000}$

अध्याय 19
Simple Interest
साधारण ब्याज

When a person borrows some money from another person or institution (bank), then the person has to pay some additional amount to repay the loan, this additional amount paid is called interest.
जब एक व्यक्ति किसी दूसरे व्यक्ति या संस्था (बैंक) से कुछ धनराशि उधार लेता है, तब व्यक्ति को ऋण या उधार चुकाने के लिए कुछ अतिरिक्त धनराशि का भुगतान करना होता है, यह अतिरिक्त भुगतान की गई राशि ब्याज कहलाती है।

Principal मूलधन

Principal is the money borrowed or deposited for a fixed period of time.
मूलधन एक निश्चित समय के लिए उधार लिया गया या जमा किया गया धन है।

Amount मिश्रधन

The sum of principal and interest is called amount.
मूलधन और ब्याज का योग मिश्रधन कहलाता है।

Rate of Interest ब्याज की दर

This is the rate at which interest is charged on the principal. It is always specified as a percentage.
यह वह दर है जिस पर मूलधन पर ब्याज लिया जाता है। इसे सदैव प्रतिशत के रूप में निर्दिष्ट किया जाता है।

Time समय

The period of time for which money is borrowed or deposited.
वह समयान्तराल जिसके लिए धन उधार या जमा किया जाता है।

Simple Interest साधारण ब्याज

If interest is calculated only on the principal amount for any time period, it is called simple interest.
यदि ब्याज की गणना किसी भी अवधि के लिए केवल मूलधन पर की जाती है, तो इसे साधारण ब्याज कहा जाता है।

1. Based on Basic Concept मूल अवधारणा पर आधारित

- Interest = Amount − Principal
  ब्याज = मिश्रधन − मूलधन
- Amount = Principal + Interest
  मिश्रधन = मूलधन + ब्याज

- $SI = \frac{P \times R \times T}{100}$

  SI → Simple Interest/साधारण ब्याज

  R → Rate of Interest/ब्याज की दर

  T → Time/समय

  or/या $R = \frac{SI \times 100}{P \times T}$ or/या $T = \frac{SI \times 100}{P \times R}$

अतिसम्भावित! यह सूत्र सभी परीक्षाओं की दृष्टि से *M. Imp* है। इस पर आधारित 1 प्रश्न परीक्षा में अवश्य पूछा जाता है।

Example **How much interest and amount will a person get on a principal of ₹ 7500 after 2 yr at 7% simple interest rate per annum?**

एक व्यक्ति को ₹ 7500 के मूलधन पर 2 वर्ष बाद 7% वार्षिक साधारण ब्याज की दर से कितना ब्याज और मिश्रधन प्राप्त होगा? *SSC (10+2) 2022*

Sol. According to the question,

$$SI = \frac{7500 \times 7 \times 2}{100} = ₹\ 1050$$

Amount = 7500 + 1050 = ₹ 8550

2. Based on Sum becomes *n* Times धनराशि के *n* गुना होने पर आधारित

- If a certain sum becomes 'n' times of itself in T yr on simple interest, then the rate per annum is

  यदि एक निश्चित राशि साधारण ब्याज पर T वर्ष में स्वयं की 'n' गुनी हो जाती है, तब

  $$SI = (n-1) \times \text{Sum/मूल धनराशि}$$

  $$R\% = \frac{n-1}{T} \times 100\%$$

  and/तथा $T = \frac{(n-1)}{R} \times 100$

परीक्षा ब्रह्मास्त्र इस टॉपिक के सभी सूत्र *SSC*, *Railway*... सभी परीक्षाओं की दृष्टि से *M. Imp* हैं। इनका निरन्तर अभ्यास करते रहिए।

- If a sum of money at a certain rate of interest becomes n_1 times in T_1 yr and n_2 times in T_2 yr, then

  यदि एक राशि एक निश्चित ब्याज की दर पर T_1 वर्षों में n_1 गुना और T_2 वर्षों में n_2 गुना हो जाती है, तब $\frac{T_1}{T_2} = \frac{n_1 - 1}{n_2 - 1}$ *M. Imp.*

- If a sum of money in a certain time become n_1 time at R_1 % and n_2 times at R_2 %, then

  यदि कोई राशि एक निश्चित समय में R_1 % पर n_1 गुना तथा R_2 % पर n_2 गुना हो जाती है, तब $\frac{R_1}{R_2} = \frac{n_1 - 1}{n_2 - 1}$ *M. Imp.*

- If there are distinct rates of interest for distinct time periods.

  यदि अलग-अलग समय अवधि के लिए ब्याज दरें अलग-अलग हों।

- If for first t_1 yr $\rightarrow R_1$, second t_2 yr $\rightarrow R_2$...

यदि पहले t_1 वर्षों के लिए ब्याज दर R_1, दूसरे t_2 वर्षों के लिए R_2..., then/तब

$$\text{SI/साधारण ब्याज} = \frac{P(R_1t_1 + R_2t_2 + ...)}{100}$$

Example **Find the rate of simple interest at which a sum of money becomes 5 times in 8 yr?**

साधारण ब्याज की वह दर ज्ञात कीजिए, जिस पर कोई धनराशि 8 वर्ष में 5 गुना हो जाती है? *RRB NTPC (CBT-I) 2021*

Sol. According to the question,

$$n = 5, T = 8 \text{ yr}$$

$$R\% = \frac{(n-1)}{T} \times 100\% = \frac{(5-1)}{8} \times 100\% = \frac{4}{8} \times 100\% = 50\% \text{ p.a.}$$

Example **In how much time will a sum of money become 4 times at 15% per annum simple interest?**

एक धनराशि 15% प्रतिवर्ष के साधारण ब्याज पर कितने समय में 4 गुना हो जाएगी? *SSC CGL 2023*

Sol. According to the question,

$$n = 4, R = 15\% \text{ p.a.}$$

$$T = \frac{(n-1)}{R} \times 100 = \frac{(4-1)}{15} \times 100 = \frac{3}{15} \times 100 = 20 \text{ वर्ष}$$

Example **If a sum of money doubles in 6 yr, when deposited in a bank at simple interest, then in how many years will that sum become 3 times?**

यदि कोई धनराशि बैंक में साधारण ब्याज पर जमा करने पर 6 वर्ष में दोगुनी हो जाती है, तो कितने वर्ष में धनराशि 3 गुना होगी? *RRC GROUP D 2021*

Sol. According to the question,

$$T_1 = 6 \text{ yr}, n_1 = 2, n_2 = 3$$

$$\therefore \quad \frac{T_1}{T_2} = \frac{n_1 - 1}{n_2 - 1} \Rightarrow \frac{6}{T_2} = \frac{(2-1)}{(3-1)} = \frac{1}{2}$$

$$\therefore \quad T_2 = \frac{6 \times 2}{1} = 12 \text{ yr}$$

Example **A sum of money becomes 3 times at the simple interest rate of 8% p.a. At what rate per cent will it become 4 times?**

एक धनराशि 8% प्रतिवर्ष की साधारण ब्याज की दर पर 3 गुना हो जाती है। किस दर से यह 4 गुना हो जाएगी?

Sol. According to the question,

$$n_1 = 3, n_2 = 4$$

$$R_1 = 8\%, R_2 = ?$$

$$\therefore \quad \frac{R_1}{R_2} = \frac{n_1 - 1}{n_2 - 1} \Rightarrow \frac{8}{R_2} = \frac{(3-1)}{(4-1)} = \frac{2}{3} \Rightarrow R_2 = \frac{8 \times 3}{2} = 12\% \text{ p.a.}$$

Example **How much interest will a person get on a principal sum of ₹ 20000 at 5% p.a. simple interest rate for 2 yr and 7% p.a. for 3 yr?**

एक व्यक्ति को ₹ 20000 के मूलधन को 2 वर्ष के लिए 5% और 3 वर्ष के लिए 7% वार्षिक साधारण ब्याज की दर से कितना ब्याज प्राप्त होगा?

Sol. According to the question,

$$\text{SI} = \frac{20000\,(5 \times 2 + 7 \times 3)}{100} = \frac{20000 \times 31}{100} = ₹\, 6200$$

3. Based on Different Rate of Interest and Different Time

भिन्न-भिन्न ब्याज दरों और भिन्न-भिन्न समय पर आधारित

- If the difference between two simple interests is x, calculated at different annual rates and times, then

  यदि दो साधारण ब्याजों के बीच अन्तर x की गणना अलग-अलग वार्षिक दरों और समय पर की जाती है, तब

$$P\ (\text{Principal/मूलधन}) = \frac{x \times 100}{(\text{Diff. of rates/ दरों का अन्तर}) \times (\text{Time/ समय})}$$

M. Imp.

- If in t yr a sum amounts to x_1 at rate of R_1 % and amounts to x_2 at rate of R_2 %. Then, the sum is given by,

  यदि कोई राशि t वर्षों में R_1 % की दर से x_1 हो जाती है और R_2 % की दर से x_2 हो जाती है, तब

$$P\ (\text{Principal/मूलधन}) = \frac{(x_1 \sim x_2)}{(R_1 \sim R_2) \times t} \times 100$$

Example **The difference between the simple interest received on a sum of money from two different bank in 2 yr is ₹ 2.5. The difference between their annual interest rate is 0.25%. Find the sum.**

दो भिन्न बैंकों से किसी राशि पर 2 वर्ष में प्राप्त होने वाले साधारण ब्याजों का अन्तर ₹ 2.5 है। उनकी ब्याज की वार्षिक दरों का अन्तर 0.25% है। राशि ज्ञात करें।

Sol. According to the question,

$$x = ₹\, 2.5$$

Difference of rates = 0.25%

Time = 2 yr

$$P = \frac{x \times 100}{\text{Difference of rates} \times \text{Time}} = \frac{2.5 \times 100}{0.25 \times 2} = 500$$

$$\therefore \quad P = ₹\, 500$$

4. Based on Installments किस्त पर आधारित

- If a sum (P) is to be deposited at the rate of $R\%$ in equal installments in t yr, then
  यदि कोई राशि P, $R\%$ प्रतिवर्ष की दर से t वर्षों में समान किश्तों में जमा करनी हो, तब

$$\text{Equal installments/समान किश्त} = \frac{100 \times P}{100T + \frac{RT}{2}(T-1)}$$ M. Imp.

Example **What annual payment will discharge a debt of ₹ 770 in 5 equal annual installments at the rate of 5% p.a. on SI?**

साधारण ब्याज पर 5% प्रतिवर्ष की दर से 5 समान वार्षिक किश्तों में ₹ 770 का ऋण किस वार्षिक भुगतान से चुकता होगा? *IBPS CLERK (PRE) 2019*

Sol. According to the question,

$$R = 5\%,\ P = ₹\,770,\ T = 5$$

$$\therefore \text{Equal installments} = \frac{100 \times P}{100T + \frac{RT}{2}(T-1)} = \frac{100 \times 770}{100 \times 5 + \frac{5 \times 5}{2} \times (5-1)}$$

$$= \frac{100 \times 770}{500 + 50} = \frac{100 \times 770}{550} = ₹\,140$$

5. Based on Some other Important Concept
कुछ अन्य महत्त्वपूर्ण अवधारणा पर आधारित

Example **A certain sum of money is lent at 5% p.a. for 2 yr, 7% p.a. for the next 3 yr and 10% p.a. after 5 yr. If the simple interest received for a period of 10 yr is ₹ 16200, then what is the sum?**

एक निश्चित राशि 2 वर्षों के लिए 5% प्रतिवर्ष, अगले 3 वर्षों के लिए 7% प्रतिवर्ष और 5 वर्षों के बाद 10% प्रतिवर्ष की दर पर उधार दी जाती है। यदि 10 वर्षों की अवधि के लिए साधारण ब्याज ₹ 16200 प्राप्त होता है, तब राशि क्या है?

Sol. According to the question,

$$SI = (5\% \times 2) + (7\% \times 3) + (10\% \times 5) = 10\% + 21\% + 50\% = 81\%$$

$$\therefore \text{ Amount} = \frac{16200}{81\%} \times 100\% = ₹\,20000$$

Example **A person divided a sum of ₹ 17200 into three parts and invested it at 5%, 6% and 9% p.a. simple interest. At the end of two years, he received the same interest in all three cases. What is amount invested at 9% p.a. interest?**

एक व्यक्ति ने ₹ 17200 की धनराशि को तीन भागों में विभाजित कर, उसे 5%, 6% और 9% वार्षिक साधारण ब्याज की दर पर निवेश किया। दो वर्षों की समाप्ति पर, उसे सभी तीनों स्थितियों में एकसमान ब्याज प्राप्त हुआ। 9% वार्षिक ब्याज पर निवेश की गई धनराशि कितनी है? *CDS 2020 (I)*

Sol. According to the question,

Let the sum invested in the ratio = $P_1 : P_2 : P_3$

$\because$ SI is constant.

$$\therefore \quad \text{Sum} \propto \frac{1}{\text{Rate\%} \times \text{Time}}$$

$$\therefore \quad P_1 : P_2 : P_3 = \frac{1}{5 \times 2} : \frac{1}{6 \times 2} : \frac{1}{9 \times 2} = 18 : 15 : 10$$

$$\Rightarrow \quad P_3 = \frac{10}{43} \times 17200 = ₹\, 4000$$

Example **₹ 9750 is invested by a man in the bank account of his two sons, whose ages are 14 yr and 18 yr in such a way that they will get equal amount at an age of 22 yr at 5% p.a. at simple interest. Find the share of younger child.**

एक आदमी द्वारा अपने दो बेटों जिनकी आयु 14 वर्ष और 18 वर्ष हैं, उनके बैंक खाते में ₹ 9750 इस प्रकार निवेश किया गया है, कि उन्हें 22 वर्ष की आयु में 5% वार्षिक साधारण ब्याज की दर से बराबर राशि मिलेगी। छोटे बेटे का हिस्सा ज्ञात करो।

Sol. According to the question,

Time remaining for elder son = 22 – 18 = 4 yr

Time remaining for younger son = 22 – 14 = 8 yr

Assume principal are P_1 and P_2 respectively,

$$P_1 + \frac{P_1 \times 5 \times 4}{100} = P_2 + \frac{P_2 \times 5 \times 8}{100}$$

$$\Rightarrow \quad \frac{120 P_1}{100} = \frac{140 P_2}{100} \Rightarrow \frac{P_1}{P_2} = \frac{140}{120}$$

$$\Rightarrow \quad \frac{P_1}{P_2} = \frac{7}{6}$$

$$\therefore \quad P_2 = \frac{6}{13} \times 9750 = ₹\, 4500$$

अध्याय 20

Compound Interest
चक्रवृद्धि ब्याज

When interest is calculated on the principal amount as well as the interest of the previous period, then it is called compound interest. This is represented by CI.

जब ब्याज की गणना मूलधन के साथ-साथ विगत् अवधि के ब्याज पर भी की जाती है, तब उसे चक्रवृद्धि ब्याज कहते हैं। इसे CI से प्रदर्शित करते हैं।

1. Based on Basic Concept मूल अवधारणा पर आधारित

- $CI = A - P = P\left[\left(1+\frac{r}{100}\right)^t - 1\right]$

  Where/जहाँ, $A \rightarrow$ Amount/मिश्रधन,

  $P \rightarrow$ Principal/मूलधन

  $r \rightarrow$ Rate/ब्याज दर, $t \rightarrow$ Time/समय

> यह सूत्र सभी परीक्षाओं SSC, Railway, CDS, UPSC (CSAT)... की दृष्टि से M. Imp है। इस पर आधारित 1 प्रश्न अवश्य पूछा जाता है।

- $A = P\left(1+\frac{r}{100}\right)^t$
- When interest is payable half-yearly, then/जब ब्याज छमाही (अर्द्धवार्षिक) देय हो, तब $R = \frac{R}{2}\%, T = 2n$

  and when interest is payable quarterly, then/और जब ब्याज तिमाही देय है, तब $R = \frac{R}{4}\%, T = 4n$

Example **What will be compound interest on ₹ 5000 for 3 yr at 10% p.a.?**

₹ 5000 पर 3 वर्षों का 10% वार्षिक दर पर चक्रवृद्धि ब्याज कितना होगा?

SSC CGL 2023

Sol. According to the question,

$P = ₹\, 5000,\ t = 3\,\text{yr},\ r = 10\%$ p.a.

$$\because \quad A = P\left(1+\frac{r}{100}\right)^n = 5000 \times \left(1+\frac{10}{100}\right)^3$$

$$= 5000 \times \frac{11}{10} \times \frac{11}{10} \times \frac{11}{10} = ₹\, 6655$$

$$\therefore \quad CI = A - P = 6655 - 5000 = ₹\, 1655$$

Example **In how much time will a sum of ₹ 10000 becomes ₹ 13310 at 20% annual compound interest compounded half-yearly?**

अर्द्धवार्षिक रूप से संयोजित 20% वार्षिक चक्रवृद्धि ब्याज की दर से कितने समय में ₹ 10000 की धनराशि ₹ 13310 हो जाएगी? *SSC (10+2) 2022*

Sol. Given, $P =$ ₹ 10000; $A =$ ₹ 13310; $r = 20\%$ p.a.

$\because$ Half-yearly rate $= \frac{20}{2}\% = 10\%$, $T = 2n$

$$\because \quad A = P\left(1 + \frac{r}{100}\right)^{2n}$$

$$13310 = 10000\left(1 + \frac{10}{100}\right)^{2n}$$

$$\Rightarrow \quad \frac{13310}{10000} = \left(\frac{11}{10}\right)^{2n} \Rightarrow \left(\frac{11}{10}\right)^{3} = \left(\frac{11}{10}\right)^{2n}$$

Compare on both sides, $2n = 3 \Rightarrow n = 1\frac{1}{2}$ yr

Example **If the interest is compounded quarterly, at what annual rate of interest will the compound interest of ₹ 32000 for 9 months be ₹ 5044?**

यदि ब्याज त्रैमासिक रूप से संयोजित हो, तो ब्याज की किस वार्षिक दर से ₹ 32000 का 9 माह का चक्रवृद्धि ब्याज ₹ 5044 होगा? *SSC CPO 2019*

Sol. Given, $P =$ ₹ 32000; CI = ₹ 5044; $R = \frac{r}{4}\%$ quarterly

and $\quad n = \frac{9}{12} \times 4 = 3$ quarter

$$\therefore \quad A = P\left(1 + \frac{R}{100}\right)^{n}$$

$$\Rightarrow \quad 37044 = 32000\left(1 + \frac{r}{4 \times 100}\right)^{3} \qquad [\because A = P + \text{CI}]$$

$$\Rightarrow \quad \frac{37044}{32000} = \left(1 + \frac{r}{400}\right)^{3} \Rightarrow \frac{9261}{8000} = \left(1 + \frac{r}{400}\right)^{3}$$

$$\Rightarrow \quad \left(\frac{21}{20}\right)^{3} = \left(1 + \frac{r}{400}\right)^{3}$$

Compare on both sides,

$$\frac{21}{20} = 1 + \frac{r}{400} \Rightarrow \frac{1}{20} = \frac{r}{400}$$

$$\Rightarrow \quad r = 20\% \text{ p.a.}$$

2. When Time is in Fractional Form जब समय भिन्नात्मक रूप में हो

- If the time is in fractional form, i.e./यदि समय भिन्नात्मक रूप में हो, अर्थात्

(Let time $= n\frac{a}{b}$/माना समय $= n\frac{a}{b}$)

Then/तब, $$A = P\left(1+\frac{r}{100}\right)^n\left(1+\frac{r\times\frac{a}{b}}{100}\right)$$

Example **What is the compound interest on a sum of ₹ 12000 for $2\frac{5}{8}$ yr at 8% p.a., when the interest is compounded annually? (rounded-off to the nearest ₹)**

₹ 12000 की राशि पर $2\frac{5}{8}$ वर्षों के लिए 8% वार्षिक दर से चक्रवृद्धि ब्याज कितना होगा, जब ब्याज को वार्षिक रूप से संयोजित किया जाता है? (निकटतम ₹ तक) *SSC CGL (TIER I) 2019*

Sol. According to the question,

$$P = ₹\ 12000,\quad r = 8\%\text{ p.a.},\quad \text{Time} = 2\frac{5}{8}\text{ yr}$$

$$\because\quad A = P\left(1+\frac{r}{100}\right)^n \times \left(1+\frac{r\times\frac{a}{b}}{100}\right) = 12000\times\left(1+\frac{8}{100}\right)^2\left(1+\frac{8\times\frac{5}{8}}{100}\right)$$

$$= 12000\times\frac{27}{25}\times\frac{27}{25}\times\frac{21}{20} = ₹\ 14696.64$$

$$\therefore\quad \text{CI} = 14696.64 - 12000 = ₹\ 2696.6 \cong ₹\ 2697$$

3. Based on Relation and Difference between SI and CI

SI और CI के बीच सम्बन्ध और अन्तर पर आधारित

- Relation between CI and SI for 2 yr

2 वर्ष के लिए चक्रवृद्धि ब्याज और साधारण ब्याज के बीच सम्बन्ध

$$\text{CI} = \text{SI}\left[1+\frac{2r}{200}\right]$$

- Difference between SI and CI at the rate of $R\%$ on amount P

किसी राशि P पर $R\%$ की दर से SI और CI का अन्तर

(i) For 2 years/दो वर्ष के लिए

$$= P\left(\frac{R}{100}\right)^2 = \frac{\text{SI}\times R}{200}$$

(ii) For 3 years/तीन वर्ष के लिए

$$= P\left(\frac{R}{100}\right)^2\left(3+\frac{R}{100}\right)$$

परीक्षा ब्रह्मास्त्र

ये सूत्र सभी परीक्षाओं की दृष्टि से महत्त्वपूर्ण हैं। इन सूत्रों के प्रयोग से समय की बचत होती है।

Example **If the simple interest on a certain sum of money for 2 yr at 7% annual interest rate is ₹ 200, then what will be the difference between compound interest and simple interest?**

यदि किसी निश्चित धनराशि पर 2 वर्षों का 7% वार्षिक ब्याज की दर पर साधारण ब्याज ₹ 200 है, तो चक्रवृद्धि ब्याज व साधारण ब्याज का अन्तर क्या होगा?

Sol. Given, $t = 2$ yr; $r = 7\%$ p.a.; SI = ₹ 200

$\therefore$ Difference between CI and SI $= \frac{\text{SI} \times R}{200} = \frac{200 \times 7}{200} = ₹\, 7$

4. Based on Sum become *n* Time धनराशि के *n* गुना होने पर आधारित

- If a sum becomes n times of itself in t yr on compound interest, then
  यदि कोई राशि चक्रवृद्धि ब्याज पर t वर्षों में स्वयं का n गुना हो जाती है, तब

$$R\% = \left[n^{\frac{1}{t}} - 1\right] \times 100\%$$

- A certain sum at CI becomes x times in n_1 yr and y times in n_2 yr, then
  चक्रवृद्धि ब्याज पर एक निश्चित राशि n_1 वर्ष में x गुना और n_2 वर्षों में y गुना हो जाती है, तब $x^{\frac{1}{n_1}} = y^{\frac{1}{n_2}}$ M. Imp

- If a certain sum at compound interest becomes A_1 in n yr and A_2 in $(n+1)$ yr, then
  यदि कोई धनराशि चक्रवृद्धि ब्याज पर n वर्षों में A_1 और $(n+1)$ वर्षों में A_2 हो जाती है, तो

  (i) Rate of compound interest/चक्रवृद्धि ब्याज की दर $= \frac{(A_2 - A_1)}{A_1} \times 100\ \%$

  (ii) Sum/धनराशि $= A_1 \left(\frac{A_1}{A_2}\right)^n$

Example **If a sum of money becomes $3\frac{3}{8}$ times on compound interest in 3 yr on the basis of annual compound interest, then what is the annual rate of interest?**

यदि कोई धनराशि वार्षिक चक्रवृद्धि ब्याज के आधार पर 3 वर्षों का चक्रवृद्धि ब्याज जोड़कर $3\frac{3}{8}$ गुनी हो जाए, तो ब्याज की वार्षिक दर कितनी है? *SSC (10+2) 2020*

Sol. Given, $n = \frac{27}{8}$ and $t = 3$ yr

$$\therefore \quad R\% = [n^{\frac{1}{t}} - 1] \times 100\% = \left[\left(\frac{27}{8}\right)^{\frac{1}{3}} - 1\right] \times 100\%$$

$$= \left[\left(\frac{3}{2}\right)^{\frac{3}{3}} - 1\right] \times 100\% = \frac{1}{2} \times 100\% = 50\%$$

Example **A sum of money doubles in 5 yr at compound interest. In how many years will it become 8 times at the same rate of interest?**

एक धनराशि चक्रवृद्धि ब्याज पर 5 वर्षों में दोगुनी हो जाती है। ब्याज की समान दर पर वह कितने वर्षों में 8 गुना हो जाएगी? *SSC CGL 2018*

Sol. Given, $x = 2, y = 8;\ n_1 = 5, n_2 = ?$

$$\therefore \quad x^{\frac{1}{n_1}} = y^{\frac{1}{n_2}}$$

$$\Rightarrow \quad 2^{\frac{1}{5}} = 8^{\frac{1}{n_2}} = (2^3)^{\frac{1}{n_2}} \Rightarrow 2^{\frac{1}{5}} = 2^{\frac{3}{n_2}}$$

Compare on both sides,

$$\frac{1}{5} = \frac{3}{n_2} \Rightarrow n_2 = 3 \times 5 = 15 \text{ वर्ष}$$

Example **A sum of money invested at compound interest becomes ₹ 800 in 3 yr and ₹ 840 in 4 yr. Find the rate of interest per annum.**

एक धनराशि को चक्रवृद्धि ब्याज पर निवेश करने पर वह 3 वर्षों में ₹ 800 और 4 वर्षों में ₹ 840 हो जाती है। ब्याज की दर प्रतिवर्ष ज्ञात कीजिए। *UK POLICE (SI) 2021*

Sol. Given, $A_1 = ₹\ 800$ and $A_2 = ₹\ 840$

$$\therefore \quad R\% = \frac{(A_2 - A_1)}{A_1} \times 100\%$$

$$= \frac{(840 - 800)}{800} \times 100\% = \frac{40}{800} \times 100\% = 5\%$$

5. When Compound Interest rates are different in each Cycle
जब चक्रवृद्धि ब्याज दर प्रत्येक चक्र में अलग-अलग हो

- If rate of interests in first and second cycle are $a\%$ and $b\%$ respectively
  यदि पहले और दूसरे चक्र में ब्याज की दर क्रमशः $a\%$ और $b\%$ हो

  Then/तब $\quad \text{CI} = P\left(a + b + \frac{ab}{100}\right)\%$

  If/यदि $\quad a = b = r\%$

  Then/तब $\quad \text{CI} = P\left(2r + \frac{r^2}{100}\right)\%$

- If rate of interest for 3 cycles is $a\%, b\%, c\%$.
  यदि 3 चक्रों के लिए ब्याज दर $a\%, b\%, c\%$ हो,

  Then/तब $\quad \text{CI} = P\left[(a + b + c) + \frac{(ab + bc + ca)}{100} + \frac{abc}{10000}\right]\%$

Example **What will be the interest on the amount of ₹ 25000 compounded annually at the rate of 4%, 5% and 6% per annum for the first, second and third year respectively?**

पहले, दूसरे और तीसरे वर्ष के लिए क्रमशः 4%, 5% और 6% प्रति वर्ष की दर से वार्षिक रूप से संयोजित ₹ 25000 की राशि पर चक्रवृद्धि ब्याज कितना होगा?

Sol. $CI = 25000 \times \left[(4+5+6) + \frac{20+30+24}{100} + \frac{120}{10000}\right]\%$

$$= 25000 \times \frac{150000 + 7400 + 120}{10000 \times 100} = 25000 \times \frac{157520}{10000 \times 100} = ₹\ 3938$$

6. Installment किस्त

- Each annual installment, for n installments/n किस्तों के लिए, प्रत्येक वार्षिक किस्त

$$= \frac{P}{\left(\frac{100}{100+r}\right) + \left(\frac{100}{100+r}\right)^2 + \ldots + \left(\frac{100}{100+r}\right)^n}$$

M. Imp

Example **A certain sum of money is repaid in two annual installments of ₹ 17640 each at 5% compound interest rate compounded annually. What was the amount borrowed?**

कुछ धनराशि वार्षिक चक्रवृद्धि ब्याज की दर से 5% चक्रवृद्धि ब्याज देकर प्रति ₹ 17640 की दो वार्षिक किस्तों में चुकाई जाती है। उधार ली गई राशि कितनी थी?

SSC SELECTION POST 2022

Sol. Given, $r = 5\%$ p.a.

Each installment = ₹ 17640 and $n = 2$

$$\therefore \text{Each installment} = \frac{P}{\left(\frac{100}{100+r}\right) + \left(\frac{100}{100+r}\right)^2}$$

$$\Rightarrow \quad 17640 = \frac{P}{\left(\frac{100}{100+5}\right) + \left(\frac{100}{100+5}\right)^2}$$

$$\therefore \quad P = 17640 \times \left[\frac{20}{21} + \frac{20 \times 20}{21 \times 21}\right] = 17640 \times \left(\frac{420+400}{441}\right)$$

$$= 17640 \times \frac{820}{441} = ₹\ 32800$$

अध्याय 21
Work and Time
कार्य और समय

According to the basic principle of work and time, time taken by a person/machine to complete a work depends on the efficiency of that person/machine.

कार्य और समय के मूल सिद्धान्त के अनुसार, किसी कार्य को किसी व्यक्ति या मशीन द्वारा लिया गया समय उस व्यक्ति/मशीन की कार्यक्षमता पर निर्भर करता है।

Basic Rules Related to Work and Time
कार्य और समय से सम्बन्धित साधारण नियम

- If a person can do a work in t day (or hour), then the work of the person in 1 day (or hour) $= \frac{1}{t}$

  यदि एक व्यक्ति किसी कार्य को t दिन (या घण्टे) में कर सकता है, तो व्यक्ति का 1 दिन (या घण्टे) का कार्य $= \frac{1}{t}$

- If a person works for 1 day (or hour) $= \frac{1}{t}$, then he will complete the work in t days (or hours).

  यदि एक व्यक्ति का 1 दिन (या घण्टे) का कार्य $= \frac{1}{t}$ है, तो वह व्यक्ति t दिन (या घण्टे) में कार्य पूरा करेगा।

- If a person does $\frac{a}{b}$ part of a work in t days, then he will complete the work in $\left(t \times \frac{b}{a}\right)$ days.

  यदि कोई व्यक्ति कार्य का $\frac{a}{b}$ भाग, t दिन में पूरा करता है, तो वह कार्य को $\left(t \times \frac{b}{a}\right)$ दिनों में पूरा करेगा।

1. Based on Basic Concept मूल अवधारणा पर आधारित

- If A completes a piece of work in x days and B completes the same work in y days, then total time taken to complete the work by A and B together

  यदि A किसी कार्य को x दिनों में पूरा करता है और B उसी कार्य को y दिनों में पूरा करता है,

  तब, A और B दोनों के द्वारा कार्य को पूरा करने में लिया गया कुल समय $= \frac{xy}{x+y}$ M. Imp

- If A alone can do a certain work in x days and A and B together can do the same work in y days, then B alone can do the same work in
यदि A अकेला किसी कार्य को x दिनों में कर सकता है और A और B मिलकर उसी कार्य को y दिनों में कर सकते हैं, तो B अकेला उसी कार्य को कर सकता है

$$=\left(\frac{xy}{x-y}\right) \text{ days/दिनों में}$$ *M. Imp*

- If A, B and C can do a piece of work in x, y and z days respectively, then total time taken to complete the work by A, B and C
यदि A, B और C किसी कार्य को क्रमशः x, y और z दिनों में कर सकता है, तब A, B और C तीनों द्वारा कार्य को पूरा करने में लिया गया समय $=\dfrac{xyz}{xy+yz+zx}$

- If A and B can do a work in x days, B and C can do the same work in y days, C and A can do the same work in z days, then total time taken when A, B and C work together
यदि A और B किसी कार्य को x दिनों में, B और C, y दिनों में तथा C और A, z दिनों में पूरा कर सकते हैं, जब A, B और C एकसाथ कार्य करें, तब लिया गया समय $=\dfrac{2xyz}{xy+yz+zx}$

समय और कार्य की मूल अवधारणा पर आधारित उदाहरण निम्नलिखित हैं

Example **Rohan can do a work in 10 days and Mohan can do the same work in 15 days. In how many days will they together complete the same work?**

रोहन अकेले एक कार्य को 10 दिनों में कर सकता है और मोहन अकेले उसी कार्य को 15 दिनों में कर सकता है। वे मिलकर उसी कार्य को कितने दिनों में पूरा करेंगे?

SSC MTS 2023

Sol. Required time $=\dfrac{xy}{(x+y)}=\dfrac{10\times 15}{(10+15)}=\dfrac{150}{25}=6$ days

Example **A, B and C can do a job working alone in 12, 16 and 24 days, respectively. In how many days they can do the job if they work together?**

A, B और C अकेले एक कार्य को क्रमशः 12, 16 और 24 दिनों में पूरा कर सकते हैं। यदि वे एकसाथ कार्य करते हैं, तो वे कितने दिनों में कार्य पूरा कर सकते हैं?

RRB NTPC 2021

Sol. Total time taken $=\dfrac{xyz}{xy+yz+zx}=\dfrac{12\times 16\times 24}{12\times 16+16\times 24+24\times 12}$

$$=\frac{16\times 24}{16+32+24}=\frac{16\times 24}{72}=\frac{16}{3} \text{ days}$$

Alternate Solution

$A \rightarrow 12$ — 4
$B \rightarrow 16$ — 3 → 48 (Total work)
$C \rightarrow 24$ — 2

$\therefore$ Required time $= \dfrac{48}{(4+3+2)} = \dfrac{16}{3}$ days

Example ***P* and *Q* can complete a work in 8 days, *Q* and *R* in 12 days and *R* and *P* in 15 days. How many days will *R* alone take to complete the work?**

P और Q एक कार्य को 8 दिनों में पूरा कर सकते हैं, Q और R 12 दिनों में और R और P 15 दिनों में पूरा कर सकते हैं। R को अकेले कार्य पूरा करने में कितने दिन लगेंगे? *SSC (10+2) 2023*

Sol. According to the question,

Work done in 1 day, $P + Q = \dfrac{1}{8}$ ...(i)

$Q + R = \dfrac{1}{12}$ ...(ii)

$R + P = \dfrac{1}{15}$ ...(iii)

Adding Eqs. (i), (ii) and (iii),

$\therefore \; 2(P+Q+R) = \dfrac{15+10+8}{120} \Rightarrow P+Q+R - \dfrac{33}{240}$

$\dfrac{1}{8} + R = \dfrac{33}{240}$ [From Eq. (i)]

So, $R = \dfrac{33}{240} - \dfrac{1}{8} = \dfrac{1}{80}$

Hence, R alone take 80 days to complete the work.

2. Based on Unitary Method ऐकिक नियम पर आधारित

- If M_1 men finish W_1 work in D_1 days, working T_1 time each day and M_2 men finish W_2 work in D_2 days, working T_2 time each day, then
यदि M_1 पुरुष W_1 कार्य को D_1 दिनों में पूरा करते हैं, प्रत्येक दिन T_1 समय कार्य करते हैं और M_2 पुरुष W_2 कार्य को D_2 दिनों में पूरा करते हैं, प्रत्येक दिन T_2 समय कार्य करते हैं, तब
$$\frac{M_1 D_1 T_1}{W_1} = \frac{M_2 D_2 T_2}{W_2}$$

यह सूत्र सभी परीक्षा के लिए *M. Imp* है। इस पर आधारित 1 प्रश्न अवश्य पूछा जाता है।

- If food is available for x days for 'A' men at a certain place and after y days B men join, then the remaining food will serve total men for
यदि एक निश्चित स्थान पर A पुरुषों के लिए x दिनों के लिए भोजन उपलब्ध है और y दिनों के बाद B पुरुष शामिल होते हैं, तब शेष भोजन कुल पुरुषों में चलेगा
$$= \frac{A(x-y)}{(A+B)} \text{ day/दिन}$$

ऐकिक नियम पर आधारित उदाहरण निम्न हैं

Example **If 6 people complete a work in 8 days, then find the number of days taken by 4 people to complete it.**

6 व्यक्ति किसी कार्य को 8 दिनों में पूरा करते हैं, तो ज्ञात कीजिए कि 4 व्यक्ति उसे कितने दिनों में पूरा करते हैं?

Sol. $M_1D_1 = M_2D_2$ [By Unitary Method]

$6 \times 8 = 4 \times D_2$

$\therefore \quad D_2 = 12$ days

Example **In a camp of 70 boys, there are food provisions for 36 days. If 50 more boys join the camp, then how many days will the food last?**

70 लड़कों के एक शिविर में, 36 दिनों के लिए भोजन की व्यवस्था है। यदि 50 और लड़के शिविर में शामिल होते हैं, तो फिर भोजन कितने दिनों तक चलेगा?

SSC (10+2) 2023

Sol. Let the food be enough for x days for 50 more boys.

$\therefore \quad 70 \times 36 = (70 + 50) \times x$

$\Rightarrow \quad x = \frac{70 \times 36}{120}$

$\Rightarrow \quad x = 21$ days

Example **30 men can complete a work in 16 days. Six day after they started working, 15 more men joined them. How many days will they now take to complete the remaining work?**

30 व्यक्ति एक कार्य को 16 दिनों में पूरा कर सकते हैं। उनके कार्य शुरू करने के छह दिन बाद, 15 अतिरिक्त व्यक्ति उनके साथ जुड़ जाते हैं। अब उन्हें शेष कार्य को पूरा करने में कितने दिन लगेंगे?

UP POLICE CONSTABLE 2024

Sol. Total work = $30 \times 16 = 480$ units

Work done in 6 days = $6 \times 30 = 180$ units

$\therefore$ Remaining work = $480 - 180 = 300$ units

Now, total number of men = $30 + 15 = 45$

$\therefore$ Required number of days = $\frac{300}{45} = \frac{20}{3}$ days

Alternate Solution

$\therefore$ Required time = $\frac{30(16-6)}{30+15} = \frac{30 \times 10}{45} = \frac{20}{3}$ days

3. Based on Efficiency कार्य कुशलता पर आधारित

- If A can do a work in x days and B is $R\%$ more efficient than A, then B alone will do the same work in $\frac{x \times 100}{100 + R}$ days.

  यदि A किसी कार्य को x दिनों में कर सकता है और B, A की तुलना में $R\%$ अधिक कुशल है, तब B अकेला उसी कार्य को $\frac{x \times 100}{100 + R}$ दिनों में करेगा।

- A can do a work in m days and B can do the same work in n days. If they work together and total wages is R, then
  A किसी कार्य को m दिनों में कर सकता है और B उसी कार्य को n दिनों में कर सकता है। यदि वे एकसाथ काम करते हैं और कुल वेतन R है,

  तब **Part of A / A का हिस्सा** $= \frac{n}{m+n} \times R$

  तथा **Part of B / B का हिस्सा** $= \frac{m}{m+n} \times R$

  Note Ratio of wages of A and B / A और B के वेतनों का अनुपात $= \frac{1}{m} : \frac{1}{n}$

- If A working alone takes x days more than A and B together and B working alone takes y days more than A and B together. Then, the number of days working together to finish a job is $\sqrt{xy}$ days.
  यदि A अकेले कार्य करते हुए A और B दोनों से x दिन अधिक लेता है तथा B अकेले कार्य करते हुए A और B दोनों से y दिन अधिक लेता है, तब A और B दोनों एकसाथ कार्य करते हुए कार्य को $\sqrt{xy}$ दिनों में समाप्त करेंगे।

कार्य कुशलता पर आधारित उदाहरण निम्न है

Example **Anu can complete a piece of work in 22 days. Shama is 60% more efficient than Anu. How many days does Shama alone take to complete the same piece of work?**

अनु एक कार्य को 22 दिनों में पूरा कर सकती है। शमा, अनु से 60% अधिक कुशल है। उसी कार्य को पूरा करने में शमा अकेले कितने दिन लेती है?

SSC (10+2) 2021

Sol. $\therefore$ Required days $= \frac{22 \times 100}{(100 + 60)} = \frac{2200}{160} = 13\frac{3}{4}$ days

Alternate Solution

$E_S : E_A = 160 : 100 = 8 : 5$

Now, $T_S : T_A = 5 : 8$

$(T_A)\ 8 \rightarrow 22$

$\therefore \quad (T_S)\ 5 \rightarrow \frac{22}{8} \times 5 = 13\frac{3}{4}$ days

4. Based on Leaving or Addng in Mid of the Work
कार्य के बीच में छोड़ने या शामिल होने पर आधारित

Example **Amit alone can complete a piece of work in 15 days and Balbir alone can do the same work in 10 days. If Amit alone works for 3 days after which Balbir joins him, then the whole work is finished in how many days?**

अमित अकेले किसी कार्य को 15 दिनों में पूरा कर सकता है और बलबीर उसी कार्य को अकेले 10 दिनों में पूरा कर सकता है। यदि अमित अकेले 3 दिनों तक कार्य करता है, जिसके बाद बलबीर उसके साथ जुड़ जाता है, तो पूरा कार्य कितने दिनों में समाप्त हुआ? *RRB NTPC 2022*

Sol. Total work = LCM of 15 and 10 = 30 units

Amit 1 day work $= \frac{30}{15} = 2$ units

Balbir 1 day work $= \frac{30}{10} = 3$ units

Amit alone work for 3 days $= 2 \times 3 = 6$ units

$\therefore$ Remaining work $= (30 - 6) = 24$ units

If they work together,

Time taken to complete the remaining work $= \frac{24}{(2+3)} = \frac{24}{5} = 4\frac{4}{5}$ days

$\therefore$ Total taken time $= 3 + 4\frac{4}{5} = 7\frac{4}{5}$ days

5. Based on Working on Alternate Days एकान्तर दिन में कार्य करने पर आधारित

Example ***A* and *B* can complete a piece of work in 10 days and 12 days, respectively. If they work on alternate days beginning with *A*, then in how many days will the work be completed?**

A और *B* एक कार्य को क्रमशः 10 दिन और 12 दिन में कर सकते हैं। यदि वे *A* से प्रारम्भ करते हैं और बारी-बारी से कार्य करें, तो कार्य कितने दिनों में पूरा हो जाएगा? *RRC GROUP D 2022*

Sol. Total work = 60 units (LCM of 10 and 12)

Work done by A in 1 day $= \frac{60}{10} = 6$ units

Work done by B in 1 day $= \frac{60}{12} = 5$ units

If they work alternate days,

Total work done in 2 days $= (6 + 5) = 11$ units

Total work done in 10 days $= 11 \times 5 = 55$ units

$\therefore$ Remaining work $= (60 - 55) = 5$ units

Now, on the 11th day, A completed remaining work.

$\therefore$ Time taken to complete the work by $A = \frac{5}{6}$ days

Thus, total time taken $= 10 + \frac{5}{6} = 10\frac{5}{6}$ days

6. Based on Men and Women पुरुषों और महिलाओं पर आधारित

Example **If 12 men or 9 women can do a piece of work in 36 days, then how many days will 16 men and 18 women take to finish the same work?**

यदि 12 पुरुष या 9 महिलाएँ किसी कार्य को 36 दिनों में पूरा कर सकते हैं, तो 16 पुरुष और 18 महिलाएँ उसी कार्य को कितने दिनों में पूरा करेंगे?

SSC CONSTABLE (GD) 2023

Sol. 12 men = 9 women

Men : Women = 3 : 4

Let total work = $12 \times 3 \times 36 = 36 \times 36$ units

Let number of days taken by 16 men and 18 women to complete the same work is D,

According to the question,

$$\therefore \quad (16 \times 3 + 18 \times 4) \times D = 36 \times 36$$

$$\Rightarrow \quad (48 + 72) \times D = 36 \times 36$$

$$\Rightarrow \quad 120 \times D = 36 \times 36$$

$$\therefore \quad D = \frac{54}{5} = 10\frac{4}{5} \text{ days}$$

Example **2 men and 7 boys can complete a work in 28 days, while 6 men and 16 boys can do the same work in 11 days. In how many days 7 men complete the same work?**

2 पुरुष और 7 लड़के एक काम को 28 दिनों में पूरा कर सकते हैं, जबकि 6 पुरुष और 16 लड़के वही काम 11 दिनों में पूरा कर सकते हैं। 7 पुरुष उसी काम को कितने दिनों में पूरा करेंगे?

Sol. (2 men + 7 boys) ×28 = (6 men + 16 boys ×11)

$\Rightarrow$ 56 men +196 boys = 66 men +176 boys

$\Rightarrow$ 10 men = 20 boys $\Rightarrow$ 1 men = 2 boys $\Rightarrow \frac{\text{Men}}{\text{Boys}} = \frac{2}{1}$

Let total work = (2 men + 7 boys) × 28 = $(2 \times 2 \times 7 \times 1) \times 28$

$= 11 \times 28$ units

$\therefore$ 7 men can complete the work $= \frac{11 \times 28}{7 \times 2} = 22$ days

अध्याय 22

Pipe and Cistern

नल और टंकी

Problems on pipe and cistern are based on the basic concept of time and work. One or more pipes are connected to a tank or cistern and are used to fill or empty the tank or cistern.

नल एवं टंकी से सम्बन्धित प्रश्न, कार्य और समय की मूल अवधारणा पर आधारित होते हैं। एक या अधिक नल एक टैंक से जुड़े होते हैं और टैंक को भरने या खाली करने के लिए प्रयुक्त होते हैं।

- **Inlet Pipe** : It fills a tank.
  इनलेट पाइप : यह टैंक को भरता है।
- **Outlet Pipe** : It empties a tank.
  आउटलेट पाइप : यह टैंक को खाली करता है।

1. Based on Basic Concept मूल अवधारणा पर आधारित

- If a pipe can fill/empty a tank in x h, then the part of tank filled/emptied in 1 h $= \frac{1}{x}$

  यदि कोई नल किसी टैंक को x घण्टे में भरता/खाली करता है, तब नल द्वारा 1 घण्टे में टैंक का भरा/खाली किया गया भाग $= \frac{1}{x}$

- Two taps A and B can fill a tank in x h and y h, respectively. If both the taps are opened together, then time taken to fill the tank

  दो नल 'A' और 'B' एक टंकी को क्रमशः x घण्टे और y घण्टे में भर सकते हैं। यदि दोनों नलों को एकसाथ खोल दिया जाए, तो टंकी को भरने में लगा समय

  $$= \left(\frac{xy}{x+y}\right) \text{ h/घण्टे}$$

- Tap A can fill a tank in x h and tap B can empty the tank in y h, then
  नल A एक टंकी को x घण्टे में भर सकता है और नल B टंकी को y घण्टे में खाली कर सकता है, तब

  Time taken to fill the empty tank/खाली टैंक को भरने में लगा समय

  $$= \frac{xy}{y-x} \quad (y > x)$$

- If $x, y, z, \ldots$ all taps are opened together, then the time required to fill/empty the tank will be
यदि $x, y, z, \ldots$ सभी नल एकसाथ खोले जाते हैं, तो टंकी को भरने/खाली करने में लगने वाला समय होगा

$$\frac{1}{x} \pm \frac{1}{y} \pm \frac{1}{z} \pm \ldots = \frac{1}{T}$$

where T, is the required time./ जहाँ T, अभीष्ट समय है।

> **Note** *Positive result shows that the tank is filling and negative result shows that the tank is getting empty.*
> *धनात्मक परिणाम दर्शाता है कि टंकी भर रही है और ऋणात्मक परिणाम दर्शाता है कि टंकी खाली हो रही है।*

नल एवं टंकी की मूल अवधारणा पर आधारित उदाहरण निम्नलिखित हैं

Example **Two pipes A and B can fill a tank in 25 h and 35 h, respectively. If both the pipes are opened simultaneously, how much time will it take to fill the tank?**

दो नल A और B एक टैंक को क्रमशः 25 घण्टे और 35 घण्टे में भर सकते हैं। यदि दोनों पाइप एकसाथ खोल दिए जाते हैं, तो टैंक को भरने में कितना समय लगेगा?

RRC GROUP D 2022

Sol. Required time $= \frac{xy}{x+y} = \left(\frac{25 \times 35}{25+35}\right) = \left(\frac{25 \times 35}{60}\right) = \frac{175}{12}$ h

Alternate Solution

$A \rightarrow 25$ (7)
$B \rightarrow 35$ (5)
175 (Total work)

$\therefore$ Required time $= \left(\frac{175}{7+5}\right) = \frac{175}{12}$ h

Example **Pipe A can fill a tank in 6 h. Pipe B can empty it in 15 h. If both the pipes are opened together, then the tank will be filled in how many hours?**

नल A एक टैंक को 6 घण्टे में भर सकता है। नल B, 15 घण्टे में खाली कर सकता है। यदि दोनों नल एकसाथ खोल दिए जाएँ, तब टैंक कितने घण्टे में भरेगा?

SSC CPO 2023

Sol. Required time $= \frac{xy}{(y-x)} (y > x) = \left(\frac{6 \times 15}{15-6}\right) = \left(\frac{6 \times 15}{9}\right) = 10$ h

Alternate Solution

$A \rightarrow 6$ (5)
$B \rightarrow 15$ (2)
30 (Total work)

$\therefore$ Required time $= \left(\frac{30}{5-2}\right) = 10$ h

Example **Pipe A and B can fill the tank in 10 h and 20 h respectively, where as pipe C can empty the tank in 8 h. If all the pipes are opened together at the same time, what is the time to fill the tank?**

नल A और B टैंक को क्रमशः 10 घण्टे और 20 घण्टे में भर सकते हैं, जहाँ नल C टैंक को 8 घण्टे में खाली करता है। यदि सभी नल एकसाथ एक ही समय पर खोल दिए जाएँ, तब टैंक कितने समय में भरता है? *IBPS CLERK (PRE) 2021*

Sol. Let total required time to fill the tank is t h.
According to the question,

$$\frac{1}{t} = \frac{1}{10} + \frac{1}{20} - \frac{1}{8} \Rightarrow \frac{1}{t} = \frac{4+2-5}{40} = \frac{1}{40} \Rightarrow t = 40\text{ h}$$

Alternate Solution

$A \to 10$ — 4
$B \to 20$ — 2 → 40 (Total work)
$C \to 8$ — −5

$$\therefore \text{ Required time} = \left(\frac{40}{4+2-5}\right) = 40\text{ h}$$

2. Based on Leakage on Tank टैंक में रिसाव पर आधारित

- If a pipe fills a tank in x h but it takes t more hours to fill due to leakage in tank. If tank is filled completely, then it will be empty in

यदि एक पाइप किसी टंकी को x घण्टे में भरता है, लेकिन टंकी में रिसाव होने के कारण उसे भरने में t घण्टे अधिक लगते हैं। यदि टंकी पूरी तरह भर जाए, तब वह टंकी खाली होगी

$$\text{Time/समय} = \frac{x(x+t)}{t}$$

SSC, Railway परीक्षा की दृष्टि से यह सूत्र बहुत महत्त्वपूर्ण है इसका अभ्यास अवश्य कीजिए।

टैंक में रिसाव पर आधारित उदाहरण निम्नलिखित है

Example **A pipe can fill a tank in 15 h. Due to a leak in the bottom, it is filled in 20 h. If the tank is full, how much time will the leak take to empty it?**

एक नल एक खाली टैंक को 15 घण्टे में भर सकता है, किन्तु तली में रिसाव के कारण यह 20 घण्टे में भरता है। यदि टैंक पूरा भरा हो, तो यह कितने समय में पूरा खाली होगा? *RRC GROUP D 2022*

Sol. Due to leak extra time to fill tank = (20 − 15) = 5 h

$$\therefore \text{ Required time} = \frac{x(x+t)}{t}$$

$$= \frac{15(15+5)}{5} = 3 \times 20 = 60\text{ h}$$

3. Based on Opening/Closing a Tap in Midway
नल को बीच में खोलने या बन्द करने पर आधारित

- Two taps A and B can fill a tank in x h and y h, respectively. If both the pipes are opened together then the time after which pipe B should be closed so that tank is full in t h.

दो नल A और B एक टंकी को क्रमशः x घण्टे और y घण्टे में भर सकते हैं। यदि दोनों पाइपों को एकसाथ खोल दिया जाता है, तब पाइप B को कितने समय बाद बन्द किया जाए कि टंकी t घण्टे में भर जाए।

$$\text{Time/समय} = y\left(1-\frac{t}{x}\right) \text{ h/घण्टे}$$ *M. Imp.*

नल को बीच में खोलने या बन्द करने पर आधारित उदाहरण निम्नलिखित है

Example **Two pipes A and B can fill a tank in 24 min and 32 min, respectively. If both the pipes are opened simultaneously, after how much time should B be closed, so that the tank is full in 18 min?**

दो पाइप A और B एक टैंक को क्रमशः 24 मिनट और 32 मिनट में भर सकते हैं। यदि दोनों पाइपों को एकसाथ खोल दिया जाता है, तो पाइप B को कितने समय बाद बन्द कर देना चाहिए, ताकि टैंक 18 मिनट में भर जाए? *SSC CPO 2023*

Sol. Required time $= \left[y\left(1-\frac{t}{x}\right)\right] = \left[32\left(1-\frac{18}{24}\right)\right]$

$$= \left[32 \times \frac{6}{24}\right] = 8 \text{ min}$$

4. Based on Efficiency of Taps नलों की क्षमता पर आधारित

Example **Pipe A fill twice as fast as pipe B and pipe B fills twice as fast as pipe C. Pipe C can fill a tank in 20 h. In how many hours will the tank fill if all the pipes are opened?**

पाइप A किसी टंकी को पाइप B की तुलना में दोगुनी गति से भरता है और पाइप B, पाइप C की तुलना में दोगुनी गति से भरता है। पाइप C उस टंकी को 20 घण्टे में भर सकता है। यदि सभी पाइपों को खोल दिया जाए, तो टंकी कितने घण्टों में भर जाएगी? *SBI PO (PRE) 2022*

Sol. The efficiency ratio of A, B and $C = 4:2:1$

$\because$ Time taken by pipe $C = 20$ h

$\therefore$ Time taken by pipe A and B is 5 h and 10 h, respectively.

Time taken by A, B and C together

$$= \frac{1}{\left(\frac{1}{5}+\frac{1}{10}+\frac{1}{20}\right)} = \frac{20}{(4+2+1)} = \frac{20}{7} \text{ h} = 2\frac{6}{7} \text{ h}$$

अध्याय 23

Speed, Time and Distance
चाल, समय और दूरी

- **Speed/चाल** The rate at which an object covers a certain distance is called the speed of that object.
  वह दर जिस पर कोई वस्तु एक निश्चित दूरी तय करती है, उस वस्तु की चाल कहलाती है।
- **Distance/दूरी** Distance is basically the length of the path between two points on a plane.
  दूरी मूलरूप से एक समतल पर दो बिन्दुओं के बीच के पथ की लम्बाई है।
- **Time/समय** Time is basically the period taken to cover a certain distance.
  समय मूलरूप से एक निश्चित दूरी तय करने में लिया गया अन्तराल है।

1. Based on Basic Concept मूल अवधारणा पर आधारित

- Distance (D) = Speed $(S) \times$ Time (T)

  दूरी = चाल × समय

  $1 \text{ km/h} = \frac{5}{18} \text{ m/s}$ and $1 \text{ m/s} = \frac{18}{5} \text{ km/h}$

  1 किमी/घण्टा $= \frac{5}{18}$ मी/से तथा 1 मी/से $= \frac{18}{5}$ किमी/घण्टा

> ये सूत्र सभी परीक्षाओं SSC, Railway, UPSC (CSAT)... की दृष्टि से M. Imp हैं। इन पर आधारित 1 प्रश्न परीक्षा में अवश्य पूछा जाता है।

Case I/स्थिति I If D is constant/यदि D अचर है

$$S \propto \frac{1}{T}$$

$$\therefore \quad \frac{S_1}{S_2} = \frac{T_2}{T_1}$$

Case II/स्थिति II If time is constant/यदि समय अचर है

$$D \propto S$$

$$\frac{D_1}{D_2} = \frac{S_1}{S_2}$$

Case III/स्थिति III If speed is constant/यदि चाल अचर है

$$D \propto T$$

$$\therefore \quad \frac{D_1}{D_2} = \frac{T_1}{T_2}$$

मूल अवधारणा पर आधारित उदाहरण निम्न हैं

Example **A car can cover a distance of 144 km in 1.8 h. In what time (in hours) will it cover double the distance when its speed is increased by 20%?**

एक कार 1.8 घण्टे में 144 किमी की दूरी तय कर सकती है। जब इसकी गति में 20% की वृद्धि की जाती है, तो यह कितने समय में (घण्टों में) दोगुनी दूरी तय करेगी? *SSC CGL 2020*

Sol. According to the question,

$$\text{Original speed} = \frac{144}{1.8} = 80 \text{ km/h}$$

Now, speed increased by 20%

$$\therefore \quad \text{New speed} = 80 \times \frac{120}{100} = 96 \text{ km/h}$$

$$\text{Required time} = \frac{2 \times 144}{96} = 3 \text{ h}$$

Example **The ratio of the speed of a bus and a car is 7 : 11. If the car covers a distance of 396 km in 6 h, what is the speed of the bus in km/h?**

एक बस और एक कार की चालों का अनुपात 7 : 11 है। यदि कार 6 घण्टे में 396 किमी की दूरी तय करती है, तो बस की चाल किमी/घण्टा में कितनी है? *RRB NTPC (CBT II) 2022*

Sol. $\text{Speed of car} = \frac{396}{6} = 66 \text{ km/h}$

According to the question,

$$\text{Speed of bus} = 66 \times \frac{7}{11} = 42 \text{ km/h}$$

2. Based on Meeting Each Other एक-दूसरे से मिलने पर आधारित

- If A and B starts walking towards each other. After meeting each other, A covered his remaining distance in t_1 time and B covered his remaining distance in t_2 time. Then, the ratio of their speed

  यदि A और B एक-दूसरे की ओर चलना प्रारम्भ करते हैं। आपस में मिलने के बाद A अपनी शेष दूरी t_1 समय में और B अपनी शेष दूरी को t_2 समय में तय करता है, तब

  उनकी गति का अनुपात $\frac{A}{B} = \sqrt{\frac{t_2}{t_1}}$ *M. Imp*

  If they meet after t h, then/यदि वे t घण्टे के बाद मिलते हैं, तब $t = \sqrt{t_1 t_2}$

Example **Two motorists travelling in opposite directions meet at some point in between. After this, they take 9 h and 16 h to reach their respective destination. What is the ratio of their speeds?**

विपरीत दिशाओं में यात्रा करने वाले दो मोटर चालक बीच में किसी बिन्दु पर मिलते हैं। इसके बाद उन्हें अपने सम्बन्धित गन्तव्य तक पहुँचने में 9 और 16 घण्टे लगते हैं। उनकी गति का अनुपात क्या है? *RRB JE 2019*

Sol. $\frac{\text{Speed of first motorist}}{\text{Speed of second motorist}} = \sqrt{\frac{16}{9}} = \frac{4}{3} = 4:3$

3. Relative Speed सापेक्ष चाल

- When two objects travel in same direction, then

  Relative speed = Difference of speeds

  जब दो वस्तुएँ एक ही दिशा में चलती हैं, तब **सापेक्ष चाल = चालों का अन्तर**

- When two objects travel in opposite direction, then

  Relative speed = Sum of speeds

  जब दो वस्तुएँ विपरीत दिशा में चलती हैं, तब **सापेक्ष चाल = चालों का योग**

Example **Piu and Titu start from the same place and at the same time and walk in opposite directions. If Piu walks at 5 km/h and Titu walks at 2 km/h, then after what time will they be 10.5 km apart?**

पीयू और टीटू एक ही जगह से एक ही समय पर विपरीत दिशाओं में चलना शुरू करते हैं। यदि पीयू 5 किमी/घण्टा की चाल से चलता है और टीटू 2 किमी/घण्टा की चाल से चलता है, तो कितने समय बाद उनके बीच की दूरी 10.5 किमी हो जाएगी? *SSC MTS 2022*

Sol. Relative speed $= (5 + 2) = 7$ km/h

$\therefore$ Required time $= \frac{10.5}{7} = 1.5$ h

4. Average Speed औसत गति

- Average speed/औसत चाल $= \frac{\text{Total distance/कुल दूरी}}{\text{Total time/कुल समय}}$ *M. Imp*

- If a man travels different distances $d_1, d_2, d_3 \ldots$ and so on in different time $t_1, t_2, t_3 \ldots$ respectively, then

  यदि एक व्यक्ति अलग-अलग दूरी $d_1, d_2, d_3 \ldots$ क्रमशः $t_1, t_2, t_3 \ldots$ समय में तय करता है, तब

  Average speed/औसत गति $= \frac{d_1 + d_2 + d_3 + \ldots}{t_1 + t_2 + t_3 + \ldots}$

- If a man travels different distances $d_1, d_2, d_3, \ldots$ and so on with different speeds $s_1, s_2, s_3 \ldots$ respectively, then
यदि एक व्यक्ति अलग-अलग दूरी $d_1, d_2, d_3 \ldots$ क्रमशः $s_1, s_2, s_3 \ldots$ चालों से तय करता है, तब

$$\textbf{Average speed/औसत गति} = \frac{d_1 + d_2 + d_3 + \ldots}{\frac{d_1}{s_1} + \frac{d_2}{s_2} + \frac{d_3}{s_3} + \ldots}$$

- If a distance is divided into n equal parts, each travelled with different speed, then
यदि एक दूरी को n बराबर भागों में विभाजित किया जाता है, प्रत्येक ने अलग-अलग गति से यात्रा की, तब

$$\textbf{Average speed/औसत गति} = \frac{n}{\frac{1}{s_1} + \frac{1}{s_2} + \frac{1}{s_3} + \ldots + \frac{1}{s_n}}$$

- If a bus travels from A to B with the speed of x km/h and returns from B to A with the speed of y km/h, then the average speed will be $\left(\frac{2xy}{x+y}\right)$ **km/h.**
यदि एक बस A से B तक x किमी/घण्टा की गति से यात्रा करती है और B से A तक y किमी/घण्टा की गति से वापस आती है, तब औसत गति $\left(\frac{2xy}{x+y}\right)$ किमी/घण्टा होगी।

अतिसम्भावित! इस सूत्र पर आधारित 1 प्रश्न परीक्षाओं में पूछा जाता है।

- Excluding stoppage, the average speed of an object is u and with stoppage its average speed is v. Then,
ठहराव को छोडकर, एक वस्तु की औसत गति u है और ठहराव के साथ इसकी औसत गति v है। तब,

The stoppage time per hour/प्रति घण्टे रूकने का समय $= \frac{u-v}{u}$

औसत गति पर आधारित उदाहरण निम्नलिखित हैं

Example **A bus travels for 3 h at 60 km/h, 4 h at 50 km/h and next 5 h at 50 km/h. Find its average speed.**
एक बस 3 घण्टे, 60 किमी/घण्टा की गति से, 4 घण्टे, 50 किमी/घण्टा की गति से और अगले 5 घण्टे, 50 किमी/घण्टा की गति से यात्रा करती है। इसकी औसत गति ज्ञात कीजिए। *SSC (10+2) 2023*

Sol. Average speed $= \left(\frac{3 \times 60 + 4 \times 50 + 5 \times 50}{3 + 4 + 5}\right)$

$= \left(\frac{180 + 200 + 250}{12}\right) = \frac{630}{12} = 52.5$ km/h

Example **A person travels at the speed of 12 km/h, 18 km/h and 36 km/h on an equilateral triangular plane. Find the average speed of the journey.**

एक व्यक्ति एक समबाहु त्रिभुजाकार तल पर 12 किमी/घण्टा, 18 किमी/घण्टा और 36 किमी/घण्टा की चाल से यात्रा करता है। यात्रा की औसत चाल ज्ञात कीजिए।

Sol. In equilateral triangle, each side are equal.

$$\text{Average speed} = \frac{3}{\left(\frac{1}{12}+\frac{1}{18}+\frac{1}{36}\right)} = \frac{3}{\frac{3+2+1}{36}}$$

$$= \frac{3\times 36}{(3+2+1)} = \frac{3\times 36}{6} = 18\text{ km/h}$$

5. Some Other Cases कुछ अन्य स्थितियाँ

- If an object travels certain distance with the speed of $\frac{A}{B}$ of its original speed and reaches its destination 't' h before or after, then the time taken by object travelling at original speed is

  यदि कोई वस्तु अपनी मूल गति की $\frac{A}{B}$ गति से निश्चित दूरी तय करती है और अपने गन्तव्य पर t घण्टे पहले या बाद में पहुँचती है, तो वस्तु द्वारा मूल गति से यात्रा करने में लिया गया समय $= \frac{A}{A \sim B} \times t$

Example **A bike is moving with $\frac{3}{5}$th of its original speed and reaches 15 min late at the destination, find the actual time taken by the bike to reach the destination.**

एक बाइक अपनी मूल चाल की $\frac{3}{5}$ भाग के साथ आगे बढ़ती है और गन्तव्य पर 15 मिनट देर से पहुँचती है, बाइक द्वारा गन्तव्य तक पहुँचने के लिए लिया गया वास्तविक समय ज्ञात कीजिए।

Sol. Required time $= \left(\frac{3}{5-3}\right) \times 15 = \frac{45}{2} = 22.5$ min

- If a man travels at the speed of S_1, then he reaches his destination t_1 time late while he reaches t_2 time before when he travels at S_2 speed, then the distance between the two places is

  यदि एक व्यक्ति S_1 गति से यात्रा करता है, तो वह अपने गन्तव्य पर t_1 समय देर से पहुँचता है जबकि वह t_2 समय पहले पहुँचता है, जब वह S_2 गति से यात्रा करता है, तो दो स्थानों के बीच की दूरी $= \frac{(S_1 \times S_2) \times (t_1 + t_2)}{S_2 - S_1}$

Example **If a man travels at the speed of 30 km/h, then he reaches his destination late by 10 min and if he travels at the speed of 42 km/h then he reaches 10 min earlier. Find the distance travelled by him.**

यदि एक व्यक्ति 30 किमी/घण्टा की चाल से यात्रा करता है, तो वह अपने गन्तव्य पर 10 मिनट देरी से पहुँचता है और यदि वह 42 किमी/घण्टा की गति से यात्रा करता है, तो वह 10 मिनट पहले पहुँच जाता है। उसके द्वारा तय की गई दूरी ज्ञात कीजिए। *ODISHA POLICE SI 2022*

Sol. Required distance $=\left[\dfrac{(30\times 42)\times\dfrac{(10+10)}{60}}{(42-30)}\right]$ km

$$=\frac{420}{12}=35\text{ km}$$

- If anyone overtakes or follows another one, then time taken to catch
  अगर कोई किसी को ओवरटेक या पीछा करता है, तो उसे पकड़ने में लगा समय

$$=\frac{\text{Distance between them / उनके बीच की दूरी}}{\text{Relative speed / सापेक्ष गति}}$$

or/या Meeting time/मिलने का समय

$$=\frac{\text{Speed of Ist}\times\text{Time}}{\text{Difference of both speeds}}\Big/\frac{\text{पहले की चाल}\times\text{समय}}{\text{दोनों चालों का अन्तर}}$$

or/या Total distance travelled to catch the thief

$$=\frac{\text{Product of speeds}\times\text{Time}}{\text{Difference of speed}}$$

चोर को पकड़ने के लिए तय की गई कुल दूरी $=\dfrac{\text{चालों का गुणनफल}\times\text{समय}}{\text{चालों का अन्तर}}$ *M. Imp*

Example **A thief is spotted by a policeman from a distance of 100 m. When the policeman starts the chase, the thief also starts running. If the speed of the thief is 8 km/h and that of the policeman is 10 km/h, then how for will the thief have to run before he is overtaken?**

एक चोर को एक पुलिसकर्मी द्वारा 100 मी की दूरी से देखा जाता है। जब पुलिसकर्मी पीछा करना शुरू करता है, तो चोर भी भागना शुरू कर देता है। यदि चोर की गति 8 किमी/घण्टा और पुलिसकर्मी की गति 10 किमी/घण्टा है, तो पुलिसकर्मी द्वारा पकड़े जाने से पहले चोर कितनी दूर भाग चुका था? *CDS 2017*

Sol. Relative speed between thief and policeman

$$= (10 - 8) = 2\text{ km/h}$$

$$= 2 \times \frac{5}{18}\text{ m/s} = \frac{5}{9}\text{ m/s} \qquad \left[\because 1\text{ km/h} = \frac{5}{18}\text{ m/s}\right]$$

$\therefore$ Time taken by policeman to catch thief $= \dfrac{100}{\frac{5}{9}} = 180$ sec

So, required distance $= \left(8 \times \dfrac{5}{18}\right) \times 180 = 400$ m

- A man covers a certain distance with x km/h and comes back with a speed of y km/h. If he takes t h to go and come back, then
एक आदमी x किमी/घण्टा से एक निश्चित दूरी तय करता है और y किमी/घण्टा की गति से वापस आता है। यदि उसे जाने और वापस आने में t घण्टे लगते हैं, तब
Distance covered (one side)/तय दूरी (एक ओर की) $= \left(\dfrac{xy}{x+y}\right) \times t$

- A man covers a certain distance with x km/h and he comes back with a speed of y km/h. If he takes t h more to come back than going, then
एक आदमी x किमी/घण्टा से एक निश्चित दूरी तय करता है और वह y किमी/घण्टा की गति से वापस आता है। यदि वह जाने से वापस आने में t घण्टे अधिक लेता है, तब
Cover distance (one side)/तय दूरी (एक ओर की)
$= \left(\dfrac{xy}{x-y}\right) \times$ (Difference between time/समयों का अन्तर)

Example **A man travel a certain distance by train at the rate of 25 km/h and walk back at the rate 3 km/h. The whole journey took 5 h 36 min. What distance did he travel by train?**

एक व्यक्ति एक निश्चित दूरी रेलगाडी से 25 किमी/घण्टा की गति से तय करता है और 3 किमी/घण्टा की गति से वापस चलता है। पूरी यात्रा में 5 घण्टा 36 मिनट का समय लगा। उसने रेलगाडी से कितनी दूरी तय की?

Sol. Covered distance $= \dfrac{25 \times 3}{25 + 3} \times \left(5 + \dfrac{36}{60}\right)$

$$= \frac{25 \times 3}{28} \times \frac{28}{5} = 15\text{ km}$$

अध्याय 24

Train रेलगाड़ी

Problems on trains are based on the basic concepts of speed, time and distance.
Here, length of trains and its different scenarios are considered.
रेलगाड़ी से सम्बन्धित समस्याएँ चाल, समय और दूरी की मूल अवधारणाओं पर ही आधारित होती हैं।
यहाँ ट्रेन की लम्बाई तथा विभिन्न स्थितियों को ध्यान में रखा जाता है।

1. When a Train passes a Pole/Stationary Man or Bridge/Platform

जब एक ट्रेन किसी खम्भे/स्थिर व्यक्ति या पुल/प्लेटफॉर्म को पार करती है

(i) When train passes a pole or stationary man, then
जब ट्रेन किसी खम्भे या स्थिर आदमी को पार करती है, तब
Covered distance/तय की गई दूरी
= Length of train/ट्रेन की लम्बाई

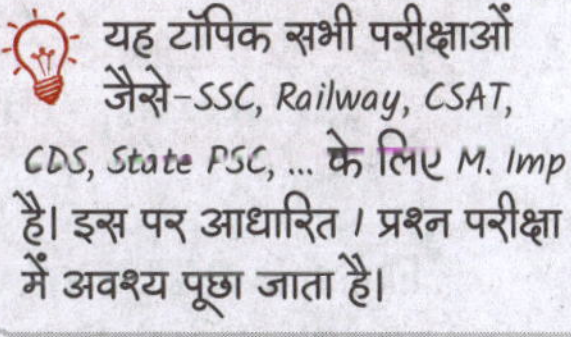
यह टॉपिक सभी परीक्षाओं जैसे-SSC, Railway, CSAT, CDS, State PSC, ... के लिए M. Imp है। इस पर आधारित 1 प्रश्न परीक्षा में अवश्य पूछा जाता है।

(ii) When train passes a bridge/platform...etc, then
Covered distance = Length of train + Length of bridge
जब ट्रेन पुल/प्लेटफॉर्म... आदि से गुजरती है, तब
तय की गई दूरी = ट्रेन की लम्बाई + पुल की लम्बाई

- If a train crosses a standing man/a pole in 't_1' sec and crosses 'p' m long platform in t_2 sec, then
यदि कोई रेलगाड़ी किसी स्थिर व्यक्ति/खम्भे को t_1 सेकण्ड में पार करती है तथा 'p' मी लम्बे प्लेटफॉर्म को t_2 सेकण्ड में पार करती है, तो

$$\text{Length of train/रेलगाड़ी की लम्बाई} = \frac{p \times t_1}{t_2 - t_1}$$

Example **A 180 m long train is moving at a speed of 90 km/h. How long will it take to pass a pole?**
एक 180 मी लम्बी रेलगाड़ी 90 किमी/घण्टा की गति से चल रही है। एक खम्भे को पार करने में यह कितना समय लेगी? *CISF HEAD CONSTABLE 2023*

Sol. Required time to pass a pole $= \left(\frac{180}{90} \times \frac{18}{5}\right) = 7.2$ sec

Example **A train of length 250 m, passes a platform of 350 m length in 50 sec. What time will this train take to pass the platform 230 m length?**

एक 250 मी लम्बी रेलगाड़ी, 350 मी लम्बे प्लेटफॉर्म को 50 सेकण्ड में पार करती है। 230 मी लम्बाई के प्लेटफॉर्म को पार करने में यह रेलगाड़ी कितना समय लेगी?

Sol. Speed of train $= \left(\frac{250+350}{50}\right) = \frac{600}{50} = 12$ m/s

$\therefore$ Required time taken $= \left(\frac{250+230}{12}\right) = \frac{480}{12} = 40$ sec

2. When a Train passes a Walking/Running Man or a Person sitting in a moving Train जब एक रेलगाड़ी, दूसरी रेलगाड़ी में चलते हुए व्यक्ति या बैठे हुए व्यक्ति को पार करती है

When a train of length L passes a person sitting in another moving train, then

जब एक L लम्बाई की रेलगाड़ी दूसरी चलती रेलगाड़ी में बैठे व्यक्ति को पार करती है, तब

Time taken to crosses/पार करने में लगा समय,

$$T = \frac{L}{\text{Relative speed / सापेक्ष चाल}}$$

M. Imp

Example **A train passes two persons who are walking in the opposite direction of the train at the rates of 4 m/s and 10 m/s in 10 sec and 8 sec, respectively. What is the speed of train?**

एक रेलगाड़ी दो व्यक्तियों जो रेलगाड़ी की विपरीत दिशा में 4 मी/से और 10 मी/से की दर से चल रहे हैं, को क्रमशः 10 सेकण्ड और 8 सेकण्ड में पार करती है। रेलगाड़ी की चाल क्या है? *RRB NTPC 2021*

Sol. Let the length and the speed of train be x m and y m/s, respectively.

$\therefore \quad x = (y+4) \times 10$ and $x = (y+10) \times 8$

According to the question,

$\Rightarrow \quad (y+4) \times 10 = (y+10) \times 8 \Rightarrow y = \frac{40}{2} = 20$ m/s

3. When a Train passes another moving Train

जब एक ट्रेन दूसरी चलती हुई ट्रेन को पार करती है

- Two trains of lengths L_1 and L_2 are moving on parallel tracks with speeds S_1 and S_2, respectively.

  L_1 और L_2 की लम्बाई की दो रेलगाड़ियाँ क्रमशः S_1 और S_2 चाल से समान्तर पटरियों पर गतिमान है।

(i) When a train passes another train in opposite direction, then
जब एक रेलगाड़ी विपरीत दिशा में दूसरी रेलगाड़ी को पार करती है, तब
Time taken to cross/पार करने में लगा समय

$$T = \frac{L_1 + L_2}{S_1 + S_2}$$

ये सूत्र सभी प्रतियोगी परीक्षा की दृष्टि से महत्वपूर्ण है। इसलिए इनका अभ्यास करें।

(ii) When a train passes another train in same direction
जब एक रेलगाड़ी उसी दिशा में दूसरी रेलगाड़ी को पार करती है
Time taken to cross/पार करने में लगा समय $T = \frac{L_1 + L_2}{S_1 - S_2}$

- If two trains of same lengths are coming from same direction and crosses a man in t_1 sec and t_2 sec, then time taken by both the trains to cross each other
  यदि समान लम्बाई की दो रेलगाड़ियाँ एक ही दिशा से आ रही हैं और एक आदमी को t_1 और t_2 सेकण्ड में पार करती हैं, तो दोनों रेलगाड़ियों द्वारा एक-दूसरे को पार करने में
  लिया गया समय $= \frac{2 \times \text{Product of time/समय का गुणनफल}}{\text{Difference of time/समय का अन्तर}}$ *M. Imp*
- If two trains of same lengths are coming from opposite directions and crosses a man in t_1 sec and t_2 sec, then time taken by both trains to cross each other
  यदि समान लम्बाई की दो ट्रेनें विपरीत दिशाओं से आ रही हैं और एक आदमी को t_1 सेकण्ड और t_2 सेकण्ड में पार करती हैं, तो दोनों ट्रेनों द्वारा एक-दूसरे को पार करने में
  लिया गया समय $= \frac{2 \times \text{Product of time/समय का गुणनफल}}{\text{Sum of time/समय का योग}}$ *M. Imp*

Example **Two trains of lengths 80 m and 90 m are moving in opposite direction at 10 m/s and 7 m/s, respectively. Find the time taken by the train to cross each other.**

दो रेलगाड़ियाँ जिनकी लम्बाई 80 मी तथा 90 मी है विपरीत दिशा में क्रमशः 10 मी/से तथा 7 मी/से से चल रही हैं। रेलगाड़ी द्वारा एक-दूसरे को पार करने में लिया गया समय ज्ञात कीजिए।

Sol. Required time $= \frac{80 + 90}{10 + 7} = \frac{170}{17} = 10$ sec

Example **Two trains of lengths 75 m and 95 m are moving in the same direction at 9 m/s and 8 m/s, respectively. Find the time taken by the faster train to cross the slower train.**

दो रेलगाड़ियाँ जिनकी लम्बाई 75 मी तथा 95 मी हैं, एकसमान दिशा में क्रमशः 9 मी/से तथा 8 मी/से से चल रही हैं। तेज रेलगाड़ी द्वारा, धीमी रेलगाड़ी को पार करने में लगने वाला समय ज्ञात कीजिए।

Sol. Required time $= \frac{75 + 95}{9 - 8} = 170$ sec

अध्याय 25

Boat and Stream

नाव और धारा

The concept of boat and stream is also based on speed, time and distance. Here, not only the speed of boat or swimmer but also the speed of current or river is taken into consideration.

नाव और धारा की अवधारणा भी चाल, समय और दूरी पर आधारित है। यहाँ न केवल नाव या तैराक की चाल बल्कि धारा या नदी की चाल को भी ध्यान में रखा जाता है।

Basic Rules Related to Boat and Stream
नाव एवं धारा से सम्बन्धित सामान्य नियम

- If speed of boat in still water $= x$ km/h and speed of current/stream $= y$ km/h, then
  यदि शान्त जल में नाव की चाल $= x$ किमी/घण्टा तथा धारा की चाल $= y$ किमी/घण्टा, तब

  (i) Downstream speed/धारा के अनुकूल चाल, $D = x + y$

  (ii) Upstream speed/धारा के प्रतिकूल चाल, $U = x - y$

  Also/साथ ही, Speed of boat/नाव की चाल $= \dfrac{D+U}{2}$

  Speed of current/धारा की चाल $= \dfrac{D-U}{2}$

  परीक्षा ब्रह्मास्त्र
  सभी परीक्षाओं की दृष्टि से ये सूत्र महत्त्वपूर्ण हैं। इन पर आधारित 1 प्रश्न अवश्य पूछा जाता है।

  (iii) If a swimmer takes same time to travel d_1 km downstream and d_2 km upstream, then
  यदि एक तैराक d_1 किमी धारा के अनुकूल और d_2 किमी धारा के प्रतिकूल यात्रा करने में समान समय लेता है, तब

  $$\frac{\text{Speed of swimmer or boat / नाव या तैराक की चाल}}{\text{Speed of stream / धारा की चाल}} = \frac{d_1 + d_2}{d_1 - d_2}$$

  (iv) A swimmer or boat travels a certain distance upstream in t_1 h, while it takes t_2 h to travel same distance downstream, then
  एक तैराक या नाव एक निश्चित दूरी को धारा के प्रतिकूल t_1 घण्टे में तय करता है, जबकि समान दूरी को धारा के अनुकूल तय करने में t_2 घण्टे का समय लेता है, तब

  $$\frac{\text{Speed of swimmer / तैराक की चाल}}{\text{Speed of stream / धारा की चाल}} = \frac{t_1 + t_2}{t_1 - t_2}$$

- Let the speed of boat is x km/h and speed of stream is y km/h. To travel d_1 km downstream and d_2 km upstream, the time is t h, then

माना नाव की चाल x किमी/घण्टा तथा धारा की चाल y किमी/घण्टा हैं, d_1 किमी धारा के अनुकूल तथा d_2 किमी धारा के प्रतिकूल तय करने में t घण्टे लगते हैं, तब

$$\frac{d_1}{x+y}+\frac{d_2}{x-y}=t$$

नाव तथा धारा पर आधारित मिश्रित उदाहरण निम्नवत् हैं

Example **The speed of stream is 3 km/h and the speed of man in still water is 5 km/h. The time taken by the man to swim 26 km downstream is**

धारा की गति 3 किमी/घण्टा तथा स्थिर जल में व्यक्ति की गति 5 किमी/घण्टा है। व्यक्ति द्वारा धारा के अनुकूल 26 किमी तैरने में लगने वाला समय ज्ञात कीजिए। ***RRB NTPC 2021***

Sol. Rate of downstream = (3 + 5) = 8 km/h

$$\therefore \quad \text{Time}=\frac{\text{Distance}}{\text{Speed}}=\frac{26}{8}=\frac{13}{4}=3\frac{1}{4}\text{ h}$$

Example **The speed of a boat in still water is 14 km/h. It goes 28 km downstream in 1 h 45 min. Find the speed of the stream.**

स्थिर जल में एक नाव की गति 14 किमी/घण्टा है। यह 28 किमी धारा के अनुकूल 1 घण्टे 45 मिनट में जाती है। धारा की गति ज्ञात कीजिए। ***RRB NTPC (CBT-II) 2022***

Sol. Assume the speed of stream be x km/h.

$\therefore$ Rate of downstream = $(x + 14)$ km/h

$$\Rightarrow \quad \frac{7}{4}=\frac{28}{(14+x)} \qquad \left[\because 1\text{ h }45\text{ min}=\frac{7}{4}\text{ h}\right]$$

$$\Rightarrow \quad 7(14+x)=28\times 4$$

$$\Rightarrow \quad 14+x=16 \Rightarrow x=16-14$$

$$\therefore \quad x=2\text{ km/h}$$

Example **The speed of boat is 6 km/h in still water and the speed of the stream is 2 km/h, if the boat takes 6 h to go to a place and come back. Then, find the distance of the place.**

शान्त जल में नाव की गति 6 किमी/घण्टा है और धारा की गति 2 किमी/घण्टा है, यदि नाव एक स्थान पर जाने तथा वापस आने में 6 घण्टे का समय लेती है, तब स्थानों के बीच की दूरी ज्ञात कीजिए। ***IBPS RRB OFFICE ASSISTANT (PRE) 2022***

Sol. Speed of boat in still water = 6 km/h

Speed of current = 2 km/h

$$\therefore \quad \frac{d}{6+2}+\frac{d}{6-2}=6 \qquad [\because \text{Assume distance} = d\text{ km}]$$

$$\Rightarrow \quad \frac{d}{8}+\frac{d}{4}=6 \Rightarrow \frac{d+2d}{8}=6$$

$$\Rightarrow \quad 3d=48 \Rightarrow d=16\text{ km}$$

$$\therefore \quad d=16\text{ km}$$

Example **A boat can travel 5 km upstream in 15 min. If the ratio of the speed of the boat in still water to the speed of the stream is 6 : 1, then how much time will the boat take to cover 19.6 km downstream?**

एक नाव धारा के प्रतिकूल 15 मिनट में 5 किमी दूरी तय कर सकती है। यदि शान्त जल में नाव व धारा की गति का अंनुपात 6 : 1 है, तब धारा के अनुकूल नाव को 19.6 किमी की दूरी तय करने में कितना समय लगेगा? *SSC MTS 2022*

Sol. Upstream speed $= 5 \times \frac{60}{15} = 20$ km/h

Speed of boat in still water : Speed of stream = 6 : 1

$\therefore$ 6 : 1 (−) (+) = 5 → 20

$\Rightarrow$ 1 → 4

$\Rightarrow$ 7 → 28

$\therefore$ Required time taken $= \frac{19.6}{28} \times 60 = 42$ min

Example **The ratio of the speed of the boat in still water to the speed of the stream is 8 : 3 and the boat can travel 22 km downstream in 24 min. Then how much time will the boat take to cover 50 km upstream?**

स्थिर जल में नाव की चाल और धारा की चाल का अनुपात 8 : 3 है और नाव धारा के अनुकूल 24 मिनट में 22 किमी का सफर तय कर सकती है, तो नाव को धारा के प्रतिकूल 50 किमी दूरी तय करने में कितना समय लगेगा? *SSC CHSL 2023*

Sol. Let the speed of boat in still water and speed of stream = $8x$ and $3x$.
Speed of boat to travel 22 km downstream in 24 min.

$$\Rightarrow (8x + 3x) = \frac{22}{\frac{24}{60}} \Rightarrow 11x = 55 \text{ km/h}$$

$$x = 5 \text{ km/h}$$

Then speed of boat in still water and speed of stream

= 40 km/h and 15km/h

Therefore, time taken by boat to cover 50 km

$$\text{Upstream} = \frac{50}{40-15} = 2\text{ h}$$

अध्याय 26

Race दौड़

In a race between two or more contestants, question are related to their winning or losing to each other by a certain distance on time. These questions are based on the concept of time, speed and distance.

दौड़ में दो या अधिक प्रतियोगियों में से किसी के भी जीतने, हारने या कितने समय, कितनी दूरी से हारने, जीतने सम्बन्धी प्रश्न होते हैं। यह प्रश्न चाल, समय व दूरी की अवधारणा पर आधारित होते हैं।

1. When A, Beats B by d m/जब A, B को d मी से हराता है,

- If in x m race, A beats B by d m. Then,
  यदि x मी दौड़ में, A, B को d मी से हराता है। तब,
  For the same time interval
  समान समय अन्तराल के लिए

$$D \propto S$$

$$\frac{D_A}{D_B} = \frac{S_A}{S_B} = \frac{x}{x-d}$$

> दौड़ टॉपिक SSC की परीक्षाओं के लिए M. Imp है। इस टॉपिक से 1 प्रश्न पूछा जा रहा है।

D_A and/और D_B → Distance covered by A and B in same interval
समान समय अन्तराल में A और B द्वारा तय दूरी
S_A and/और S_B → Speed of A and B/A और B की चाल

- If in a race of length L, the time taken by A and B be t_A and t_B $(t_B > t_A)$, then the distance (d) by which A beats B.
  यदि लम्बाई L की दौड़ में, A और B द्वारा लिया गया समय t_A और t_B $(t_B > t_A)$ हो, तो दूरी (d) जिससे A, B को हराता है

$$= \left(\frac{L}{t_B}\right)(t_B - t_A)$$ *M. Imp*

Example **In a race of 1200 m, Ram can beat Shyam by 200 m or by 20 sec. What must be the speed of Ram?**

1200 मी की दौड़ में राम, श्याम को 200 मी या 20 सेकण्ड से हरा सकता है। राम की चाल क्या होगी? **SSC CPO 2022**

Sol. Distance covered by Ram = 1200 m
Distance covered by Shyam = 1000 m

$$\text{Speed of Shyam} = \frac{200}{20} = 10 \text{ m/s}$$

Now, $\frac{D_{Ram}}{D_{Shyam}} = \frac{S_{Ram}}{S_{Shyam}}$

$\Rightarrow$ $\frac{1200}{1000} = \frac{S_{Ram}}{10} \Rightarrow S_{Ram} = 12\text{ m/s}$

$\therefore$ Speed of Ram = 12 m/s

Example **In a 500 m race, A reaches the final point in 28 sec and B reaches in 35 sec. By how much distance does A beat B?**

500 मी की दौड़ में, A अन्तिम बिन्दु पर 28 सेकण्ड में पहुँचता है और B अन्तिम बिन्दु पर 35 सेकण्ड में पहुँचता है। A, B को कितनी दूरी से हराता है?

DSSSB 2022

Sol. Required distance $= \left[\frac{500}{35}(35-28)\right] = \frac{500 \times 7}{35} = 100\text{ m}$

2. When A Gives Start of b m to B / जब A, B को b मी की शुरुआत देता है

- If in a race of length L m, A gives start of b m to B and start of 'c' m to C, then B gives start to C

  यदि लम्बाई L मी की दौड़ में, A, B को b मी की शुरुआत देता है और C को c मी की शुरुआत देता है, तो B, C को शुरुआत देता है,

$$= L\left(\frac{c-b}{L-b}\right)$$

Example **In a 1 km race, A gives B a start of 120 m and C a start of 160 m. They all finish the race together. How much start can B give to C in order to finish the same race together?**

1 किमी की दौड़ में A, B को 120 मी की बढ़त देता है और C को 160 मी की बढ़त देता है। वे सभी एकसाथ दौड़ समाप्त करते हैं। समान दौड़ को एकसाथ समाप्त करने के लिए B, C को कितनी बढ़त दे सकता है?

Sol. $\because$ 1 km = 1000 m

$\therefore$ Required start $= 1000\left(\frac{160-120}{1000-120}\right)$

$= \left(1000 \times \frac{40}{880}\right) = 45\frac{5}{11}\text{ m}$

- If in a race of length L m, A gives start of distance 'd' to B and still beats him by time t, then B's speed

  यदि L मी लम्बाई की दौड़ में, A, B को दूरी 'd' की शुरुआत देता है और फिर भी उसे t समय से हरा देता है, तो B की गति

$$S_B = \frac{L-d}{\left(\frac{L}{S_A} + t\right)}$$

Example **A and B participated in a 150 m race. A gives B a start of 15 m and still beats him by 9 sec. If A runs at a speed of 12 km/h, find the speed of B in km/h?**

A और B ने 150 मी की एक दौड़ में भाग लिया। A, B को 15 मी आगे से दौड़ शुरू करने देता है और फिर भी उसे 9 सेकण्ड से हरा देता है। यदि A, 12 किमी प्रति घण्टे की गति से दौड़ता है, तो B की गति किमी प्रति घण्टा में ज्ञात कीजिए।

Sol. Speed of $A = 12\text{ km/h} = 12 \times \frac{5}{18}\text{ m/s}$ $\left[\because 1\text{ m/s} = \frac{18}{5}\text{ km/h}\right]$

$$\text{Speed of } B = \frac{150 - 15}{\left(\frac{150}{12 \times \frac{5}{18}} + 9\right)}$$

$$= \frac{135}{(45+9)} = \frac{135}{54}\text{ m/s} = \left(\frac{135}{54} \times \frac{18}{5}\right)\text{ km/h}$$

$$= 9\text{ km/h}$$

3. Based on Circular Motion वृत्तीय गति पर आधारित

- When two persons A and B are running around a circular track of length L m with speeds S_1 m/s and S_2 m/s in the same direction. Time taken by A to complete the race is T_1 and time taken by B to complete the race is T_2.

  जब दो व्यक्ति A और B, L मी लम्बाई के एक वृत्ताकार ट्रैक के चारों ओर एक ही दिशा में S_1 मी/से और S_2 मी/से की गति से दौड़ रहे हैं। A द्वारा दौड़ पूरी करने में लिया गया समय T_1 है और B द्वारा दौड़ पूरी करने में लिया गया समय T_2 है।

  (i) Time taken to meet each other for the first time

  पहली बार एक-दूसरे से मिलने में लगा समय $= \frac{L}{S_1 - S_2}$ sec/सेकण्ड

  (ii) If they run in opposite direction, then time taken to meet each other for first time

  यदि वे विपरीत दिशा में दौड़ रहे हैं, तब

  पहली बार एक-दूसरे से मिलने में लगा समय $= \left(\frac{L}{S_1 + S_2}\right)$ sec/सेकण्ड

  (iii) The time in which they meet each other at the starting point

  वह समय जिसमें वे प्रारम्भिक बिन्दु पर एक-दूसरे से मिलते हैं।

  $= \text{LCM}\left(\frac{L}{S_1}, \frac{L}{S_2}\right)$ sec/सेकण्ड *M. Imp*

Example **In a circular race of 2500 m, a man and a woman starts running from a point in opposite directions with a speed of 37 km/h and 35 km/h, respectively. After how much time from the start of the race will they meet for the first time?**

2500 मी की एक वृत्ताकार दौड़ में एक पुरुष और एक महिला एक बिन्दु से विपरीत दिशा में क्रमशः 37 किमी/घण्टा और 35 किमी/घण्टा की गति से दौड़ना शुरू करते हैं। दौड़ शुरू होने के कितने समय बाद वे पहली बार मिलेंगे? *SSC (10+2) 2022*

Sol. Relative speed between the man and woman

$$= 37 + 35 = 72 \text{ km/h}$$

$$= 72 \times \frac{5}{18} \text{ m/s} = 20 \text{ m/s} \qquad \left[\because 1 \text{ km/h} = \frac{5}{18} \text{ m/s}\right]$$

$\therefore$ Time taken to meet for the first time $= \frac{2500}{20} = 125$ sec

Example **In a race around a circular cycling track of 75 km, two cyclists are riding at a speed of 30 km/h and 25 km/h. After what time (in hours) will they meet at the starting point of their journey?**

75 किमी के वृत्ताकार साइकिल ट्रैक के चारों ओर एक दौड़ में दो साइकिल चालक 30 किमी/घण्टा और 25 किमी/घण्टा की गति से साइकिल चला रहे हैं। कितने समय (घण्टे) के बाद वे उस बिन्दु पर मिलेंगे, जहाँ से उन्होंने अपनी यात्रा शुरू की थी? *SSC CGL (TIER- I) 2023*

Sol. Time taken by each to cover one round $\frac{75}{30}$ and $\frac{75}{25}$, respectively.

$$\therefore \text{ Required time} = \text{LCM of}\left(\frac{75}{30} \text{ and } \frac{75}{25}\right) = \text{LCM of}\left(\frac{5}{2}, \frac{3}{1}\right) = 15 \text{ h}$$

अध्याय 27

Statistics सांख्यिकी

Statistics is simply defined as the study of data. Statistics deals with the analysis and computation of numerical data.

सांख्यिकी को सरल शब्दों में आँकड़ों के अध्ययन के रूप में परिभाषित किया जाता है। सांख्यिकी संख्यात्मक आँकड़ों के विश्लेषण और गणना से सम्बन्धित है।

Data आँकड़े

Data is a group of information which describe the properties and quantitive nature of variable.

आँकड़े सूचनांओं का एक समूह है, जो चर की गुणात्मक तथा मात्रात्मक प्रवृत्ति का वर्णन करते हैं।

Properties of Data आँकड़ों के गुण

1. **Lower Limit** The figure or number on the left side of a class, is called its lower limit

   निम्न सीमा एक वर्ग के बाएँ ओर की संख्या या अंक, इसकी निम्न सीमा कहलाती है।

2. **Upper Limit** The figure or number on the right side of a class, is called its upper limit.

   उच्च सीमा एक वर्ग के दाएँ ओर की संख्या या अंक, इसकी उच्च सीमा कहलाती है।

3. **Class-Interval** Class intervals are the groups in which all observations are divided.

   वर्ग-अन्तराल वर्ग अन्तराल वे समूह हैं, जिनमें सभी प्रेक्षण विभाजित होते हैं।

4. **Class Mark** It is the mid-point of the class interval.

   $$i.e.\ \textbf{Class mark} = \frac{\textbf{Lower limit} + \textbf{Upper limit}}{2}$$

   वर्ग-माध्य (चिन्ह) यह वर्ग अन्तराल का मध्य-बिन्दु होता है।

   $$\text{अर्थात् } \textbf{वर्ग-माध्य} = \frac{\textbf{निम्न सीमा} + \textbf{उच्च सीमा}}{2}$$

5. **Frequency of an observation** The number of times an observation occurs, is called its frequency.

   प्रेक्षण की बारम्बारता एक प्रेक्षण की बारी-बारी से आने की संख्या, इसकी बारम्बारता कहलाती है।

6. **Class size** Difference between the upper limit and lower limit of class interval is called class size.
वर्ग अन्तराल के उच्च सीमा और निम्न सीमा के अन्तर को वर्ग माप कहते हैं।
Class size/वर्ग माप = Upper limit (उच्च सीमा) − Lower limit (निम्न सीमा)

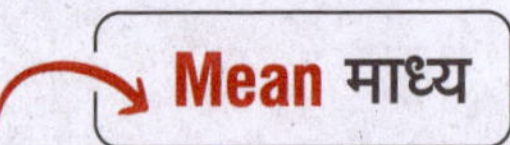

Mean is the average of all the numbers in the given data.
माध्य दिए गए आँकड़ों की सभी संख्याओं का औसत है।

1. **Mean** $(\bar{x}) = \dfrac{\textbf{Sum of all observations}}{\textbf{Total number of observations}}$

माध्य $(\bar{x}) = \dfrac{\textbf{सभी प्रेक्षणों का योग}}{\textbf{प्रेक्षणों की कुल संख्या}}$

माध्य टॉपिक सभी परीक्षाओं (SSC, Railway, CSAT IAS) की दृष्टि से M. Imp है। इस पर आधारित 1 प्रश्न परीक्षा में अवश्य पूछा जाता है।

Example **The mean of five observations x, $x-1$, $x-2$, $x-3$ and $x-4$ is 20, then the mean of the first two observations is**
यदि 5 आँकड़ों $x, x-1, x-2, x-3$ और $x-4$ का माध्य 20 है, तब पहले दो आँकड़ों का माध्य है *RRB NTPC 2021*

Sol. Mean $= \dfrac{\text{Sum of all observations}}{\text{Total observation}}$

$$\Rightarrow \quad 20 = \frac{x + x - 1 + x - 2 + x - 3 + x - 4}{5}$$

$$\Rightarrow \quad 5x - 10 = 100$$

$$\Rightarrow \quad x = \frac{110}{5} = 22$$

Mean of first two observations $= \dfrac{x + x - 1}{2} = \dfrac{22 + 22 - 1}{2}$

$$= \frac{43}{2} = 21.5$$

2. **Mean of Ungrouped Data/अवर्गीकृत आँकड़ों का माध्य**
If $x_1, x_2, x_3, \ldots x_n$ are n observations, then
यदि $x_1, x_2, x_3, \ldots x_n$, n प्रेक्षण हैं, तब

(i) **Mean/माध्य** $(\bar{x}) = \dfrac{x_1 + x_2 + x_3 + \ldots + x_n}{n} = \dfrac{1}{n}\sum_{i=1}^{n} x_i$

(ii) **Mean/माध्य** $(\bar{x}) = A + \dfrac{1}{n}\sum_{i=1}^{n} d_i$

Where, A = assumed mean and $d_i = x_i - A$
जहाँ, A = काल्पनिक माध्य तथा $d_i = x_i - A$

3. Mean of Grouped Data/वर्गीकृत आँकड़ों का माध्य

If $x_1, x_2, x_3, \dots x_n$ are n observations with respective frequencies $f_1, f_2, f_3, \dots f_n$, then

यदि $x_1, x_2, x_3, \dots x_n$, n प्रेक्षण हैं तथा $f_1, f_2, f_3, \dots f_n$, इनकी संगत बारम्बारताएँ हैं, तब

(i) **Mean/माध्य** $(\bar{x}) = \dfrac{f_1x_1 + f_2x_2 + f_3x_3 + \dots + f_nx_n}{f_1 + f_2 + f_3 + \dots + f_n} = \dfrac{\sum_{i=1}^{n} f_i x_i}{\sum_{i=1}^{n} f_i}$

Example **If the arithmetic mean of the following data is 21.5, then value of k will be**

यदि दिए गए आँकड़ों का गणितीय माध्य 21.5 है, तब k का मान होगा

UP TGT 2021

x	5	15	25	35	45
f	6	4	3	k	2

Sol. Arithmetic mean $= \dfrac{\sum f_i x_i}{\sum f_i}$

$$\Rightarrow \quad 21.5 = \frac{f_1x_1 + f_2x_2 + f_3x_3 + f_4x_4 + f_5x_5}{f_1 + f_2 + f_3 + f_4 + f_5}$$

$$\Rightarrow \quad 21.5 = \frac{5 \times 6 + 15 \times 4 + 25 \times 3 + 35 \times k + 45 \times 2}{6 + 4 + 3 + k + 2}$$

$$\Rightarrow \quad 21.5 = \frac{30 + 60 + 75 + 35k + 90}{15 + k}$$

$$\Rightarrow \quad 322.5 + 21.5k = 255 + 35k$$

$$\Rightarrow \quad 13.5k = 67.5$$

$$\therefore \quad k = 5$$

(ii) **Mean/माध्य** $(\bar{x}) = A + \dfrac{\sum_{i=1}^{n} f_i d_i}{\sum_{i=1}^{n} f_i}$

Where/जहाँ, $d_i = x - A$

Note *If class intervals are given, then firstly find class mark* (x). / *यदि वर्ग अन्तराल दिए हों, तब पहले वर्ग माध्य* (x) *ज्ञात करते हैं।*

Example **The score in maths of 50 students of a class is given in the following table. Calculate the mean score.**

दी गई तालिका में किसी वर्ग के 50 छात्रों के गणित के अंक दिए गए हैं। माध्य अंक की गणना कीजिए।

Class-Interval/वर्ग-अन्तराल	Frequencies/बारम्बारता
0-10	2
10-20	11
20-30	28
20-30	6
40-50	3

SSC (10+2) 2022

Sol.

Class-Interval	Frequencies (f_i)	Class-marks (x_i)	$f_i x_i$
0-10	2	$\frac{0+10}{2}=5$	10
10-20	11	$\frac{10+20}{2}=15$	165
20-30	28	$\frac{20+30}{2}=25$	700
30-40	6	$\frac{30+40}{2}=35$	210
40-50	3	$\frac{40+50}{2}=45$	135
Total	50		1220

Here, $\Sigma f = 50$ and $\Sigma f_i x_i = 1220$

$$\therefore \text{ Mean } (\bar{x}) = \frac{\Sigma f_i x_i}{\Sigma f_i} = \frac{1220}{50} = 24.4$$

Median माध्यिका

Median is defined as the middle most or the central value of the variable in a set of observations, when the observations are arranged either in ascending order or descending order of their magnitudes.

माध्यिका को अवलोकनों के समूह में चर के सबसे मध्य या केन्द्रीय मान के रूप में परिभाषित किया जाता है, जब अवलोकनों को उनके परिमाण के आरोही क्रम या अवरोही क्रम में व्यवस्थित किया जाता है।

1. Median of Ungrouped Data/अवर्गीकृत आँकड़ों की माध्यिका

Suppose there are n number of observations. Then, is arranged the data in ascending or descending order.

माना प्रेक्षणों की संख्या n है, तब इन्हें आरोही या अवरोही क्रम में व्यवस्थित करते हैं।

(i) If n is odd, then

Median = Value of $\left(\frac{n+1}{2}\right)$th term

यदि n विषम है, तब

माध्यिका = $\left(\frac{n+1}{2}\right)$ वें पद का मान *M. Imp*

> **अतिसम्भावित!**
> माध्यिका पर आधारित 1 प्रश्न SSC, Railway आदि महत्वपूर्ण परीक्षाओं में अवश्य पूछा जाता है।

Example **The median of the following data will be**

नीचे दिए गए आँकड़ों की माध्यिका होगी

32, 25, 33, 27, 35, 29 and 30 *SSC CGL (TIER II) 2023*

Sol. 32, 25, 33, 27, 35, 29 and 30.

Ascending order,

25, 27, 29, 30, 32, 33, 35

n = total number of observations = 7 (odd)

$$\text{Median} = \left(\frac{n+1}{2}\right)\text{th term} = \left(\frac{8}{2}\right)\text{th term}$$

$$= 4\text{ th term} = 30$$

(ii) If n is even, then

$$\textbf{Median} = \frac{\text{Value of } \frac{n}{2}\text{th term} + \text{Value of } \left(\frac{n}{2}+1\right)\text{th term}}{2}$$ *M. Imp*

यदि n सम है, तब

$$\text{माध्यिका} = \frac{\frac{n}{2}\text{ वें पद का मान} + \left(\frac{n}{2}+1\right)\text{ वें पद का मान}}{2}$$

Example **The nominal marks obtained by students who appeared in a test are given below. Find the median marks of the students.**

एक परीक्षा में बैठने वाले छात्रों द्वारा प्राप्त प्राप्तांक नीचे दिए गए हैं। छात्रों के माध्यिका अंक ज्ञात कीजिए।

Marks/अंक	11	20	8	16	29	22
Number of Students/छात्रों की संख्या	14	5	11	9	3	8

Sol.

Marks	Students	Cumulative Frequency
8	11	11
11	14	25
16	9	34
20	5	39
22	8	47
29	3	50

Here, $N = 50 \quad \therefore \frac{N}{2} = 25\text{th}, \frac{N}{2} + 1 = 26\text{th}$

$$\text{Median} = \frac{(25\text{th} + 26\text{th})\text{ term}}{2} = \frac{11 + 16}{2} = 13.5$$

2. Median of Grouped Data/वर्गीकृत आँकड़ों की माध्यिका

Firstly we find commulative frequencies of all the classes and find median class. Then,

पहले हम सभी वर्गों की संचयी बारम्बारताएँ ज्ञात करते हैं तथा माध्यिका वर्ग ज्ञात करते हैं, तब

$$\textbf{Median/माध्यिका} = l + \frac{\frac{N}{2} - C_f}{f} \times h$$ *M. Imp*

Where, l = Lower limit of median class/माध्यिका वर्ग की निम्न सीमा

C_f = Commulative frequency of class preceding the median class/माध्यिका वर्ग से पहले वर्ग की संचयी बारम्बारता

f = Frequency of median class/माध्यिका वर्ग की बारम्बारता

h = Class size/वर्ग माप

Example **100 surnames were randomly picked up from a local telephone directory and the frequency distribution of the number of letters from the English alphabets in the surnames was obtained as follows.**

100 उपनामों को एक स्थानीय टेलीफोन निर्देशिका से यादृच्छिक रूप से उठाया गया और उपनामों में अंग्रेजी वर्णमाला के अक्षरों की संख्या का आवृत्ति वितरण निम्नानुसार प्राप्त किया गया।

Number of Letters अक्षरों की संख्या	Numbers of Surnames उपनामों की संख्या
1-4	6
4-7	30
7-10	40
10-13	16
13-16	4
16-19	4

The median number of letters in the surnames is
उपनामों में अक्षरों की माध्यिका संख्या है *RRC GROUP D 2022*

Sol.

Class-Interval	Frequency (f_i)	Cumulative Frequency
1-4	6	6
4-7	30	36
7-10	40	76
10-13	16	92
13-16	4	96
16-19	4	100

Total frequency $= \Sigma f_i = N = 100$

So, $\frac{N}{2} = 50$, it lies in (7 – 10) interval

$\therefore$ Median class = 7 – 10

$$l = 7, h = 10 - 7 = 3, cf = 36, f = 40$$

$$\text{Median} = l + \left(\frac{\frac{N}{2} - cf}{f}\right) \times h = 7 + \left(\frac{50 - 36}{40}\right) \times 3 = 7 + 1.05 = 8.05$$

Mode is that value of variable, which occurs most frequently or we can say that the observation which has highest frequency is the mode.
बहुलक वह मान है, जो अधिकांशतः और हम कह सकते हैं जो आँकड़ों में सबसे ज्यादा आवृत्ति के साथ आया हो।

1. **Mode of Ungrouped Data/अवर्गीकृत आँकड़ों का बहुलक**
   Mode is that value which is repeated maximum number of times.
   बहुलक वह मान है, जिसकी सबसे अधिक बार पुनरावृत्ति हुई है।

Example **Find the mode of the following data of students ages.**
दिए गए छात्रों की आयु का बहुलक ज्ञात कीजिए।
16, 17, 15, 17, 16, 15, 14, 14, 13, 17, 13, 12, 12, 16, 10, 14, 17, 10, 11
RRC GROUP D 2022

Sol. Most frequent age of given students = 17 (4 times)

$\therefore$ Mode = 17

2. Mode of Grouped Data/वर्गीकृत आँकड़ों का बहुलक

$$\text{Mode/बहुलक} = l + \left(\frac{f_1 - f_0}{2f_1 - f_0 - f_2}\right) \times h$$ M. Imp

Where/जहाँ,

l = Lower limit of modal class
बहुलक वर्ग की निम्न सीमा

h = Size of the class interval
वर्ग अन्तराल की माप

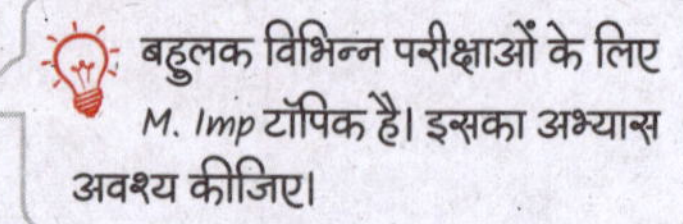

f_0 = Frequency of the class preceding the modal class
बहुलक वर्ग से पहले वर्ग की बारम्बारता

f_1 = Frequency of the modal class/बहुलक वर्ग की बारम्बारता

f_2 = Frequency of the class succeeding the modal class
बहुलक वर्ग से बाद के वर्ग की बारम्बारता

Example **A group of students as class IX conducted a survey of the number of family members in a household for 40 houses in a locality, which resulted in the given frequency table. What is the mode of the data?**

कक्षा 9 के छात्रों के एक समूह ने एक क्षेत्र के 40 घरों में परिवार के सदस्यों की संख्या के लिए एक सर्वेक्षण किया, जिसके परिणामस्वरूप दी गई बारम्बारता सारणी प्राप्त हुई। आँकड़ों का बहुलक क्या है?

Family Size परिवार का आकार	Number of Families परिवारों की संख्या
2-4	10
4-6	8
6-8	12
8-10	6
10-12	4

RRC GROUP D 2022

Sol. Maximum class frequency = 12

So, modal class = 6 − 8

Here, $l = 6, h = 8 - 6 = 2, f_1 = 12, f_0 = 8, f_2 = 6$

By formula for grouped data,

$$\text{Mode} = l + \left(\frac{f_1 - f_0}{2f_1 - f_0 - f_2}\right) \times h = 6 + \left(\frac{12 - 8}{24 - 8 - 6}\right) \times 2$$

$$= 6 + 0.8 = 6.8$$

- **Relation between Mean, Median and Mode**
माध्य, माध्यिका और बहुलक में सम्बन्ध

Mode = 3 Median − 2 Mean
बहुलक = 3 माध्यिका − 2 माध्य

M. Imp

Example **For a sample data, mean = 60 and median = 48. For this distribution, the mode is**
एक आँकड़े के लिए, माध्य = 60 और माध्यिका = 48 है। इस वितरण के लिए बहुलक है *SSC CGL (TIER II) 2023*

Sol. Given, mean = 60, median = 48
Mode = 3 Median − 2 Mean = $3 \times 48 - 2 \times 60 = 144 - 120 = 24$

Range is the difference of maximum and minimum values of data.

Range = Maximum value − Minimum value

परास, अधिकतम आँकड़े और न्यूनतम आँकड़े के मान का अन्तर होता है।

परास = अधिकतम मान − न्यूनतम मान

Mean Deviation माध्य विचलन

1. **Mean Deviation for Ungrouped Data/अवर्गीकृत आँकड़ों का माध्य विचलन**
Let $x_1, x_2, \ldots x_n$ be n observations, then
माना $x_1, x_2, \ldots x_n$, n प्रेक्षण हैं, तब

$$\text{Mean deviation/माध्य विचलन} = \frac{\sum_{i=1}^{n} |x_i - \bar{x}|}{n}$$

Where/जहाँ, $\bar{x} \to$ Mean/माध्य

2. **Mean Deviation for Grouped Data/ वर्गीकृत आँकड़ों का माध्य विचलन**
Let $x_1, x_2, \ldots x_n$ be n observations and $f_1, f_2, \ldots f_n$ are their corresponding frequencies. Then,
माना $x_1, x_2, \ldots x_n$, n प्रेक्षण तथा $f_1, f_2, \ldots f_n$ इनकी संगत बारम्बारताएँ हैं। तब,

$$\text{Mean deviation/माध्य विचलन} = \frac{\sum_{i=1}^{n} f_i |x_i - \bar{x}|}{N}$$

Where/जहाँ, $\bar{x} \to$ Mean/माध्य

$$N = \sum_{i=1}^{n} f_i = \text{Total frequency/कुल बारम्बारता}$$

1. **Variance of Ungrouped Data/अवर्गीकृत आँकड़ों का प्रसरण**

   Variance of n observations $x_1, x_2, \ldots, x_n$ is

   n प्रेक्षणों $x_1, x_2, \ldots, x_n$ का प्रसरण $\sigma^2 = \dfrac{\sum_{i=1}^{n}(x_i - \bar{x})^2}{n}$

   Where/जहाँ, σ^2 = variance/प्रसरण, $\bar{x}$ = mean/माध्य

2. **Variance of Grouped Data/वर्गीकृत आँकड़ों का प्रसरण**

   Let $x_1, x_2, \ldots x_n$ be n observations and $f_1, f_2, \ldots f_n$ are their corresponding frequencies, then

   माना $x_1, x_2, \ldots x_n$, n प्रेक्षण हैं तथा इनकी संगत बारम्बारताएँ $f_1, f_2, \ldots f_n$ हैं, तब

   Variance/प्रसरण, $\sigma^2 = \dfrac{\sum_{i=i}^{n} f_i (x_i - \bar{x})^2}{n}$ Where/जहाँ, $\bar{x}$ = Mean/माध्य

Example **Find the variance of the following data.**
दिए गए आँकड़ों का प्रसरण ज्ञात कीजिए।
6, 7, 5, 9, 12, 15

Sol. Mean $(\bar{x}) = \dfrac{6+7+5+9+12+15}{6} = \dfrac{54}{6} = 9$

Variance $= \dfrac{\Sigma (X - \bar{x})^2}{n} = \dfrac{(9-6)^2 + (9-7)^2 + (9-5)^2 + 0^2 + 3^2 + 6^2}{6}$

$= \dfrac{9+4+16+9+36}{6} = \dfrac{74}{6} = \dfrac{37}{3}$

- **Standard Deviation/मानक विचलन**

  Standard Deviation/मानक विचलन

  $(SD) = \sqrt{\sigma^2} = \sigma$

  Where/जहाँ, σ^2 is variance / प्रसरण है।

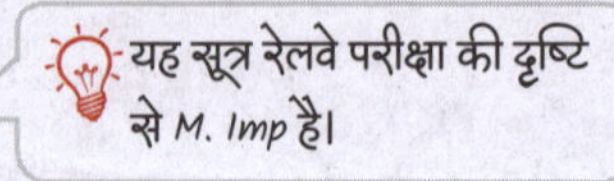

Example **Calculate the standard deviation for the following data. / दिए गए आँकड़ों का मानक विचलन ज्ञात कीजिए।**
4, 7, 9, 10, 15 *RRB NTPC 2021*

Sol. Mean, $\bar{x} = \dfrac{4+7+9+10+15}{5} = \dfrac{45}{5} = 9$

Here, $n = 5$

$SD = \sqrt{\dfrac{\Sigma (X - \bar{x})^2}{n}} = \sqrt{\dfrac{(4-9)^2 + (7-9)^2 + (9-9)^2 + (10-9)^2 + (15-9)^2}{5}}$

$= \sqrt{\dfrac{25+4+1+36}{5}} = \sqrt{\dfrac{66}{5}} = 3.63$

Graphical Representation of Data
आँकड़ों का चित्रमय निरूपण

1. Bar Graphs/दण्ड आरेख

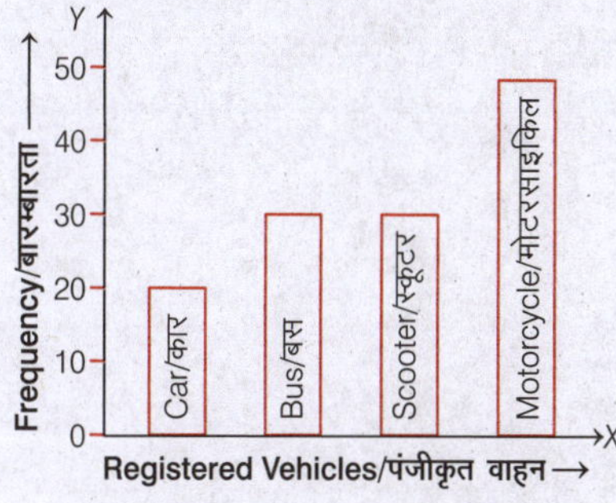

2. Histograms/आयत चित्र

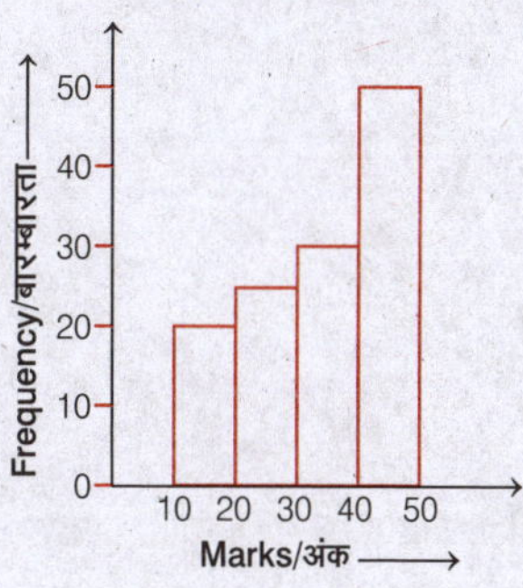

3. Frequency Polygon/आवृत्ति बहुभुज

A frequency polygon is a line graph of class frequency plotted against class mid-point. It can be obtained by joining the mid-points of the top of the rectangles in the histogram.

एक आवृत्ति बहुभुज, वर्ग आवृत्ति का रेखा ग्राफ निरूपण है, जिसे वर्ग आवृत्ति के मध्य-बिन्दु के विरुद्ध अंकित किया जाता है।

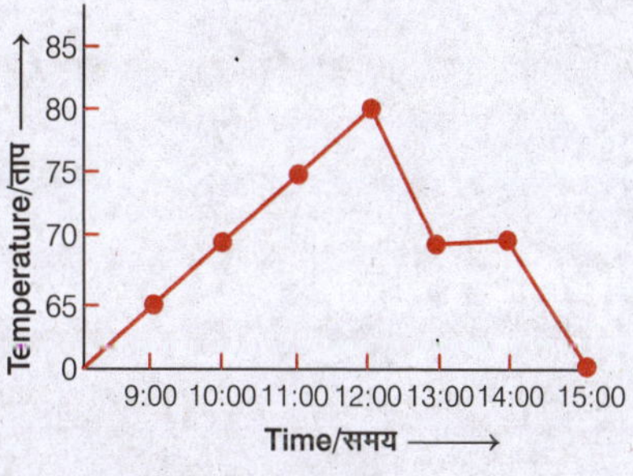

4. Circle Graph (Pie-chart)/वृत्त चित्र (पाई-चार्ट)

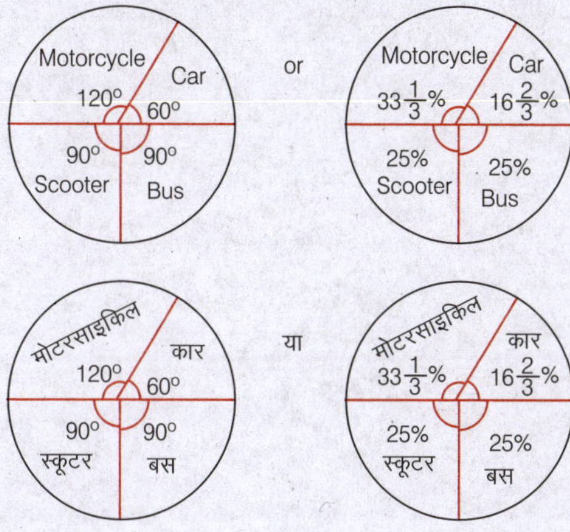